C0-AZS-543

STRUCTURE AND FUNCTION OF PLANT ROOTS

Developments in Plant and Soil Sciences

Volume 4

Also in this series:

1. J. Monteith and C. Webb, eds., Soil Water and Nitrogen in Mediterranean-type Environments. Proceedings based on a Workshop organized by the International Center for Agricultural Research in the Dry Areas, held in Aleppo, Syria, January 1980.
   ISBN 90-247-2406-6
2. J. C. Brogan, ed., Nitrogen Losses and Surface Run-off from Landspreading of Manures, Proceedings of a Workshop in the EEG Programme of Coordination of Research on Effluents from Livestock, held at The Agricultural Institute, Johnstown Castle Research Centre, Wexford, Ireland, May 20–22, 1980.
   ISBN 90-247-2471-6
3. J. D. Bewley, ed., Nitrogen and Carbon Metabolism. Symposium on the Physiology and Biochemistry of Plant Productivity, held in Calgary, Canada, July 14–17, 1980.
   ISBN 90-247-2472-4

Series ISBN 90-247-2405-8

Structure and Function of Plant Roots

Proceedings of the 2nd International Symposium, held in Bratislava, Czechoslovakia, September 1–5, 1980

edited by

R. BROUWER

O. GAŠPARÍKOVÁ

J. KOLEK

B.C. LOUGHMAN

Chapters 23, 28, 29, 34, 37, 38, 40, 45, 47, 51, 52, 53, 55, 59, 65, 66, 69, 72, 74, 76
reprinted from *Plant and Soil* Vol. 63, No. 1 (1981)

1981

MARTINUS NIJHOFF / DR W. JUNK PUBLISHERS
THE HAGUE / BOSTON / LONDON

Distributors:

for the United States and Canada

Kluwer Boston, Inc.
190 Old Derby Street
Hingham, MA 02043
USA

for all other countries

Kluwer Academic Publishers Group
Distribution Center
P.O. Box 322
3300 AH Dordrecht
The Netherlands

Library of Congress Cataloging in Publication Data

Main entry under title:

Structure and function of plant roots.

(Developments in plant and soil sciences; v. 4)
'Organised by the Slovak Academy of Sciences' – Pref.
Includes index.
1. Roots (Botany) – Congresses. I. Brouwer, Rienk. II. International Symposium on Structure and Function of Roots (2nd : 1980 : Bratislava, Czechoslovakia) III. Slovenská akadémia vied. IV. Series.
QK644.S83 582'.010428 81-16800

ISBN 90-247-2510-0 AACR2
ISBN 90-247-2405-8 (series)

PRINTED IN THE NETHERLANDS

Preface

The 2nd International Symposium on Structure and Function of Roots organised by the Slovak Academy of Sciences was held in Bratislava in early September 1980 and continued the theme of its 1971 predecessor at Tatranska Lomnica.

The symposium started with 'structural characteristics of roots in relation to the process of growth'. The approach was notable for the way in which it brought together cytologists, electron microscopists and physiologists to discuss problems of common interest and set the scene for a multidisciplinary approach to the remainder of the Symposium. Metabolic aspects of roots under normal conditions and after being exposed to stress received considerable attention, together with interesting contributions concerned with responses of roots to gravity.

The session on water movement provided a useful balance between theoretical papers on the one hand and the results of the use of new experimental approaches to the problem on the other. The Symposium continued with the examination of factors affecting the transport of ions into and across roots, particular attention focussing on metabolic aspects concerned with delivery of ions to the shoot. The problems associated with coordinating the separate functions of the root were introduced by a group of papers, again supported by contributions in which new experimental methods were described. These led to an integrated approach to the correlation of root function with the well-being of the plant as a whole, which formed the final day of the Symposium, and it was particularly noticeable from the trend of the discussion that a number of the participants had benefitted from the contact with other disciplines during their week's stay in Bratislava.

The Organising Committee had not intended to publish the proceedings, but during the Symposium the feeling emerged that the high quality of many of the papers merited a permanent record, particularly if relatively rapid publication could be achieved. The Committee felt that it would be a good opportunity to present a broad spectrum of international contributions including a number from Eastern Europe that might not otherwise be readily available.

These points were presented to the full Symposium and it was decided to aim for manuscript submission by early December 1980. With the cooperation of all concerned, the texts of delivered papers limited to seven text pages and poster contributions to three pages were in the hands of the Editors by the end of the year. A deliberate decision was taken to allow all Symposium contributors to have space in the published volume if the work merited it, and as a consequence the stringent restraints on space meant that authors were required to present contributions that were overly concise by normal Symposium standards. Nevertheless, it was felt that the opportunity to include information from such a wide range of research groups from so many countries would perhaps provide a volume the general usefulness of which outweighed the enforced brevity of the contributions. We hope that some flavour of the Symposium as felt by the participants comes through to the reader even though some of the contributions contain only the key points of poster presentations with a minimum of text.

The facilities provided by the Comenius University were outstanding and the hospitality shown to the visitors by the Institute of Experimental Biology and Ecology of the Slovak Academy will long be remembered. It will be the wish of many that the 3rd Symposium takes place long before the lapse of the nine-year interval that divided the first two.

THE EDITORS

Contents

II. Metabolism of roots

III. Transport phenomena

A. Water transport

IV. Functional integrity of the root system

V. Responses of roots to stress

VI. Interaction between roots and shoots

List of Contributors

Aljochina, Natalia D., Plant Physiology Department, Biology Faculty, Moscow State University, 117 234 Moscow, USSR

Barlow, P. W., Agricultural Research Council, Letcombe Laboratory, Wantage, Oxon, OX12 9JT, England

Beffa, R., Institut de Biologie et de Physiologie Végétales, Place de la Riponne 6, 1005 Lausanne, Suisse

Beneš, K., Institute of Exp. Botany, 166 30 Praha 6, Ke Dvoru 16/15, Czechoslovakia

Bogemans, G. J., Laboratorium voor Plantenfysiologie Vrije Universiteit Brussel, Paardenstraat 65, B1640 St.Genesius-Rode, Belgium

Bolyakina, Y. P., K. A. Timiriazev Institute of Plant Physiology, Academy of Sciences, Botanicheskaya 35, 127 296 Moscow, USSR

Boone, F. R., Soil Tillage Laboratory, Agricultural University Wageningen, The Netherlands

Bowling, D. J. F., Department of Botany, University of Aberdeen, St.Machar Drive, Aberdeen AB9 2UD Scotland, U.K.

Brouwer, R., Botanical Laboratory, State University of Utrecht, Lange Nieuwstraat 106 – 3512 PN Utrecht, The Netherlands

Bujtás, Claire, Department of Plant Physiology, Eötvös University, H-1445, Budapest, P.O.B. 324, Hungary

Černohorská, Jana, Plant Physiology Dept., Charles University, Viničná 5, 128 44 Praha 2, Czechoslovakia

Čiamporová, Milada, Slovak Academy of Sciences, Institute of Experimental Biology and Ecology, Dúbravská cesta 14 885 34 Bratislava, Czechoslovakia

Chanson, A., Institut de Biol. et de Physiol. Végétales, Place de la Riponne 6, 1005 Lausanne, Suisse

Chloupek, O., Plant Breeding Station, 664 43 Želešice near Brno, Czechoslovakia

Considine, J. A., Horticultural Research Institute, Department of Agriculture, Burwood Highway, Knoxfield, Australia, 3180

Cseh, Edith, Department of Plant Physiology, Eötvös University H-1445, Budapest, POB 324, Hungary

Czajkowska, E., Warsaw Agricultural University Dep. of Plant Biology, Rakowiecka str. 26/30 02–528 Warsaw, Poland

Dainty, J., Department of Botany, University of Toronto, Toronto, Ontario, Canada M5S 1A1

Danilova, M. F., Komarov Botanical Institute of the Academy of Sciences of the U.S.S.R. Leningrad U.S.S.R. 197022

Dejaegere, R., Lab. Plant Physiology, Vrije Univ. Brussels 65, Paardenstr., 1640 St-Genesius-Rode, Belgium

Delegher, V., Lab. Plant Physiology, Vrije Univ. Brussels 65, Paardenstr., 1640 St-Genesius-Rode, Belgium

Dvořák, M., Plant Physiology Dept., Charles University, Viničná 5, 128 44 Praha 2, Czechoslovakia

El Bassam, N., Institute of Crop Science and Plant Breeding, Federal Research Center of Agriculture, Braunsschweig – Völkenrode (FAL), Bundesallee 50, D-3300 Braunschweig, FRG

Erdei, L., Biol. Research Centrum, H-6701 Szeged, POB 521, Hungary

Felipe, M. R., Institute de Edafologia y Biologia Vegetal, Serrano, 115, Madrid /6/, Spain

Filippenko, V. N., Institute of General and Inorganic Chemistry of the Academy of Sciences of the USSR, Moscow 117071, Leninskii Prospekt, 31, USSR

Foy, C. D., Institut für Nutzpflanzenforschung – Pflanzenernährung, Lentzeallee 55–57, D 1000 Berlin 33

François, G., Laboratorium voor Plantenfysiologie, Vrije Universiteit Brussel, Paardenstraat 65, B1640 St. Genesius-Rode, Belgium

Frič, F., Institute of Experimental Biology and Ecology of the Slovak Academy of Sciences, 885 34 Bratislava Czechoslovakia

Gašparíková, Otília, Institute of Experimental Biology and Ecology of the Slovak Academy of Sciences, 885 34 Bratislava, Czechoslovakia

Grinieva, Galina, M., K. A. Timiriazev Institute of Plant Physiology USSR Academy of Sciences, 127296 Moscow, Botanicheskaja 35, USSR

Hadačová, Věra, Institute of Experimental Botany, Czechoslovak Academy of Sciences, Ke dvoru 15, 166 30 Praha 6, Czechoslovakia

Hecht-Buchholz, Charlotte, Institut für Nutzpflanzenforschung – Pflanzenernährung, Lentzeallee 55–57, D 1000 Berlin 33

Herdová, Zuzana, Institute of Experimental Biology and Ecology, Slovak Academy of Sciences, 885 34 Bratislava, Dúbravská 14, Czechoslovakia

Holobradá, Margita, Institute of Experimental Biology and Ecology, Slovak Academy of Sciences, 885 34 Bratislava, Dúbravská 14, Czechoslovakia

Ivanov, V. B., Institute of General and Inorganic Chemistry, Academy of Sciences, Moscow 117071, Leninskii Prospekt 31, USSR

Janas, K., Department of Plant Cytology and Cytochemistry and Laboratory of Plant Growth Substances, Institute of Physiology and Cytology, University of Lódź, Banacha 12/16, 90–237 Lódź, Poland

Janáček, K., Kleinová, M., Laboratory for Cell Membrane Transport, Institute of Microbiology CSAV, Vídeňská 1083, 142 20 Praha 4 Krč, Czechoslovakia

Ješko, T., Institute of Experimental Biology and Ecology of the Slovak Academy of Sciences, 885 34 Bratislava, Czechoslovakia

Kadej, Agneszka, Kadej, F., Central Laboratory and Institute of Biology of the Marie Curie-Skodowska University, Plac M. C. Skodowska 3, 30–031 Lublin, Poland

Keltjens, W., Department of Soil Science and Plant Nutrition, Agricultural University, De Dreijen 3, 6703 BC Wageningen, The Netherlands

Kenzhebaeva, S. S., Plant Physiology Department, Biology Faculty, Moscow State University, 117234 Moscow, USSR

Khavkin, E. E., Siberian Institute of Plant Physiology and Biochemistry, USSR Academy of Sciences, Irkutsk 33, P.O. Box 1243, USSR 664033

Kholodova, V. P., K. A. Timiriazev Institute of Plant Physiology, Academy of Sciences, Botanicheskaya 35, 117 234 Moscow, USSR

Kleinová, Marie, Laboratory for Cell Membrane Transport, Institute of Microbiology ČSAV, Vídeňská 1083 142 20 Praha 4 Krč, Czechoslovakia

Kluikova, A. I., Plant Physiology Department, Biology Faculty, Moscow State University, 117234 Moscow, USSR

Kolek, J., Institute of Experimental Biology and Ecology SAV 885 34 Bratislava, Czechoslovakia

Kononowicz, A. K., Department of Plant Cytology and Cytochemistry and Laboratory of Plant Growth Substances, Institute of Physiology and Cytology, University of Lódź, Banacha 12/16, 90–237 Lódź, Poland

Kononowicz, H., Department of Plant Cytology and Cytochemistry and Laboratory of Plant Growth Substances, Institute of Physiology and Cytology, University of Lódź, Banacha 12/16, 90-237 Lódź, Poland

Kozinka, V., Institute of Experimental Biology and Ecology of CBEV SAV, 885 34 Bratislava, Dúbravská 14, Czechoslovakia

Kramer, D., Institut für Botanik, Fachbereich Biologie, Technische Hochschule Darmstadt, D-6100 Darmstadt, FRG

Krasavina, Marina S., K. A. Timiriazev Institute of Plant Physiology Botanicheskaya 35, 127 296 Moscow, USSR

Kubica, Š., Institute of Experimental Biology and Ecology SÀV, Dúbravská cesta 14, 885 34 Bratislava, Czechoslovakia

Kurkova, Elena B., K. A. Timiriazev Institute of Plant Physiology, Academy of Sciences of the USSR, Moscow, 127 296, USSR

Lambers, H., School of Agriculture and Forestry, University of Melbourne, Parkville, Vic. 3052, Australia

Loughman, B. C., Department of Agricultural Science, University of Oxford, Parks Road, Oxford OX1 3PP UK

Lux, A., Jr., Department of Plant Physiology, Comenius University, Odborárske nám. 5, 886 04 Bratislava, Czechoslovakia

Luxová, Mária, Slovak Academy of Sciences, Institute of Experimental Biology and Ecology, 885 34 Bratislava, Czechoslovakia

Lyalin, O. O., Institute of Agrophysics, Leningrad, USSR

Markov, E. Y., K. A. Timiriazev Institute of Plant Physiology, USSR Academy of Sciences Moscow, USSR 127106

MacLeod, R. D., Department of Plant Biology, University of Newcastle upon Tyne NE1 7RU, England

Martyn, G. I., N. G. Kholodny Institute of Botany, Ukrainian Academy of Sciences, Kiev, USSR, 252601

Meshcheriakov, A. B., K. A. Timiriazev Institute of Plant Physiology, Academy of Sciences, Moscow, USSR

Michalov, J., Institute of Experimental Biology and Ecology of Slovak Academy of Sciences, Dúbravská cesta 14, 885 34 Bratislava, Czechoslovakia

Miller, D. M., Agriculture Canada Research Centre, London, Ont., Canada N6A 5B7

Minarčic, P., Institute of Experimental Biology and Ecology of the Slovak Academy of Sciences, 885 34 Bratislava, Dúbravská 14, Czechoslovakia

Mishutina, Natalya E., K. A. Timiriazev Institute of Plant Physiology, Academy of Sciences, 127276 Moscow, Botanicheskaya 35, USSR

Mistrík, I., Institute of Experimental Biology and Ecology of Slovak Academy of Sciences, Dúbravská 14, 885 34 Bratislava, Czechoslovakia

Murín, A., Laboratory of Karyology, Institute of Molecular and Subcellular Biology, Comenius University, Bratislava, Czechoslovakia

Musatenko, Ljudmila I., Institute of Botany Ukr.SSR.Acad.Sci. Department of Plant Physiology, Repina 2, 252601 Kiev – 4, GSP USSR

Neirinckx, L., Lab.Plant.Physiology, Vrije Univ. Brussels 65, Paardenstr., 1640 St-Genesius-Rode, Belgium

Nicholas, D. J. D., Departments of Agricultural Biochemistry and Botany, University of Adelaide, Adelaide, 5001, Australia

Oaks, Ann, Biology Department, McMaster University, Hamilton, Ontario, Canada L8 S4KI

Obroucheva, Natalia V., K. A. Timiriazev Institute of Plant Physiology USSR Academy of Sciences, 127106 Moscow, Botanicheskaya 35, USSR

Olszewska, Maria J., Department of Plant Cytology and Cytochemistry and Laboratory of Plant Growth Substances, Institute of Physiology and Cytology, University of Lódź, Banacha 12/16, 90–237 Lódź, Poland

Orlova, M. S., K. A. Timiriazev Institute of Plant Physiology, Academy of Sciences, Irkutsk 33, P.O. Box 1243, USSR 664033

Paulech, C., Institute of Experimental Biology and Ecology, Slovak Academy of Sciences, 885 34 Bratislava, Dúbravská 14, Czechoslovakia

Pilet, P. E., Institut de Biol. et de Physiol. Végétales, Place de la Riponne 6, 1005 Lausanne, Suisse

Pospíšil, F., Institute of Experimental Botany, Czechoslovak Academy of Sciences, Na Karlovce 1, 160 00 Praha 6 Czechoslovakia

Pozuelo, J. M., Instituto de Edafologia y Biologia Vegetal, Madrid, Spain

Priehradný, S., Institute of Experimental Biology and Ecology, Slovak Academy of Sciences, 885 34 Bratislava, Dúbravská 14, Czechoslovakia

Procházka, S., Department of Botany and Plant Physiology, University of Agriculture, Zemědelská 1, 662 65 Brno, Czechoslovakia

Przemeck, E., Institute of Agricultural Chemistry of the Georg – August University Göttingen, von Siebold-Strasse 6, 3400 Göttingen, FRG

Pšenáková, Tatiana, Institute of Experimental Biology and Ecology of Slovak Academy of Sciences, Dúbravská cesta 14, 885 34 Bratislava, Czechoslovakia

Richards, D., Horticultural Research Institute, Department of Agriculture, Burwood Highway, Knoxfield, Victoria, Australia, 3180

Rodchenko, Oktyabrina, P., Siberian Institute of Plant Physiology and Biochemistry, Siberian Branch, USSR Academy of Sciences, P.O.Box 1243, Irkutsk 33, USSR 664033

Sahulka, J., Institute of Experimental Botany of the Czechoslovak Academy of Sciences, Ke dvoru 16/15, 166 30 Praha 6, Vokovice, Czechoslovakia

Sarič, M. R., Faculty of Natural Sciences, Institute of Biology, Novi Sad, Yugoslavia

Schrader, B., Institute of Agricultural Chemistry of the Georg–August University Göttingen, von Siebold-Strasse 6, 3400 Göttingen, FRG

Shevyakova, Nina I., K. A. Timiriazev Institute of Plant Physiology Academy of Sciences, 127276 Moscow, Botanicheskaya 35, USSR

Sindelářová, M., Institute of Experimental Botany, Czechoslovak Academy of Sciences, Na Karlovce 1, 160 00 Praha 6, Czechoslovakia

Škorič, D., Faculty of Agriculture, Institute of Field and Vegetable Crops, Novi Sad, Yugoslavia

Smith, F. A., Department of Agricultural Biochemistry and Botany, University of Adelaide, Adelaide, 5001, Australia

Smith, Sarah E., Department of Agricultural Biochemistry and Botany University of Adelaide, Adelaide 5001, Australia

Snijder, J., University of Groningen, Department of Plant Physiology, P.O.Box 14, 9750 AA Haren (Gr.). The Netherlands

Sokolova, S. V., K. A. Timiriazev Institute of Plant Physiology, 127 276 Moscow, Botanicheskaya 35, USSR

Spek, Louise Y., Botanical Laboratory, State University of Utrecht, Lange Nieuwstraat 106, 3512 PN Utrecht, The Netherlands

Starck, Zofia, Warsaw Agricultural University, Dept. of Plant Biology, Rakowiecka str. 26/30, 02–528 Warsaw, Poland

Stassart, J. M., Laboratorium voor plantenfysiologie, Institut voor molekulaire biologie, Vrije Universiteit Brussel, Paardenstraat 65, B-1640 Sint Genesius Rode, Belgium

Sytnik, K. M., N. G. Kholodny Institute of Botany, Ukrainain Academy of Sciences, Kiev, USSR, 252 601

Szaniawski, R. K., Department of Botany, Technical University, D-8 München 2, Arcisstrasse 21, FRG

Tánczos, O. G., Department of Plant Physiology, Biol. Center, P.O.Box 14, 9750 AA Haren (Gr), The Netherlands

Thompson, A., Department of Plant Biology, University of Newcastle upon Tyne, Newcastle upon Tyne, NE1 7RU, England

Tikhaya, Nataliya I., K. A. Timiriazev Institute of Plant Physiology, Academy of Sciences, 127 276 Moscow, Botanicheskaya 35, USSR

Troughton, A., Welsh Plant Breeding Station, Aberystwyth, SY23 2EB, U.K.

Vakhmistrov, D. B., K. A. Timiriazev Institute of Plant Physiology, Academy of Sciences, 127276 Moscow, Botanicheskaya 35, USSR

Vartanian, Nicole, Phytron, CNRS, 91190 Gif-sur-Yvette, Ecologie végétale, Université de Paris XI, Orsay 91405, France

Veen, B. W., Centre for Agrobiological Research, P.O. Box 14, 6700 AA Wageningen, The Netherlands

Vető, F., Biophysical Inst. Med. Univ., H-7643 Pécs, Szigeti u.12, Hungary

Vizárová, Gabriela, Institute of Experimental Biology and Ecology, Slovak Academy of Sciences, 885 34 Bratislava, Dúbravská cesta 14, Czechoslovakia

Wiedenroth, E. M., Humboldt-University Berlin, Section of Biology GDR – 1040 Berlin, Invalidenstr. 43

Zaitseva, Maria G., K. A. Timiriazev Institute of Plant Physiology, Academy of Science Moscow, Botanicheskaya 35, 127 276 Moscow, USSR

Zeleneva, I. V., Siberian Institute of Plant Physiology and Biochemistry, USSR Academy of Sciences, Irkutsk 33, P.O. Box 1243 USSR 664033
Zholkevich, V. N., K. A. Timiriazev Plant Physiology Institute, Academy of Sciences of the USSR, 127 276 Moscow, Botanicheskaya 35, USSR

I. GROWTH PROCESSES, STRUCTURAL CHARACTERISTICS

1. Cellular organization of root growth

VICTOR B. IVANOV

Institute of General and Inorganic Chemistry, Academy of Sciences of the USSR, Moscow, U.S.S.R.

Slowing down of cell proliferation in the basal part of the meristem may be caused either by lengthening of mitotic cycles or withdrawal of a fraction of the cells from the mitotic cycle. This alternative is hard to resolve by conventional methods, e.g., from the curves of labelled mitoses after a short-term incubation of the roots in ^{3}H-thymidine. In the root any cell is shifted farther from the tip due to the growth of the cells that are closer to the tip. Therefore the term 'cycle duration at a certain distance from the tip' is meaningless since by plotting the curves of labelled mitoses at various distances from the tip we compare different cells at different times, whereas the idea of this method is to watch the same cells over a long period. These difficulties made it necessary to use other means for the solution of this alternative [1, 2] (Fig. 1).

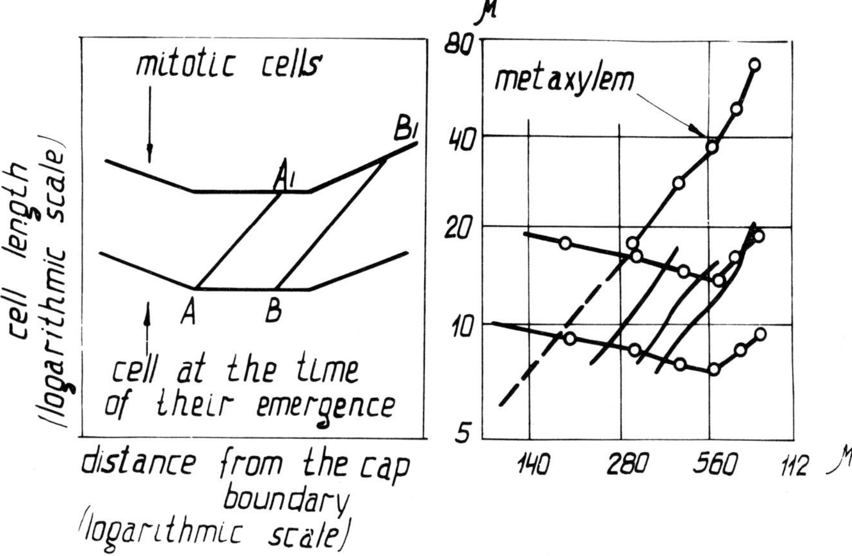

Fig. 1. Estimation of mitotic cycle duration by analysis of mitotic cell lengths. Metaxylem is a marker of cell growth without mitoses [1].

R. Brouwer et al. (eds.), Structure and Function of Plant Roots, 3–7.

From the changes in the lengths of dividing cells along the root we have established first of all at what distance from the root tip the cells under consideration had arisen as the result of the previous mitoses. Then we calculate within what time they enter the next mitosis, i.e., to determine cycle duration along the length of the meristem.

The measurements on roots of *Zea mays, Triticum aestivum, Allium cepa* have indicated that the cycle duration changes slightly along the meristem above the quiescent centre and a small number of adjacent cells. Retardation of cell proliferation in the second half of the meristem is due to the withdrawal of some cells from the mitotic cycle. In various tissues, despite the different sizes of their cells, cycle duration is almost the same.

The analysis of lengths of mitotic and interphase cells shows that in the meristem above the quiescent centre there are no cells which are quiescent and which do not divide for a longer time than duration of two mitotic cycles [3, 4].

The cessation of division and the beginning of elongation are controlled by different mechanisms, although in roots these two processes almost coincide under normal conditions. The moment when the elongation of various cells in the meristem begins does not change after irradiation, although mitoses stop completely. During the most part of the growth period after irradiation the growing parts of the roots consist of two fragments: the zone where the cell length changes little with the distance from the root tip, and the zone where it increases rapidly (Fig. 2).

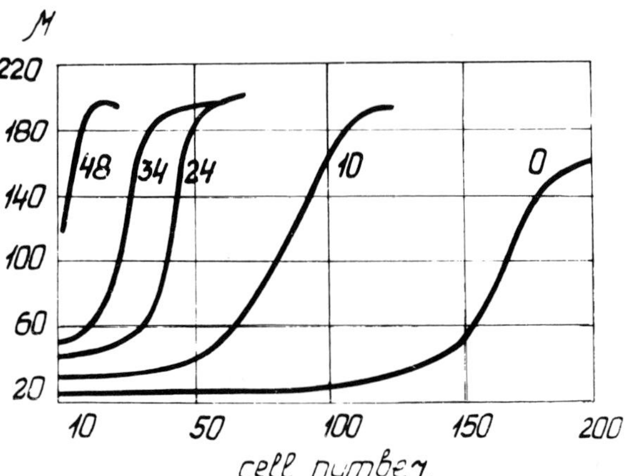

Fig. 2. Relationship between average cell length and number of cells in longitudinal rows of corn root cortical cells at 0, 10, 24, 34, 48 h after irradiation (100 kR) [4].

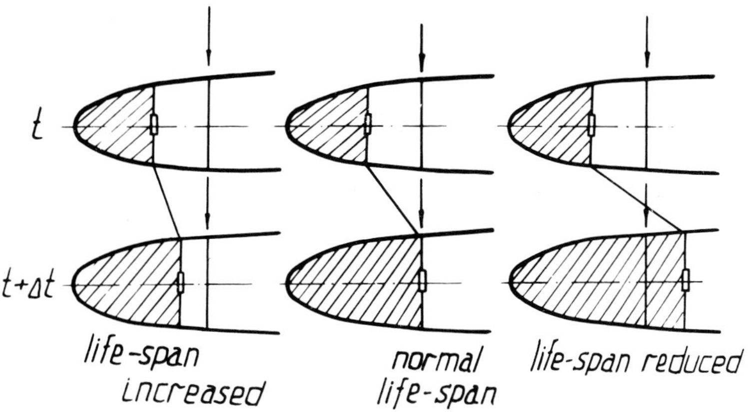

Fig. 3. Life-span of cells in meristem [4].

We have suggested the estimation of the life-span of cells in the meristem as a tool for analysis of mechanisms regulating transition of cells to elongation [4]. The change in the rate of transition to elongation may be accounted for by two reasons (Fig. 3). Firstly, the life-span of cells in the meristem may vary. Secondly, the number of cells that arose during this period may vary as a result of divisions of the cells beginning to elongate.

One approach to help evaluate the effects of a factor on the life-span of cells in

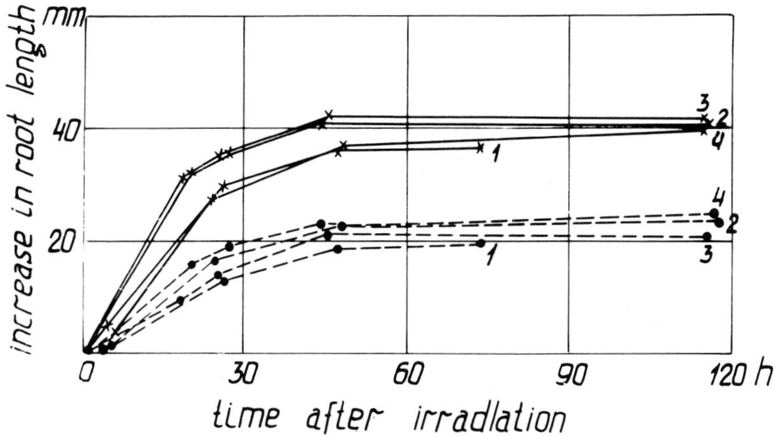

Fig. 4. Growth of irradiated (10 kR) corn seedling roots in distilled water (solid line) and in cycloheximide (1 μg/ml) (broken-line). Each curve represents mean values for 10 roots in independent experiments [4].

the meristem is the analysis of its action on irradiated roots, and the roots in which divisions are suppressed by inhibitors or on decapitated roots. Various inhibitors, while affecting the length of the mature cells and slowing down the growth of the irradiated roots, had no effect on the duration of growth of the latter, neither did they affect the time by which one and the same portion of the cells of the irradiated root had completed elongation (Fig. 4).

Determination of the life-span of cells in the meristem for the majority of cells and its independence on the rate of cell proliferation provides automatic deceleration of transition of cells to elongation, cell division in the meristem being inhibited. Therefore, the treatment of the root with various selective inhibitors of mitoses before the whole meristem is exhausted slows down the root growth by decreasing the rate of transition of cells to elongation. At first the rate of root growth does not decrease, but some time later it starts to slow down. This retarded inhibition of root growth can be used as a test for selection of inhibitors with the most selective antimitotic effect [5].

The decapitation experiments demonstrate that removal of the quiescent centre and of a certain portion of the cells above it may entail regeneration of a new quiescent centre and further elongation of the root axis. Such roots react to irradiation in the same way as non-decapitated roots, although the response of the cells of the quiescent centre and the main part of the meristem to irradiation is

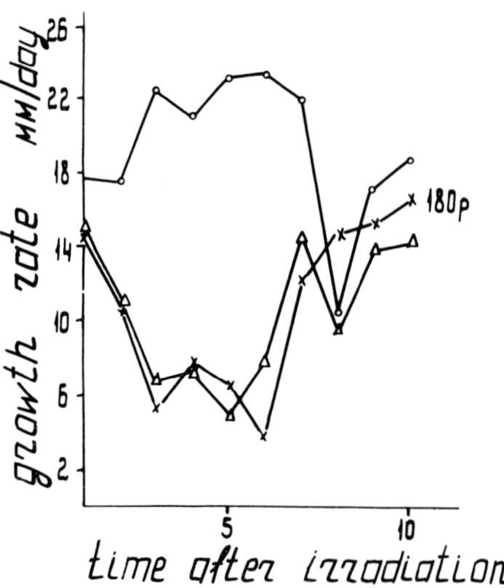

Fig. 5. Growth-curves of roots of *Vicia faba* seedling. ○ - control, X - irradiated (180 kR) roots, △ - irradiated (180 kR) and decapitated roots [5].

different (Fig. 5) [6]. These experiments show that in the root there is a mechanism determining the behaviour of the cell in relation to the state of more apical cells.

REFERENCES

1. Balodis, V. A. and Ivanov, V. B. 1970 Proliferation of root cells in the basal part of meristem and apical part of elongation zone. Tsitologia, 12, 983–992 (in Russian).
2. Ivanov, V. B. 1968 A new method for estimation of the mitotic cycle duration and the probability of cell entering the mitosis. Tsitologia 10, 770–773 (In Russian).
3. Ivanov, V. B. 1971 Critical size and transition to cell division. 1. Sequence of transition to mitosis of sister cells in the root tip of maize seedlings. Sov. J. Develop. Biol. 2, 524–535.
4. Ivanov, V. B. 1974 Cellular basis of plant growth. 'Nauka' Moscow.
5. Ivanov, V. B. and Larina, L. P. 1980 Effect of the quiescent center removal on the growth of irradiated roots. Proc. Acad. Sci. U.S.S.R. 252, 506–508 (in Russian).

2. Growth region of the primary root of maize (*Zea mays* L.)

MÁRIA LUXOVÁ

Slovak Academy of Sciences, Institute of Experimental Biology and Ecology, 385 34 Bratislava, Czechoslovakia

The root of maize has been used by many research workers as an object for studies of the structure and morphogenesis, of physiology and biochemistry of its functions as well as of the problems of uptake and transport. This cooperation directed at a single object causes a faster advance and, at the same time, provides a source of mutual confrontation and supplementation of knowledge. This paper presents the results of the growth analyses of the maize root (hybrid CE 380), summarized [3, 4] and completed with the aim of characterizing the growth region of the root growing under defined conditions. The seeds were allowed to germinate in darkness at 25 °C on sheets of moist filter paper in a vertical position. Primary roots that had grown 60 ± 5 mm within 72 h were used for analysis.

The following growth parameters hold for the analysed roots (Figs. 1, 2): in the longitudinal section the root cap measures 0.4–0.5 mm; the extent of the growth region is approximately 7 mm; the meristematic region represents the distal quarter of it. The unequal elongation along the growth region is characterized by low values in the region of the apical meristem, with a basipetally continuing rise of the elongation growth rate up to the attainment of the maximum followed by slowing down until complete cessation. At an average increment of 1.44 mm h^{-1} in the first six hours of the third day of the experiment, the meristematic region participates only to the extent of 5–6 per cent in root elongation. The elongation of the third and fourth millimetre section from the root cap junction is most intensive.

Cell division is asynchronous and does not stop at one level for the different types of tissues in the meristematic region. The prolonged mitotic activity of the epidermis is characteristic of the apical meristem of the maize root. Cell division ceases in the cortex centripetally, and in the central cylinder centrifugally, so that the inner cortical layers and the peripheral region of the central cylinder take the longest time to divide. The last mitoses occur here at a distance of approximately 1500 μm, the cells dividing transversely 5–6 times. The mitotic activity of the central metaxylem is the shortest, its cells stop dividing at a distance of approximately 400 μm after having passed 3–4 cell cycles. They grow further endo-

R. Brouwer et al. (eds.), Structure and Function of Plant Roots, 9–14.

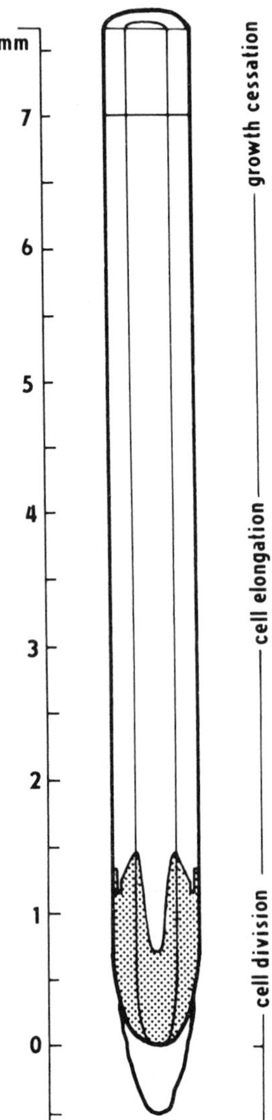

Fig. 1. Diagram of longitudinal section of a root tip of maize illustrating the extent of the growth region. Apical meristem is indicated by dotting.

Fig. 2. Longitudinal section of the maize root apex.

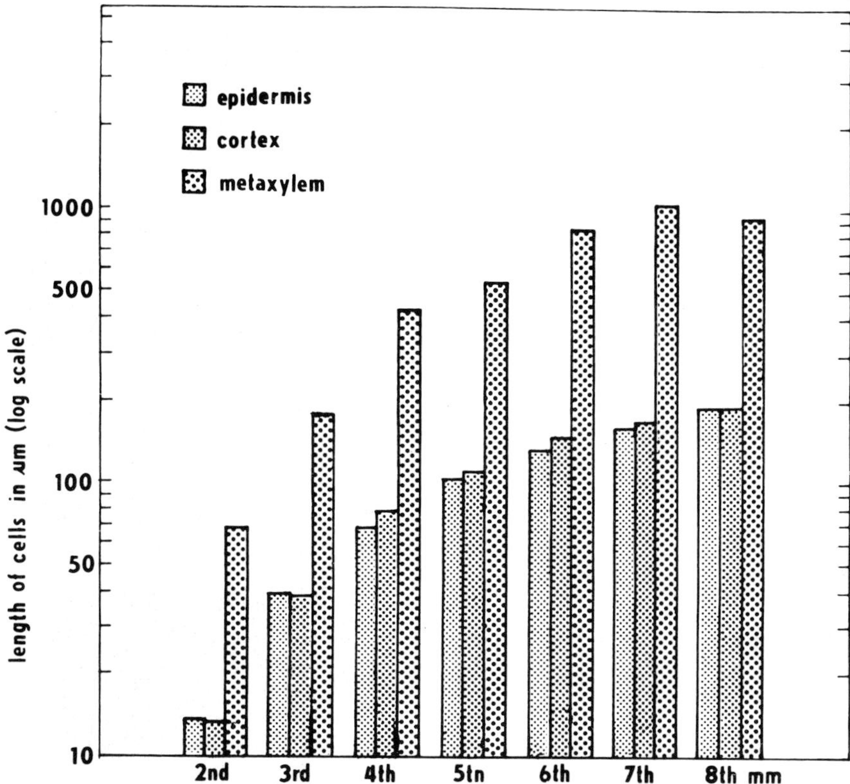

Fig. 3. Average length of cells of the epidermis, the cortex (fifth layer from the periphery) and of the vessel members of the central metaxylem in successive 1 mm long sections, expressed by use of logarithmical scale.

mitotically and become highly polyploid [2, 5]. Within the growth region of the root an equilibrium must be maintained between cell division and cell elongation: the more often the cells divide transversely, the shorter they are. The increase of the average length of the epidermal and cortical cells and of vessel members of the central metaxylem along the growth region is presented in Fig. 3. The values have been obtained by measuring all cells in the longitudinal cell columns along the growth region of six roots, beginning 1 mm from the root cap junction. Successive 1 mm long sections of longitudinal cell columns are used in the calculation. Epidermal and cortical cells proliferating longer elongate only 40 to 60 times, attaining an average length of approximately 190 µm, whereas members of the central metaxylem with short proliferating times elongate up to 150 times and attain an average length of approximately 1000 µm.

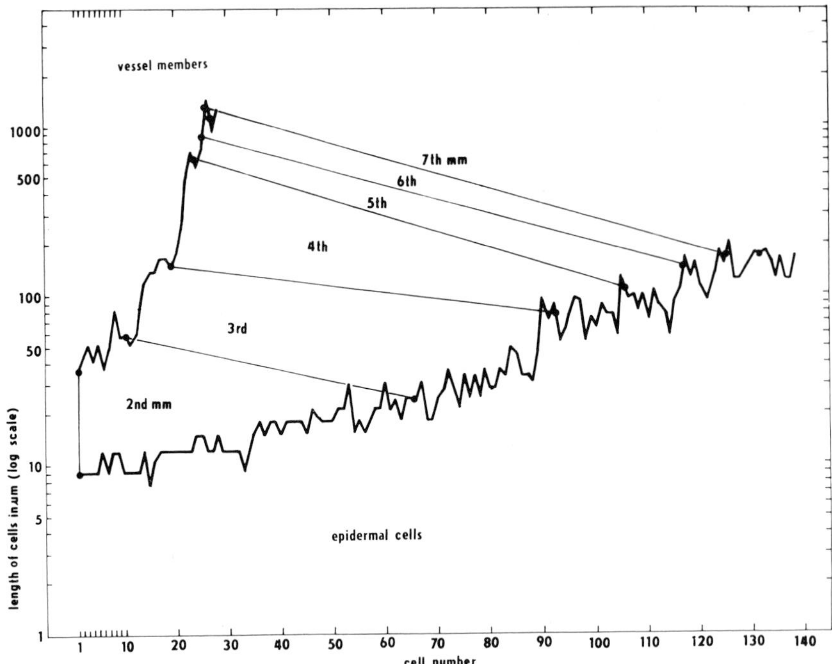

Fig. 4. Fluctuating length of successive cells of the epidermis and of the central metaxylem in the longitudinal cell column. The number of cells in successive 1 mm long sections is indicated.

Graphical demonstration of length of successive cells in longitudinal cell columns demonstrates the fluctuating and even oscillating character of the curves for each type of tissue (Fig. 4). Cytological analyses have demonstrated that the differences in length of cells within the specific tissue type can be caused by:

- their position in the packet of cells originating during successive cell cycles, the end cells in the packet are usually narrower and longer;
- differences in proliferating capacity of some cells;
- unequal cell division;
- change in the polarity of cell division, in which the transverse cell division is substituted by a longitudinal one;
- differentiated growth of the individual cell with a difference in the elongation rate of its proximal and distal part, caused by the growth gradient along the growth region [3].

The changes which occur in the course of ontogenesis are not considered since

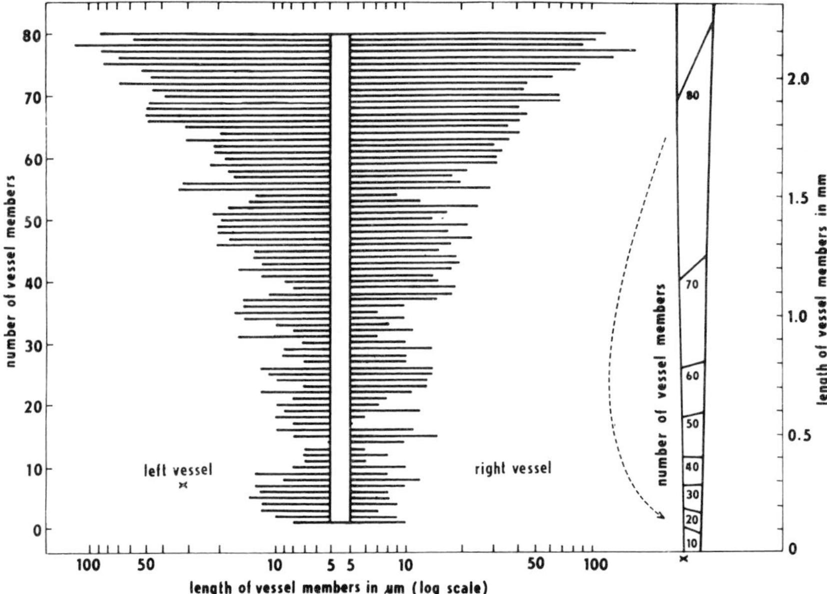

Fig. 5. Lengths of successive vessel members of two opposite vessels which point to the existence of spiral nutations of the root. For analysis two vessels of the central metaxylem documented in Fig. 2 were used.

the analysed roots are in the initial stage of development. As subsequent analyses have shown, however, the cell length is connected also with the existence of spiral nutations of the root. These were observed with roots of maize by use of time-lapse film technique [1]. As a result of the spiral nutations the cells of the same type and of the same sequence differ in length (Fig. 5).

The final length of the root cell is the result of a series of processes representing the transition of the growing cell from its origin up to the completion of its elongation. The growth processes within the limits of a single cell are subordinated to the regulatory mechanism(s) controlling the growth of the root as a whole.

REFERENCES

1. Erickson, R. O. and Sax, K. B. 1956 Elemental growth rate of the primary root of Zea mays. Proc. Amer. Phil. Soc. 100, 487–498.
2. List, A., Jr. (1963) Some observations on DNA content and cell and nuclear volume growth in the developing xylem cells of certain higher plants. Amer. J. Bot. 50, 320–329.

3. Luxová, M. 1979 The cytological analysis of causes affecting the root cell length. Biol. Plant 21, 355–360.
4. Luxová, M. 1980 Kinetics of maize root growth (in Ukr.). Ukr. bot. j. 37, 68–72.
5. Swift, H. 1950 The constancy of deoxyribose nucleic acid in plant nuclei. Proc. Nat. Acad. Sci. U.S.A. 36, 643–654.

3. Regulation of the mitotic cycle in seedling roots of *Vicia faba* by environmental factors

AUGUSTÍN MURÍN

Laboratory of Karyology, Institute of Molecular and Subcellular Biology, Comenius University, Bratislava, Czechoslovakia

The growth of a plant body is a result of reproduction, enlargement and differentiation of cells. Although cell enlargement and differentiation are partly independent of cell division, nevertheless, they soon stop if there is no supply of new cells by the mitotic cycles. Thus, if we want to regulate the growth of plants, we should know how to regulate the mitotic cycle.

Many factors regulate the mitotic cycle and for the sake of simplicity we can include them into two groups. In the first group there are the endogeneous factors that regulate the mitotic cycle in a range characteristic of individual species, varieties, organs, tissues and cell generations. For instance, among flowering plants the shortest mitotic cycle time has been found to be 6.3 h in the root tips of *Helianthus annuus* [1], while the longest one is about 120 h in the root tips of *Trilium grandiflorum* [3]. Other endogeneous factors regulate the mitotic cycle during ontogenesis and at different stages of development the duration of mitotic cycles varies to a great extent. This variation can affect different organs and tissues as well as cell generations in the same tissue. For instance, the mitotic cycle in initial cells of the root tip of *Vicia faba* can last about 120 h, whereas their derivative cells in the 4th, 5th and later generations can have a cycle time of about 12 h [8].

The second group of factors that regulate the mitotic cycle significantly are the exogeneous ones. Some experimental data concerning the primary effects of the most essential factors of the environment, i.e. temperature, water and oxygen are presented. These changes evidently influence the rate of cell reproduction, enlargement and differentiation, and the range of their regulatory effects as important.

Seedlings of *Vicia faba* at the 5th day of germination and with primary roots of 6–8 cm in length were used. At this stage of development the seedlings are supplied with nutrients by the cotyledons, so the experimental conditions are simple and the results conclusive.

The results have shown that among the three environmental factors tested, temperature has the strongest effect on the mitotic cyle. Changing the temperature from 25 °C to 3 °C, for instance, slows down the mitotic cycle from 12 h up

R. Brouwer et al. (eds.), Structure and Function of Plant Roots, 15–18.

Table 1. Temperature and mitotic cycle

°C		h: min
35	cca	24:00
30		12:29
25		11:56
20		22:21
13		34:02
8		90:19
3		264:00

to 264 h ([5], Table 1). In this range of temperature both the duration of mitosis and interphase appear to change proportionally [6]. The same may be said about proportional changes of G_1, S and G_2 phase of the interphase [2]. Along with the changes in the rate of cell reproduction proportional changes appear also in root growth.

Another environmental factor that regulates the mitotic cycle is water. Changes in the amount of water in cells can change their rate of reproduction and elongation greatly. One method of changing the water content in meristematic and elongating cells is to adjust the osmotic potential of the culture medium. In this connection a question arises as to the values of osmotic potential at which the cells can get the minimum of water for growth and reproduction and what is the duration of one mitotic cycle.

The experiments with polyethylene glycol 4000 as an osmoticum have shown that the optimum osmotic potential at which reproduction and elongation proceed with the highest rate is within the range 0–0.1 MPa. Above this level the increase in osmotic potential causes a decrease in the rate of both processes ([7], Table 2). The decrease in the rate of these processes is not proportional and cell elongation is more sensitive than cell reproduction. At the highest tested osmotic

Table 2. Osmotic potential, water availability and mitotic cycle

MPa	h: min
0	12:23
−0.39	12:57
−0.79	14:50
−1.27	23:03

potential of —1.27 MPa, for instance, the rate of root growth decreased 6 fold, while the mitotic cycle time increased less than twice. This finding suggests that there is a possibility of regulating cell elongation by increased osmotic potential of the medium. It is also interesting that loss of water by the meristematic cells by plasmolysis at an osmotic potential of —1.27 MPa stops the mitotic cycle only for a period of about 6 h and then the cells are able to recover and renew the cycle, although, at a slower rate. We can conclude therefore that changes in the water content of meristematic cells regulate the mitotic cycle much less than temperature and within the range of maximum and minimum water content the cycle time is prolonged about twice.

The third vital factor of the environment is oxygen. Although the physiological and biochemical function of oxygen is well known, nevertheless, its primary effect on cell reproduction and elongation is not clear. According to López-Sáez et al. [4] the mitotic cycle in the root tips of *Allium cepa* appears to be strongly dependent on the oxygen content of the medium. If the oxygen content of the air supply used for aeration decreased from 20% to 2% then the mitotic cycle time increased about four times.

Media with different oxygen content were tested and the results show (Table 3) that in the roots grown in wet sawdust or in aerated distilled water the mitotic cycle time is similar and indicates the optimal rate. If the distilled water was not aerated the oxygen content at the beginning of treatment was about 10 mg mg/l, and the mitotic cycle was prolonged up to 115 per cent compared with the control. Greater prolongation of the mitotic cycle to 126% resulted when the roots were grown in boiled distilled water where the oxygen content was 2 mg/l of water. As we could not decrease substantially the oxygen content in water by a longer period of boiling, we could not determine the minimum at which the cycle time is the longest. These results suggest that even when the changes of oxygen content in a medium are large, the changes in the current mitotic cycle are only small. It agrees with an older observation of Nabokich [see 9] who found that seedlings rich in carbohydrates can grow for 24–48 h even without oxygen.

On the basis of results given above we can conclude that the essential environ-

Table 3. Oxygen and mitotic cycle

Medium:	h:min
Wet sawdust	12:13
Aerated dist. water	12:23
Dist. water (10 mg O_2/l)	14:13
Boiled dist. water (2 mg O_2/l)	15:34

mental factors regulate the mitotic cycle within different ranges and the widest range is attributed to the temperature.

REFERENCES

1. Burholt, D. R. and Van 't Hof, J. 1971 Quantitative thermal-induced changes in growth and cell population kinetics of *Helianthus* roots. Amer. J. Bot. 58, 386–393.
2. González-Fernández, A., Gimenéz-Martín, G., Diez, J. L., de la Torre, C. and López, J. F. 1971 Interphase development and beginning of mitosis in the different nuclei of polynucleate homocaryotic cells. Chromosoma (Berl.) 36, 100–111.
3. Grant, C. J. 1965 Chromosome aberrations and the mitotic cycle in Trilium root tips after X-radiation. Mutation Res. 2, 247–262.
4. López-Sáez, J. F., González-Bernáldez, F., González-Fernández, A. and Garcia Ferrero, G. 1969 Effect of temperature and oxygen tension on root growth, cell cycle and cell elongation. Protoplasma 67, 213–221.
5. Murín, A. 1966 The effect of temperature on the mitotic cycle and its time parameters in root tips of Vicia faba. Naturwissenschaften 53, 312–313.
6. Murín, A. 1967 Einfluss der Temperatur auf die Mitose und den Mitosezyklus. Acta Fac. Natur. Univ. Comen, Botanica 14, 83–120.
7. Murín, A. 1979 Effects of high osmotic potential of a medium on mitotic cycle in roots of Vicia faba L. Biol. Plant 21, 345–350.
8. Murín, A. and Luxová, M. 1978 The sequence and duration of mitotic cycles in the apical root meristem of Vicia faba L. Biol. Plant 20, 293–298.
9. Němec, B. and Pastýrik, L. 1956 Všeobecná botanika. Bratislava, 858 pp.

4. Maturing seed root growth and metabolism

LUDMILA I. MUSATENKO

N. G. Kholodny Institute of Botany, Ukrainian Academy of Sciences, Kiev 252601, U.S.S.R.

Embryo organs from the maturing seeds of a dwarf French bean (*Phaseolus vulgaris* L.) were used [1]. Detailed study of growth of the hypocotyl-embryo root axis using anatomical preparations has shown that the final size of the root cap (about 250 μ) is achieved and its growth is completed by the 20th day, that of the root by the 34th and of the hypocotyl by the 40th day after anthesis. It should be noted that while the size of the embryo root increases (3-fold) during early maturation and stabilizes by the 30th day (1.75 mm), the size of the hypocotyl gradually increases during the whole maturation period.

Cellular analysis of the embryo axis growth has shown that the embryo root and hypocotyl grow due to cell division, whereas no elongation is observed. During the whole period of the growth of the embryo axis the maximum size of parenchymatous cells does not exceed 19–20 μ. After cell division, daughter cells in the distal ends of the embryo axis do not achieve the size of the mother cells which results in the smaller cell size in these regions. The cessation of embryo axis cell division coincides in time (until the 40th day) with the cessation of growth of its organs.

Active metabolic processes are characteristic of the intensive embryo root growth during seed formation and they result in more than 10-fold increase of the fresh and dry mass of the root embryo. During the same time the number of cells in the embryo axis increases about 5-fold.

The free amino acid content of the root meristem depends upon the growth processes in it: the content increases during intensive growth and decreases during seed dehydration, which is probably caused by the synthesis of reserve proteins and the transition to dormancy [2].

Protein and nucleic acid content also increases, finally stabilizing during the last stages of maturation.

The examination of native RNAs has shown that the proportion of high polymeric RNAs to low polymeric ones in the root increases as seed maturation proceeds.

The distribution of RNP particles from the extracts of embryo roots of seeds of different maturity on a sucrose gradient made it possible to detect, through

R. Brouwer et al. (eds.), Structure and Function of Plant Roots, 19–21.

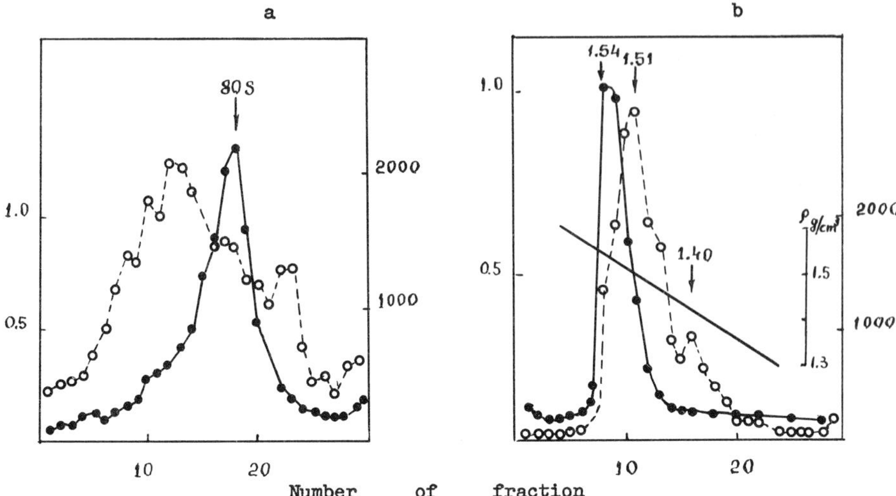

Fig. 1. Density gradient distribution in sucrose (15–60 per cent, a) and CsCl (b) of cytoplasmic RNPs from embryo root of maturing seed (35 days after anthesis). Left and filled circles – absorbance at 260 nm; right and open circles – radioactivity.

measuring optical density, a distinct monoribosomal peak (Fig. 1a). ^{3}H-uridine-labeled newly synthesized RNA is primarily localized in the pre-ribosomal zone corresponding to the polysomes throughout seed maturation and on the 30th day after anthesis a component is present in the postribosomal zone which probably represents free informosomes.

UV absorption of CsCl gradients shows mainly ribosomes with buoyant density 1.53–1.54 g/cm^3. Radioactivity content predominates in the component with buoyant density 1.50–1.52 g/cm^3 which is typical for polysomes. The component with buoyant density 1.40–1.43 g/cm^3 which is typical for free informosomes appears 35–37 days after anthesis (Fig. 1b).

Sedimentation and density analysis show that active synthesis of different RNP particles is maintained in the root almost until the final stages of seed development.

Thus, active cell division and intensive metabolism before the final stages of maturation determine the formation of a complex system providing the specificity of growth and metabolic transformations during early stages of seed germination.

REFERENCES

1. Martyn, G. I., Musatenko, L. I., Nesterova, A. N. and Berestetsky, V. A. 1980 Morphophysiological characteristic of bean seed embryo at the embryonal period of development. Ukr. Bot. Zh. (U.S.S.R.) 37, 11–13.
2. Sytnik, K. M., Bogdanova, T. L. and Musatenko, L. I. 1980 Amino acids of ornithine cycle during ripening and germinating of seeds. Ukr. Bot. Zh. (U.S.S.R.) 37, 7–10.

5. Development of the mature root growth pattern in the course of seed germination

NATALIE V. OBROUCHEVA

K. A. Timiriazev Institute of Plant Physiology of the USSR Academy of Sciences, Moscow 127106, U.S.S.R.

Seedling development in the course of seed germination begins with growth of the radicle pole of the embryo resulting in radicle emergence on the seed surface and on its immediate contact with soil moisture. The ensuing development of the seedling depends on the rate of primary root differentiation and achievement of maximal growth rate. Only after the root has provided the young seedling with water does the growth and development of above-ground organs begin and seedlings emerge.

The structure of the radicle in the mature seeds varies from a group of initial cells to a promeristem containing files of epidermal, cortical and stelar cells. Very seldom, for example, in seeds of *Phaseolus vulgaris,* the radicle consists not only of meristem but of a short primordial elongation zone as well.

In dicotyledonous plants the radicle is small while the hypocotyl occupies the major part of embryonic axis. It is very often difficult to distinguish a morphological border line between embryonic root and hypocotyl and evaluate correctly the root length. To overcome this difficulty various anatomical features have to be taken into account. The most reliable identification of boundaries between hypocotyl, radicle and epicotyl was achieved by observing the sequence of successive segmentation divisions during embryogenesis [10].

Emergence of the radicle and the ensuing root growth are considered here in terms of cell division and elongation. This trend of growth analysis, so-called cellular analysis, was initiated by Burström [2]. He suggested the measurement of the rate of cell elongation by following cells entering the zone of elongation and leaving it, and by converting the cell number longitudinally into time units.

Cellular analysis of growth involves the estimation of rates of cell division and cell elongation and their contribution to growth rate [7, 12]. The sequential measurements of cell length in a longitudinal cortical file from the initial cell up to the upper end of the root has permitted us to establish the number of meristematic, elongating and mature cells. The calculation of cell balances step by step in the course of radicle growth is the only way to estimate the cell-doubling time [8] and the duration of cell elongation [11] during germination.

R. Brouwer et al. (eds.), Structure and Function of Plant Roots, 23–27.

RADICLE EMERGENCE

In the seeds of *Quercus robur, Aesculus hippocastanum* and *Pisum sativum* the radicle emergence is the result of hypocotyl elongation. In imbibed pea seeds the hypocotyl is the first to elongate, and radicle growth begins 9 h later when it has already protruded. In seeds of *P. vulgaris* the initiation of elongation in the hypocotyl and radicle coincide probably due to the primordial zone of elongation in these radicles. The growth of the hypocotyl in dicotyledonous seeds provides root protrusion. In seeds of *Zea mays* coleorhiza is the first to elongate, the radicle protruding later through it. In all seeds investigated root growth begins with the basal radicle cells gradually increasing in length and cell elongation is the process initiating and providing root emergence. (Fig. 1, 2)

INITIATION OF CELL DIVISIONS

The first mitotic figures appear in pea roots 3 h later than the onset of root elongation. In *A. hippocastanum* [9], *Vicia faba* [13] and some cereals [4, 6] the division of cells also commences in rather long roots, after the onset of elongation.

In seeds of *Lactuca sativa* [3] and *P. vulgaris* [14] the start of cell proliferation accompanies the start of cell elongation and no mitotic activity was observed prior to the initiation of elongation [1]. Meristematic cells of radicles are predominantly at the G_1-phase and must double their DNA and pass the G_2-phase

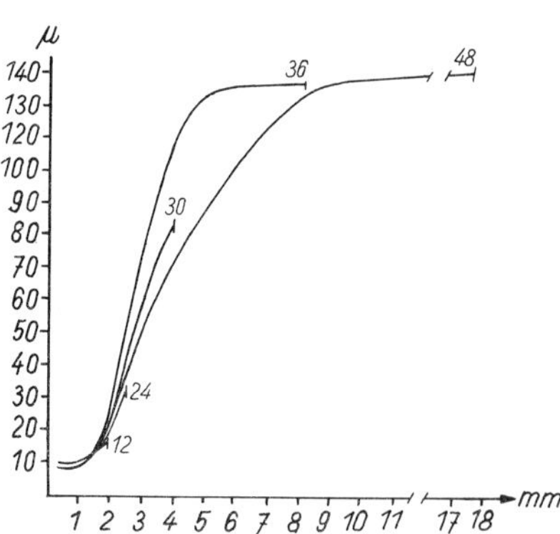

Fig. 1. Cortical cell length (μ) as a function of the distance from root tip (mm) after 12, 24, 30, 36 and 48 h imbibition of *Zea mays* seeds. The vertical lines denote the root-mesocotyl boundary.

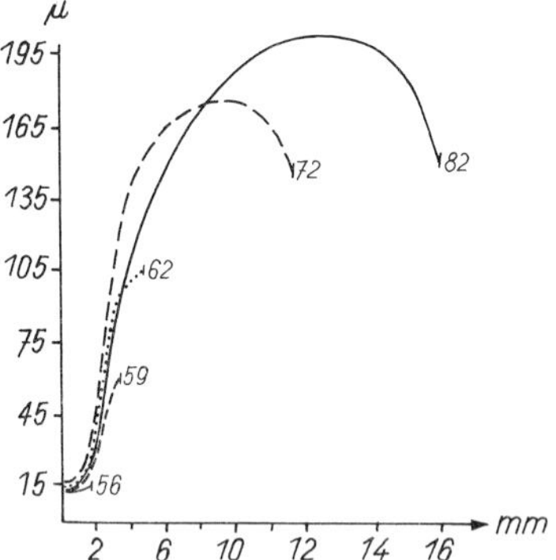

Fig. 2. Cortical cell length (μ) as a function of the distance from root tip (mm) after 56, 62, 72 and 82 h-long imbibition of *Pisum sativum* seeds. The vertical lines denote the root-hypocotyl boundary.

before the first mitoses appear. This sequence of events preceding the mitotic activity explains the often observed delay in cell proliferation as compared to the onset of elongation.

INDEPENDENT INITIATION OF CELL ELONGATION AND CELL DIVISION

Independent beginning of elongation and proliferation was demonstrated on the so-called γ-plantlets [5] emerging from seeds irradiated with high doses of γ-rays. γ-plantlets grow for some time without any mitosis, only as the result of cell elongation. In non-irradiated radicles the independent control of the onset of these two growth processes is also evident, because 1) in radicles of some plants mitotic activity develops in already elongating roots; 2) proliferation initiates in the apical half of the radicle while cell elongation begins from its basal end. The basal cells commence to elongate without preceding mitotic divisions and independent regulation of cell elongation and proliferation initiations is a plausible assumption. During embryogenesis the radicle grows mainly by cell multiplication and its initiation in the imbibed seeds can be considered as a resumption of the mitotic activity while the initiations of elongation represents an innovation in the imbibed radicles requiring an independent control mechanism.

The principle growth process during seed germination is cell elongation and it provides rapid root penetration into the soil.

DEVELOPMENT OF THE ZONAL PATTERN OF ROOT GROWTH

Zone of elongation

The elongation of the radicle begins from its basal end. The basal cells often are slightly longer than meristematic cells and do not divide at all during germination. These cells are ready programmed to elongate during the maturation of the seed. They are ready to begin elongation under favourable water and temperature conditions.

The basal cells gradually increase in length (Fig. 1). It takes 36 h for the most basal cells of *Z. mays* to finish their elongation and become $130 \pm 15\,\mu$ long mature cells thus completing the development of the zone. Later, more cells enter the zone and its length increases up to 7 ± 0.2 mm but the length of the mature cells is almost the same. The pattern of the formation of the zone is similar to one described for *A. hippocastanum* [9].

Another pattern was observed with the radicles of *Q. robur* [8] and *P. sativum* (Fig. 2). The most basal radicle cells elongate and the first mature cells remain

34 μ long in *Q. robur* and 150 μ long in *P. sativum* while the cells later initiating the elongation grow up to 120 μ and 200 μ correspondingly.

Meristem

In contrast to basal cells which are destined to elongate the apical cells of the radicles are destined to proliferate. They represent about 35 per cent of the total number of radicle cells in *Q. robur*, about 50 per cent in *A. hippocastanum* and 75 per cent in *P. sativum* and *Z. mays*. Such radicle meristems are not long enough and the role of the first divisions appears to be to increase the size of the meristem, the number of meristematic cells and proliferative pool. The mitotic activity of the developed meristem results then in the transition of cells from the meristem to the zone of elongation. Meristem formation is clearly demonstrated by observations on *Q. robur* radicles, the size of meristem increases from 0.4 to 1 mm, and number of cells in a cortical file from 36 to 100 [8]. This process is less evident in *A. hippocastanum* and *Z. mays* radicles, while in pea radicles the meristem does not increase in length and meristematic cells at once replenish the pool of cells capable of elongating. This is the reason why the acceleration of pea radicle growth is accomplished earlier than that of other radicles investigated.

The development of a zonal pattern of growth needs about 16 h after onset of root growth in pea, 24 h in maize and 96 h in oak. Thus the meristem and the zone of elongation appear with almost an invariable number of cells.

RATES OF CELL ELONGATION AND CELL DIVISION

Zonal pattern development coincides with an acceleration of elongation and proliferation in the radicles. As to elongation, the cells grow faster and to greater length. In *Z. mays* seedling roots the cells elongate for 8–13 h instead of 36 h, and cell-doubling time is 10 h instead of 30–40 h at the very beginning. In oak roots also elongation and proliferation occur in 20–40 mm roots two times faster then in the emerged radicles [8]. The acceleration of both processes appears to correlate with the development of hormonal systems thus providing the high rate of root growth specific for the seedlings.

Only a zonal pattern of growth organization can provide rapid root growth via the necessary coordination between the meristem rapidly producing new cells and elongation zone channelling them to rapid size increase. As the result of the zonal structure the high rate of root growth can be maintained for a long time. On the contrary, the organs lacking zonal organization of growth [7] can neither provide rapid growth nor maintain it.

REFERENCES

1. Bewley, J. D. and Black, M. 1978 Physiology and biochemistry of seeds in relation to germination, Vol. 1. Berlin: Springer.
2. Burström, H. 1941 On formative effects of carbohydrates on root growth. Bot. Notiser 3, 310–334.
3. Evenary, M., Klein, S., Anchori, H. and Feinbrun, N. 1957 The beginning of cell division and cell elongation in germinating lettuce seeds. Bull. Res. Council Israel 6D, 33–37.
4. Grif, V. G. 1956 On a possibility of mitotic activity in plant cells under negative temperatures. Dokl. AN S.S.S.R. (Rep. Acad. Sci. U.S.S.R.) 108, 734–737.
5. Haber, A. H. 1968 Ionizing radiations as research tools. Ann rev. plant. physiol. 19, 463–489.
6. Ikenberry, G. J. 1966 Reactivation of the dormant embryo and the initiation and development of buds and leaves in Avena sativa. Iowa State Coll. J. Sci. 41, 7–23.
7. Ivanov, V. B. 1974 Cellular approach to plant growth. Moscow: Nauka.
8. Kovalev, A. G. and Obroucheva, N. V. 1976 Characterization of the initial part of the sigmoid curve of the root growth. Fiziol rast (Sov. plant physiol.) 23, 340–345.
9. Kovalev, A. G. and Obroucheva, N. V. 1977 Cellular analysis of the S-curve of the root growth. Peculiarities of cell division and elongation in the Aesculus hippocastanum roots. Ontogenes (Sov. develop. biol.) 8, 397–405.
10. Kuras, M. 1978 Activation of embryo during rape seed germination. Structure of embryo and organization of root apical meristem. Acta Soc. Bot. Polon. 47, 65–82.
11. Obroucheva, N. V. 1965 Physiological characterization of growing root cells. Moscow: Nauka.
12. Obroucheva, N. V. and Kovalev, A. G. 1979 On physiological interpretation of sigmoid growth curves of plant organs. Fiziol rast (Sov. plant physiol.) 26, 1029–1042.
13. Rogan, P. G. and Simon, E. W. 1975 Root growth and the onset of mitosis in germinating *Vicia faba*. New phytol. 74, 273–275.
14. Sytnik, K. M., Martin, G. I. and Musatenko, L. I. 1977 Cytological analysis of embryo root growth. Dokl. AN U.S.S.R. (Rep. Acad. Sci. Ukr. S.S.R.) 9B, 851–853.

6. Structural and functional aspects of roots of germinating seeds

LUDMILA I. MUSATENKO, OTÍLIA GAŠPARÍKOVÁ,* GENNADY I. MARTYN and KONSTANTIN M. SYTNIK

N. G. Kholodny Institute of Botany, Ukrainian Academy of Sciences, Kiev 252601, U.S.S.R.

Some results of the study of embryo root structure and metabolism during seed germination are presented in this paper. They may help to elucidate the mechanisms responsible for both dormant seeds transition to active functioning and their growth and differentiation.

A dwarf French bean (*Phaseolus vulgaris* L.) was used. The embryo axis is relatively well differentiated, its length being about 5 mm. Primary leaves, hypocotyl, epicotyl, and embryo root with a root cap are readily discernible. Germination time may be divided into three periods [1]. The first ons is about 5–6 h and is associated with intensive water uptake and increase in weight and size of the embryo axis. Examination of the intact and excised embryo axis give evidence that the duration of this period is determined by the ability of the testa to allow penetration of water inside the seeds. It is evident that uptake of water by the embryo axis occurs primarily due to physical laws, as analogous results were obtained in experiments with non-viable seeds impaired by fixation solution or heat (120 °C for 1 h). The second period lasts for about 13 h. There is almost no change in either weight or size of the embryo. The third period, that of increase in size and weight of the embryo axis, rupture of testa, root emergence, is completed by seedling formation. It should be noted that hypocotyl is the first to grow during germination; it constitutes most of the embryo axis (up to 4 mm) and also provides root emergence. The hypocotyl growth occurs mainly due to cell elongation, cell division in it is found exclusively in conducting bundles. A certain sequence in cell elongation in the hypocotyl was also established: an increase in size of the cells is at first detected just near the root – hypocotyl borderline, then – further up the hypocotyl. The growth of the hypocotyl ceases after the cells have achieved their final size and its length is determined by the extent of cell elongation whereas the number of cells is not changed. Until 24 h of germination no quantitative change occurs in cells of the embryo root which is about 1.5 mm long and consists of two zones – meristem (0.5 mm) and elongation (1.0 mm).

* Institute of Experimental Biology and Ecology of CBEV SAV, Bratislava, Czechoslovakia

R. Brouwer et al. (eds.), Structure and Function of Plant Roots, 29–33.

During the first 24 h of germination practically no change occurred in either the number of the cells in a row or in their length in the embryo root, therefore zone and root sizes remain unchanged [3]. Growth of the embryo root is initiated, as a rule, in seeds with emerged root by almost simultaneous initiation of cell division and cell elongation 24 h after soaking, and already 8 h later the length of the meristem and elongation zones doubles. After 36 h of germination mature cells appear. After 44 h of germination the length of the meristematic zone equals 1.4 mm, thenceforth remaining virtually unchanged, and equilibrium is achieved between the number of cells formed by cell division in the meristem and those which have begun to elongate. By 72 h the length of the elongation zone stabilizes and is about 5 mm. At this time exponential growth is completed and linear growth of the root begins, that is the number of cells formed in the meristem, those which began to elongate, and those which ceased to grow are, for some time, roughly proportional. The stationary growth occurs without any change both in meristem size and elongation zones. We presumed that the size changes observed in the meristematic zone between 24 and 32 as well as 40 and 44 h of germination may be due to partial, natural synchronization of cell division occurring at this period.

Twenty-four hours after soaking, intensive root growth begins and results in 50-fold increase of its original length by the end of the third day and the number of cells is more than tripled. Intensive synthetic processes that accompany and cause such an active root growth lead to considerable accumulation of fresh and dry matter, mainly due to mature and elongation cells. Thus, for example, the dry weight of meristematic cells is 2–3 times less than cells of the elongation zone.

A certain qualitative and quantitative specificity of free amino acid accumulation in root growth zones during early germination has also been established [2]. The conversions of amino acids of the ornithine cycle are of major importance for the early stages of development. They accumulate in seeds in great amounts and then, during germination, break down to ammonia, thus creating a pool of reduced nitrogen which is consumed by the seedling for new syntheses. It can also be suggested that the functioning of the ornithine cycle during seed germination is associated with the formation of proline (via ornithine) which is oxidized to hydroxyproline, an obligatory protein component of cell walls associated with formation and growth of new cells.

During early germination, in the root growth zones considerable amounts of arginine (60–80% of total amino acids), serine, citrulline and glutamic acid accumulate. Towards the end of the formation of growth zones free amino acid content, especially arginine, rapidly declines, while specific weight of glutamic acid increases markedly. Glutamic acid, being of high reactive power, readily interacts with arginine and other amino acids and thus converts into glutamine – the major form of reduced nitrogen – on the one hand, or – it may be involved in proline formation: via glutamine semialdehyde and pyrroline-5-carboxylic acid. Thus, the amino acids of the ornithine cycle, and glutamic acid though having the same functions, are differently used in the plant at different stages: during early germination the ornithine cycle is active and later on, glutamic acid appears to be the main donor of reduced nitrogen.

The fact that reduced nitrogen and free amino acids are necessary for the biosynthetic processes in growing cells is proved by intensive protein synthesis which results in the 3–4-fold increase of protein content in the hypocotyl, root and its growth zones within 72 h of germination.

Biosynthetic processes in growing cells are also closely related to synthesis and accumulation of nucleic acids, particularly of RNA. The amount of RNA accumulated in the root during seed maturation remains unchanged in the course of the first 24 h of germination, but increases in the hypocotyl after imbibition. Between the first and the third day of germination RNA content increases both in the hypocotyl and in the root. Meanwhile, the growth zones in the embryo root are characterized by stable, rather high RNA and DNA content which is obviously due to the reproductive function of the meristem and active elongation during the development and formation of the seedling.

The synthesis of RNA was also examined in embryos of dry seeds 6 h after

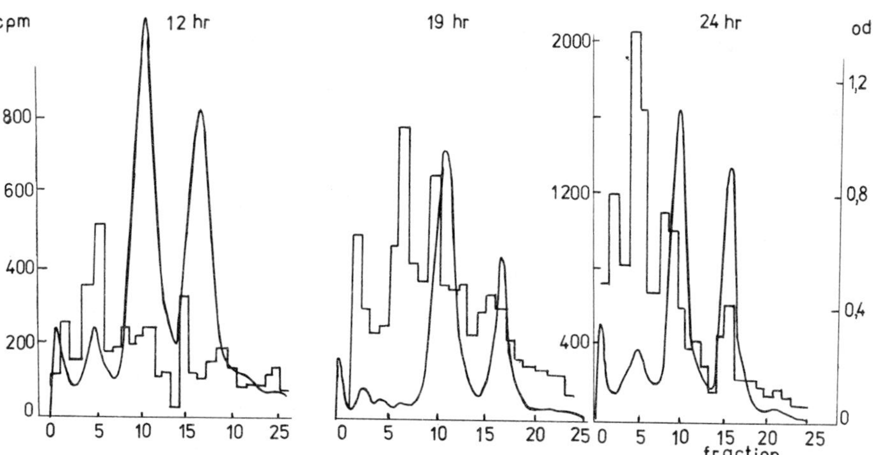

Fig. 1. Profiles of high molecular RNA fraction isolated from root of germinating French bean seed. Left – radioactivity, right – absorbance at 260 nm. The radioactivity pattern is represented as an histogramme.

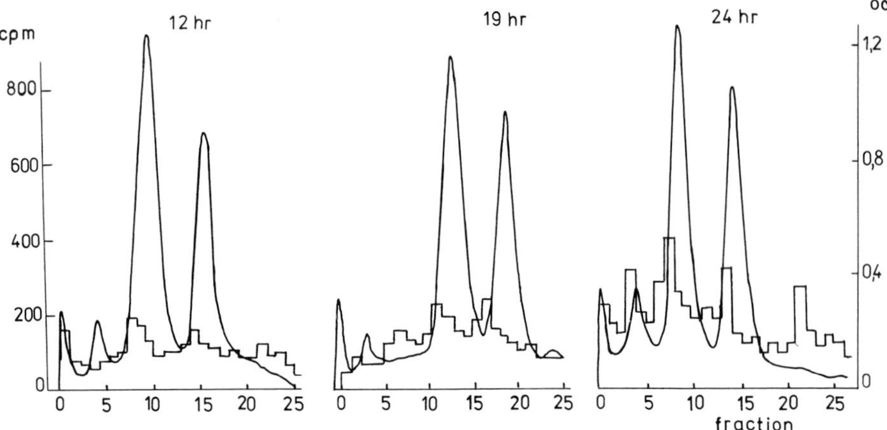

Fig. 2. Profiles of high molecular RNA fraction isolated from hypocotyl of germinating French bean seed. See Fig. 1.

soaking (at the end of the imbibition period); 12 and 19 h (in the middle and at the end of the lag-period); and 24 h after soaking (root growth due to mitotic activity). Embryos were excised from seeds germinated at 25 °C on wetted filter paper, then – during the last hour of the experiment – were incubated on ^{3}H-5-uridine solution (3.7 M Bq/ml, sp. act. 1.073 G Bq/mM) with streptomycine (50 mkg/ml), then rinsed in water, and divided into organs.

The experimental data which characterize RNA synthesis in germinating seed show that incorporation of label into the embryo is detectable already within 12 h of germination and at the subsequent stages it increases to a maximum at the 24 h. The total radioactivity of ^{3}H-5-uridine incorporation into the root is much greater than in the hypocotyl after 19 and 24 h of germination.

Polyacrylamide (2.5 per cent) gel electrophoresis of RNA has demonstrated that the rate of rRNA synthesis in the root (Fig. 1) is greater than in the hypocotyl (Fig. 2). Especially high labeling is detected in a high molecular weight precursor of rRNA at the first day of germination. It is suggested that mitotic divisions in the root are associated with the *de novo* rRNA synthesis, the latter was not detected in hypocotyls whose growth proceeded by elongation.

Thus, cell analysis of root and hypocotyl growth and dynamics of nitrogen containing compounds in them makes it possible to confirm some previous conclusions concerning both growth and metabolic specificity during early stages of seed germination.

REFERENCES

1. Martyn, G. I., Musatenko, L. I. and Sytnik, K. M. 1976 Morphology of the germ axis early stages of Phaseolus seeds germination. Dokl. AN Ukr. S.S.R., Ser. 'B' (U.S.S.R.) N 11, 1042–1045.
2. Sytnik, K. M., Bogdanova, T. L. and Musatenko, L. I. 1980 Amino acids of ornithine cycle during ripening and germinating of seeds. Ukr. Bot. Zh. (U.S.S.R.) 37, 7–10.
3. Sytnik, K. M., Martyn, G. I. and Musatenko, L. I. 1977 Cytology analysis of the germ root growth. Dokl. AN Ukr. S.S.R., Ser. 'B' (U.S.S.R.) N 9, 854–857.

7. Cell proliferation during the development of lateral root primordia

RONALD D. MACLEOD and ANNE THOMPSON

Dept. of Plant Biology, University of Newcastle upon Tyne, Newcastle upon Tyne NE1 7RU, England

Outside the apical meristem, the most prominent group of proliferating cells in roots are those representing stages in the development of lateral root primordia. Thus, of the 231 500 cells in meristems, excluding the pericycle, in the apical 9.8 cm of roots of intact plants of *Zea mays,* 125 000 are confined to the apical meristem of the primary [3] and 106 500 to developing lateral root primordia [data derived from 18]. Unlike the single apical meristem, however, lateral root anlagen occur as numerous groups of initial cells along the primary and not as one discrete cluster. Moreover, lateral root primordia form a different type of meristematic system to that found at the root apex. Thus, root apical meristems are open systems, in the sense that cells are continuously lost at their proximal and distal margins. Also, once established, they do not fluctuate in cell number [23]. In contrast, since no cells are lost from developing lateral root primordia, they increase in cell number from the time of their initiation until secondary root emergence.

Each anlage is initiated close to the proximal boundary of the root apical meristem by the division of a relatively small number of cells [15, 20]. Such primordia increase in size and cell number as the apical meristem of the primary grows away from the point of initiation. Mean values for the overall period of primordium development in the experiments reported here were about 5.8, 3.5, 5.7 and 4.2 days in *Vicia, Pisum, Zea* and *Phaseolus* respectively [18].

CYTOKINETIC INVESTIGATIONS

Investigations of cell proliferation in developing lateral root primordia in *Pisum* and *Vicia* have been hampered by the fact that exogenously supplied substances such as colchicine [5, 8] or tritiated thymidine (^{3}H-TdR) [6, 7, 10, 11, 13] cannot be used to monitor the passage of cells into mitosis or the period of deoxyribonucleic acid (DNA) synthesis (S), or, indeed through the mitotic cycle as a whole.

This problem was eventually overcome by measuring rates of increase in cell number over known intervals of time in anlagen of different ages. Serial trans-

R. Brouwer et al. (eds.), Structure and Function of Plant Roots, 35–42.

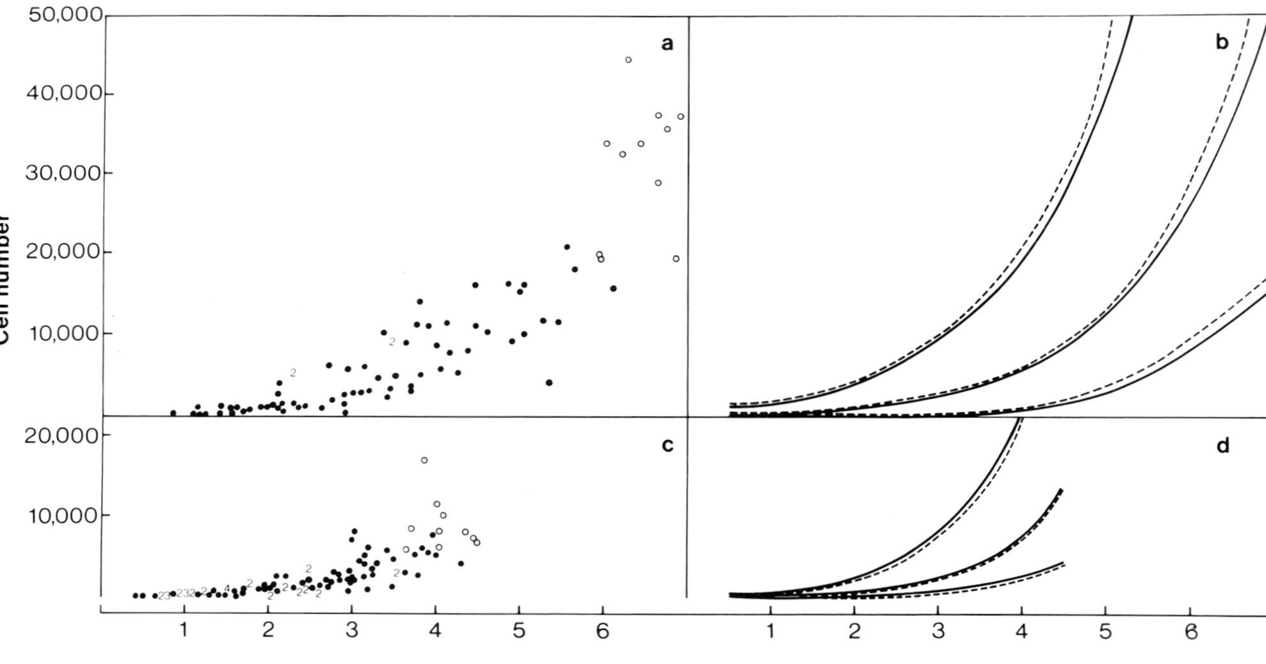

Fig. 1. Scatter diagrams in which cell number of primordia (●) and newly emerged lateral roots (○) are plotted against age of primary root tissue (days) in primaries of intact plants of *Vicia faba* (a) and *Phaseolus vulgaris* (c). The figures in these plots refer to the number of primordia or NE at each site. The curves and the 95% confidence limits corresponding to each of the former diagrams are presented for primordia only (– – – – –) and for primordia together with NE (———) for both *Vicia* (b) and *Phaseolus* (d).

verse sections were made of primary roots from the apex proximally along the root to the zone of lateral emergence. Median longitudinal sections of primordia and newly emerged laterals (NE) were then examined and, by dividing total anlage or NE volume by mean cell volume, values were obtained for anlage or NE cell number [9]. The age of the tissue in which each median longitudinal section was found was established by dividing the distance at which that section occurred from the apex of the primary root by the rate of root elongation.

Scatter diagrams of cell number against time were then drawn for the roots of intact plants of the four species investigated. Examples of such diagrams are presented for *Vicia* (Fig. 1a) and *Phaseolus* (Fig. 1c). Computer analysis of the data was carried out to find that general equation which best described increase in cell number with time. In each case, this proved to be the equation describing exponential increase in cell number.

i.e.

$$N_t = N_0 e^{rt},$$

where N_t and N_0 were the numbers of cells at times t and 0 h respectively, e the exponential function, r the rate of cell production (cells $cell^{-1}$ h^{-1}) and $t = (t - 0)h$. The curves showing the mean rate of increase in cell number (r) calculated from this equation are presented for the overall period of development of both primordia and primordia together with NE in primary roots of intact plants of *Vicia* (Fig. 1b) and *Phaseolus* (Fig. 1d). r was then used to calculate mean cell doubling time (Td) for the overall period of primordium development.

Thus,

$$\frac{\ln 2}{r} = \text{Td [19]},$$

Thus, Td for this period in roots of intact plants of *Pisum, Zea, Vicia* and Phaseolus was 13.9, 17.3, 22.4 and 14.8 h respectively.

Previous studies on cell proliferation during anlage development in the roots of intact plants of *Vicia, Pisum, Zea* and *Phaseolus* have shown, however, that Td is not constant as each primordium increases in cell number [14]. In the investigations reported here, the total period of anlage development was subdivided into intervals [see 19] and the rate of cell production, and thus Td, over each interval calculated, increase in cell number being assumed to be exponential as before.

Td clearly changed in each species as the primordia increased in cell number.

Table 1. Cell doubling time (Td) and the proportion of proliferating cells (Pf) in primordia and newly emerged laterals (NE) in the primary root of intact plants of *Phaseolus vulgaris*. Results are presented for the overall developmental period studied and at intervals over this time.

Age of site (days) in which primordia and NE occur.	Td (h)	Pf
Overall	14.75	0.46
0.5–1 to 1–1.5	6.73	1.00
1–1.5 to 1.5–2	10.04	0.67
1.5–2 to 2–2.5	12.84	0.52
2–2.5 to 2.5–3	138.60	0.05
2.5–3 to 3–3.5	9.90	0.68
3–3.5 to 3.5–4*	33.00	0.20
3.5–4* to 4–4.5*	23.90	0.28

* includes NE

Thus, for example, in *Phaseolus*, the youngest primordia seen in sections of roots of intact plants had a Td of 6.7 h. This lengthened as the primordia increased in cell number to reach 138.6 h, at which time a cavity began to develop in the adjacent cortical tissue of the primary root. Td then decreased to 9.9 h (Table 1). Td changed in exactly the same way over this period of anlage development in *Vicia, Pisum* and *Zea* [18, 19]). Further increase in primordium cell number was accompanied by a lengthening of Td up to around the time of lateral emergence when a second decrease was apparent.

If it is assumed that the youngest primordia investigated here consist solely of proliferative cells, as has been reported elsewhere [2, see also 19], the proliferative fraction (Pf) will be 1. Thus, mean cycle time (Tm) of the cycling cells in these primordia will be equal to cell doubling time (Td). There is published evidence (see Discussion in [19]) that Tm is fairly constant throughout root apical meristems, the only exceptions being the cells of the quiescent centre and, sometimes, the cap initials [4]. Assuming Tm also to be of constant duration in root primordia over the period between anlage inception and lateral emergence, any change that takes place in Td can only result from a change in size of the Pf.

If Td over the earliest stages of anlage development investigated is equated with Tm at any time during primordium development, then Pf can be determined either over this developmental period as a whole or at times during it by substituting the appropriate value for Td in the equation – Thus, in roots of

$$\frac{\text{Tm}}{\text{Td}} = \text{Pf}$$

intact plants of *Phaseolus*, for example, if a constant 6.73 h is assumed for Tm, Pf will decrease from 1 in the smallest anlagen examined to 0.05 around the time a cavity begins to form in the adjacent cortical tissues of the primary, before rising again to 0.68 following the development of a cortical lacuna. Pf subsequently decreased again before rising to 0.28 around the time of lateral emergence (Table 1). This pattern of change was also found during lateral root anlage development in *Zea* [18], *Pisum* and *Vicia* [19].

The pattern in which Td and Pf change during primordium development in the roots of intact plants of the four species investigated seems to be correlated with the formation of a cortical lacuna in the adjacent tissues of the primary root and later with the development of a vascular connection between the emerging lateral and the stele of the primary i.e. with sources of substances like sucrose [12] which are essential if cells are to continue to cycle [217].

ANLAGE DEVELOPMENT IN CULTURED ROOTS

Excised roots are known to have different rates of elongation to roots of intact plants [17]. Similarly, cell proliferation during lateral root primordium development in excised roots may be different from that found in intact ones of the same species. Since it was proposed to try to manipulate anlage inception, increase in cell number and emergence as lateral roots in cultured material, the same type of cytokinetic investigation was carried out on developing primordia in excised roots as had been carried out on such primordia in roots of intact plants.

Table 2. Period (days) between primordium initiation and lateral emergence, cell doubling time (Td) in hours over this period and the number of cells in emerging laterals (*) or in the largest primordia present after 6 days in culture ($^{+}$) in excised roots and in those of intact plants of *Phaseolus*, *Zea*, *Pisum* and *Vicia*.

Species	Nature of root	Period of development	Number of cells	Td (h) for overall period of anlage development
Phaseolus	Intact	4.75	9 100*	14.8
	Excised	4.20	4 185*	63.0
Zea	Intact	4.51	9 560*	17.3
	Excised	5.70	3 415*	21.7
Pisum	Intact	3.50	11 500*	13.9
	Excised	6.00[1]	2 690^{+}	57.8[1]
Vicia	Intact	5.80	29 000*	22.4
	Excised	6.00[1]	5 315^{+}	–[2]

[1] data for overall period of primordium development in culture.
[2] a value could not be accurately determined in this case.

During the initial stages of these investigations, it was demonstrated that, whereas laterals emerged from excised roots of *Zea* and *Phaseolus,* which were 1 cm long at the time of excision, such emergence did not take place from similar roots of *Pisum* and *Vicia*. Primordium formation still took place in the latter roots however, but at a slower rate than normal.

It is evident (Table 2) from a comparison of the interval of time between anlage inception and emergence, the number of cells attained at the time of emergence, or at the end of a 6 day culture period, and from the values obtained for Td for the overall period of primordium development investigated, that primordia in excised roots increase in cell number much more slowly than those in roots of the corresponding intact plants.

It was possible to follow the changes which took place in Td and Pf in primordia developing in excised roots of *Zea* and *Pisum* but not, because of the scattered nature of the data obtained, in similar roots of *Vicia* (fig. 2a) or *Phaseolus* (Fig. 2b). In the former two species Td and Pf changed in a similar fashion to that reported in primordia in the roots of intact plants.

In excised roots of *Pisum* no lateral emergence took place even after 20 days in culture. Moreover, in these roots, anlage cell number eventually levelled off at a figure about one quarter of that reached at the time of lateral emergence from the roots of intact plants (Table 2). This arrest was accompanied by a rapid decline in MI and an increase in the combined proportion of nuclei in G1 and G2 (pre- and post-synthetic interphase) i.e. it seems that cells in these primordia go out of cycle and halt in interphase in those phases in which cell arrest takes place following the generation of a stationary phase in apical meristems of excised roots grown in the absence of any carbohydrate source of energy [22].

Several points can be made from these comparisons of lateral root primordium development in excised roots and those of intact plants.

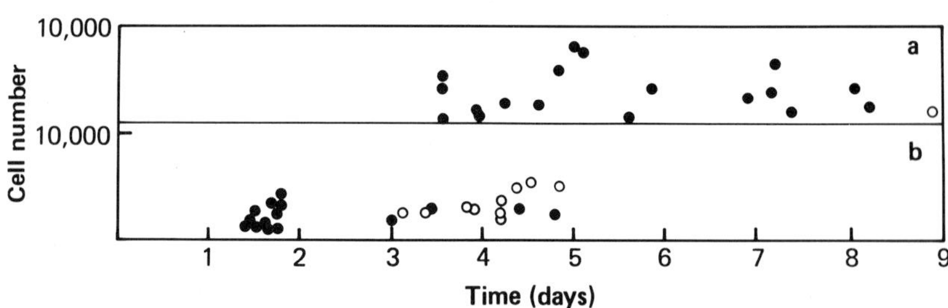

Fig. 2. Scatter diagrams in which cell number of primordia (●) and newly emerged lateral roots (○) are plotted against age of primary root tissue in excised roots of *Vicia* (a) and *Phaseolus* (b).

(1) Primordium development is multiphasic as has been previously reported [1, 16]. The results of the investigations reported here suggest that there are at least three phases –
   (a) inception and an initial increase in cell number;
   (b) further increase in cell number;
   (c) lateral emergence.

   All three stages are found in the anlagen developing in roots of intact plants while only the first phase occurs in primordia in excised roots of *Pisum* and *Vicia* and the first and last in primordia in such roots of *Zea* and *Phaseolus*, i.e. the second phase is not a prerequisite for normal lateral emergence.

(2) From the observed correlation between the formation of a lacuna rich in carbohydrates in the primary root cortex next to developing primordia and the changes which take place in Td and Pf in these anlagen as they increase in cell number it can be inferred that the contents of such cavities are necessary for the maintenance of proliferative activity in the latter meristems, at least until mature vascular tissue differentiates and connects the emerging lateral to the stele of the primary.

REFERENCES

1. Blakely, L. M., Rodaway, S. J., Hollen, L. B. and Croker, S. G. 1972 Control and kinetics of branch root formation in cultured root segments of *Haplopappus ravenii*. Plant Physiol. 50, 35–42.
2. Clowes, F. A. L. 1961 Apical meristems, 1st Edn. Oxford: Blackwell.
3. Clowes, F. A. L. 1972 Non-dividing cells in meristems. Chromosomes Today 3, 110–117.
4. Clowes, F. A. L. 1975 The cessation of mitosis at the margins of a root meristem. New Phytol. 74, 263–271.
5. Davidson, D. P. 1965 A differential response to colchicine of meristems of roots of Vicia faba. Ann. Bot. 29, 253–264.
6. Davidson, D. and MacLeod, R. D. 1968 Heterogeneity in cell behaviour in primordia of Vicia faba. Chromosoma 25, 470–474.
7. Hummon, M. R. 1962 The effects of tritiated thymidine incorporation on secondary root production by *Pisum sativum*. Amer. J. Bot. 49, 1038–1046.
8. MacLeod, R. D. 1976a The development of lateral root primordia in Vicia faba L and their response to colchicine. Ann. Bot. 40, 551–562.
9. MacLeod, R. D. 1976b Growth of lateral root primordia in Vicia faba L. New Phytol. 76, 143–151.
10. MacLeod, R. D. and Davidson, D. 1968 Delayed incorporation of ^{3}H-thymidine by primordial cells. Chromosoma 24, 1–9.
11. MacLeod, R. D. and Davidson, D. 1970 Incorporation of ^{3}H-deoxynucleosides: changes in labelling indices during root development. Canad. J. Bot. 48, 1659–1663.
12. MacLeod, R. D. and Francis, D. 1976 Cortical cell breakdown and lateral root primordium development in Vicia faba L. J. exp. Bot. 27, 922–932.
13. MacLeod, R. D. and McLachlan, S. M. 1975 Tritiated-thymidine labelled muclei in primordia and newly emerged lateral roots of Vicia faba L. Ann. Bot. 39, 535–545.

14. MacLeod, R. D. and Thompson, A. 1979 Development of lateral root primordia in Vicia faba, Pisum sativum, Zea mays en Phaseolus vulgaris: Rates of primordium formation and cell doubling times. Ann. Bot. 44, 435–449.
15. McCully, M. E. 1975 The development of lateral roots. In: Torrey, J. G. and Clarkson, D. T. (eds.), The development and function of roots, pp. 105–124. New York: Academic.
16. Pecket, R. C. 1957 The initiation and development of lateral meristems in the pea root. I. The effect of young and of mature tissue. J. exp. Bot. 8, 172–180.
17. Scadeng, D. W. F. and MacLeod, R. D. 1976 The effect of sucrose concentration on cell proliferation and quiescence in the apical meristem of excised roots of Pisum sativum L. Ann. Bot. 40, 947–955.
18. Thompson, A. 1980 The development of lateral root primordia in four species of angiosperm. PhD thesis, University of Newcastle upon Tyne, England.
19. Thompson, A. and MacLeod, R. D. 1981 Increase in size and cell number of lateral root primordia in the primary of intact plants and in excised roots of Pisum sativum and Vicia faba. Amer. J. Bot. (in press).
20. Torrey, J. G. 1965 Physiological bases of organisation and development in the root. Handb. Pflanzenphysiol. 15(1), 1256–1327.
21. Van't Hof, J. 1968 Control of cell progression through the mitotic cycle by carbohydrate provision. I. regulation of cell division in excised plant tissue. J. Cell. Biol. 37, 773–778.
22. Van't Hof, J. and Kovacs, C. J. 1972 Mitotic cycle regulation in the meristem of cultured roots: the principal control point hypothesis. In: Miller, M. W. and Kuehnert, C. C. (eds.), The dynamics of meristem cell populations, pp. 15–32. New York: Plenum.
23. Webster, P. L. and MacLeod, R. D. 1980 Characteristics of root apical meristem cell population kinetics: a review of analyses and concepts. Env. exp. Bot. 20, 335–358.

8. Argyrophilic nuclear structures in root apices

PETER W. BARLOW

Agricultural Research Council, Letcombe Laboratory, Wantage, Oxon, OX12 9JT, England

The general structure of the plant cell nucleus is quite well known at the level of resolution afforded by both the light and electron microscope [3]. Nuclei of root meristems, in particular, are favoured for structural studies since they are so readily available and because they are also so obviously active in the genetic and metabolic activities that result in mitosis, cell growth and cell differentiation. Micrographs showing the appearance of chromatin, nucleolus and nuclear envelope in meristematic cells are familiar illustrations in most books on plant cytology. However, patient searching of electron microscope sections, and the use of new cytochemical techniques, can reveal structures within the nucleus that are less familiar, and whose significance or function is unknown. The aim of this contribution is to draw attention to two unfamiliar nuclear components and to present data relating to their behaviour in different regions of the root apex.

MATERIALS AND METHODS

Roots 2–4 cm long were removed from seedlings of *Pisum sativum* and *Zea mays* and were fixed and impregnated with $AgNO_3$ according to [7]. For light microscopy the silver-impregnated apices were embedded in wax and sectioned longitudinally at 16 μm. After dewaxing, the sections were mounted under a coverslip in Canada Balsam. For electron microscopy the apices were embedded in resin and gold-silver thin sections prepared.

RESULTS

Silver impregnation of interphase nuclei clearly reveals, in addition to the nucleolus, two other types of argyrophilic object. These latter are less familiar to cytologists and have no generally accepted names. One type of object appears as a projection (up to 1 μm in length or diameter) from the nucleolus (Fig. 1) and is probably identical to the karyosome described by Hyde [2]. Although 'karyosome' is not a particularly suitable term it will be used here. The other type of object is a small spherical body (0.25–1 μm diameter) that lies in the nucleoplasm (Fig. 2). For the present this will be called 'argyrophilic intranuclear body' (AIB).

R. Brouwer et al. (eds.), Structure and Function of Plant Roots, 43–47.

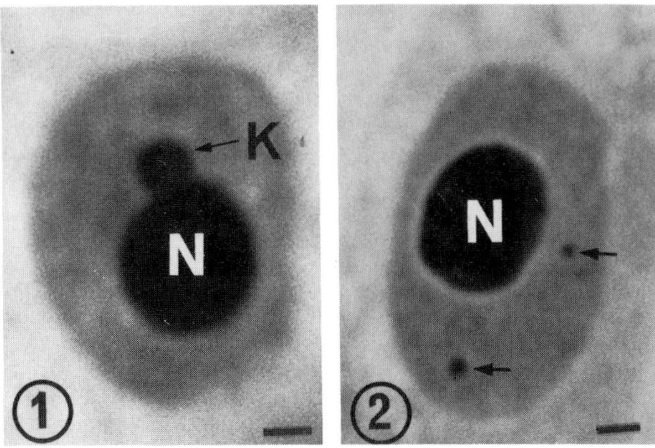

Fig. 1. A nucleus in the quiescent centre of a root of *Pisum* with a karyosome (K) attached to the nucleolus (N). Scale bar equals 1 μm.
Fig. 2. A nucleus of *Zea* with two AIBs (arrowed) lying in the nucleoplasm. The nucleolus (N) appears to lack a karyosome. Scale bar equals 1 μm.

In thick (16 μm) sections many nuclei are uncut and so the number of karyosomes and AIBs per nucleus can be counted with the light microscope. Their numbers vary according to the region within the root (Table 1). Of all the regions, both meristematic and non-meristematic, that were analysed in apices of *Z. mays* the quiescent centre has the highest mean number of AIBs; in contrast,

Table 1. The frequency with which karyosomes are seen attached to nucleoli, and the mean number of AIBs in nuclei in different regions of the root apex of *Pisum sativum* and *Zea mays*.

		Per cent nuclei with the following number of karyosomes:			Mean number of AIBs per nucleus
Species	Region of apex	0	1	2	
P. sativum	Quiescent centre	54.9	40.2	4.9	1.5
	Stele 280–430 μm*	98.9	1.1	0.5	6.5
	Stele 1200 μm*	59.8	34.3	5.9	2.7
Z. mays	Quiescent centre	86.2	12.3	1.5	2.2
	Stele 280–430 μm*	63.6	33.1	3.3	1.2
	Stele 1200 μm*	44.0	48.0	8.0	0.5

* Refers to the distance from the cap-quiescent centre junction. The number of nuclei scored in each region was between 120 and 200.

karyosomes are observed here the least frequently. The latter are most frequently seen in non-meristematic, maturing tissue. In apices of *P. sativum* it is nuclei of the actively dividing region that have the most AIBs and in which karyosomes are least frequent; in the quiescent centre the converse holds.

In order to account for the different frequencies of AIBs and karyosomes in different regions of the apex a more detailed study was made. It was supposed that the frequencies to the two objects vary across interphase and that nuclear volume may be used as an indicator of nuclear age. Thus, small nuclei are presumed to be in early interphase, and large nuclei to be in late interphase, closer to entering mitosis. Volumes of about 100 nuclei were calculated and sorted into order of increasing size (i.e. youngest to oldest). It was consistently found that in various regions of the meristem AIB frequency was highest in early interphase and fell as nuclei proceeded towards mitosis. A karyosome, however, was rarely if ever associated with nucleoli in early interphase, but in late interphase one or two karyosomes were frequently seen. In prophase nuclei the karyosome was occasionally observed to be detached from the nucleolus.

Thin sections of silver impregnated nuclei, when viewed in the electron

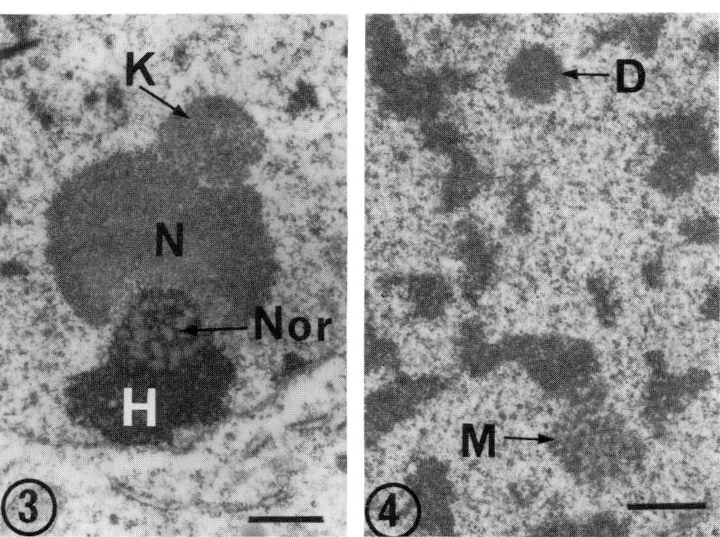

Fig. 3. Electron micrograph of a nucleous (N) with a karyosome (K) within a root cap nucleus of *Zea*. Note that the nucleolar organizer region (Nor) and its heterochromatin (H) are quite distinct from the karyosome. Scale bar equals 0.5 μm.

Fig. 4. Electron micrograph of a micropuff (M) and a dense body (D) within a meristematic nucleus of *Pisum*. (Micrograph kindly provided by Dr. E. G. Jordan and J. Pacy.) Scale bar equals 0.5 μm.

microscope, do not allow much information to be gained concerning the fine structure of either karyosome or AIB. The silver deposit within these objects obscures their structure and also the impregnation procedure does not preserve nuclear structure well. Therefore, it is necessary to use sections of conventionally fixed and stained nuclei and then attempt to correlate the structures so revealed with those seen in silver impregnated material. Although this approach is indirect, there seems to be little doubt that the karyosome corresponds to a body with a fibrillar texture that lies embedded on the surface of the nucleolus (Fig. 3). The counterpart of the AIB is less certain. Two structures are seen in thin sections of interphase nuclei that could be the AIB; one is a loose fibrillar structure known as a 'micropuff' [4, 8], the second is a more tightly structured fibrillo-granular body (Fig. 4). The micropuffs are generally held to be centromeres [4] and it seems unlikely that they are the counterparts of the AIBs seen in the light microscope since the number and location of the AIBs seen in interphase does not correspond with that expected of centromeres. Therefore, it is more likely that the dense bodies (Fig. 4) are the AIBs.

DISCUSSION

Silver was regularly used at the turn of this century as an agent for staining nuclear and nucleolar structure, particularly in nerve tissue [9]. Interest in silver staining has recently revived because silver has been found to have a high affinity for sites of ribosomal gene activity on chromosomes which, in turn, may be related to the strongly argyrophilic nature of certain nucleolar proteins [1, 5]. The method of silver impregnation used in the present study is one that has given information on nucleolar organisation in plants [6, 7, 10], but it can also be used to vizualise, in addition to the nucleolus, two other nuclear components called here karyosome and AIB. Although the present data relate to karyosomes and AIBs in *Pisum* and *Zea*, these two bodies are seen within the nuclei of a number of other plant species (P. W. Barlow, unpublished) and are probably a standard feature of plant nuclei in general.

The karyosome seems to be a product of the nucleolus. In meristematic cells the karyosome is inferred to grow out from, or upon, the nucleolar surface during interphase and then presumably it disperses during prophase when the nucleolus also disperses. The subsequent fate of the karyosome material is unknown. Karyosomes are also present in nuclei of non-meristematic cells, but whether they go through a sequence of growth and dispersion, or whether they are static fixtures on the nucleolus is not known.

It has been suggested [8] that the AIBs are pieces of nucleolar material that have been released from that organelle and which are in transit to the cytoplasm. Another possibility is that in meristematic cells they are fragments of prenucleolar material that have failed to be included in the nucleolus as it reforms in telophase.

The disappearance of AIBs from the nucleus during late interphase coincides with the appearance of the karyosome. At first sight this might suggest an interrelationship between these two structures. But this could be a fortuitous occurrence and the two structures may, in fact, have independent modes of development.

The relative ease with which karyosomes and AIBs can be demonstrated in plant material means that it should be possible to discover more about their role in the life of the cell.

REFERENCES

1. Hubbell, H. R., Rothblum, L. I. and Hsa, T. C. 1979 Identification of a silver binding protein associated with the cytological silver staining of actively transcribing nucleolar regions. Cell Biol. Internat. Reps. 3, 615–622.
2. Hyde, B. B. 1967 Changes in nucleolar ultrastructure associated with differentiation in the root tip. J. Ultrastruct. Res. 18, 25–54.
3. Jordan, E. G., Timmis, J. N. and Trewavas, A. J. 1980 The plant nucleus. In: Tolbert N. E. (ed.), Biochemistry of plants, vol. I. A comprehensive treatise: The plant cell, pp. 489–588. New York: Academic Press.
4. Lafontaine, J. G. and Luck, B. T. 1980 An ultrastructural study of plant cell (Allium porrum) centromeres. J. Ultrastruct. Res. 70, 298–307.
5. Lischwe, M. A., Smetana, K., Olson, M. O. J. and Busch, H. 1979 Proteins C23 and B23 are the major nucleolar silver staining proteins. Life Sci. 25, 701–708.
6. Moreno Díaz de la Espina, S., Risueño, M. C., Fernández-Gómez, M. E. and Tandler, C. J. 1976 Ultrastructural study of the nucleolar cycle in meristematic cells of Allium cepa. J. Microscop. biol. cell 25, 265–278.
7. Risueño, M. C., Fernández-Gómez, M. E. and Giménez-Martín, G. 1973 Nucleoli under the electron microscope by silver impregnation. Mikroskopie 29, 292–298.
8. Risueño, M. C., Moreno Díaz de la Espina, D., Fernández-Gómez, M. E., and Giménez-Martín, G. 1978 Nuclear micropuffs in Allium cepa cells. I. Quantitative, ultrastructural and cytochemical study. Cytobiol 16, 209–223.
9. Růžička, V. 1899 Zur Geschichte und Kenntnis der feineren Struktur der Nukleolen centraler Nervenzellen. Anat. Anz. 16, 557–563.
10. Stockert, J. C., Fernández-Gómez, M. E., Giménez-Martín, G. and López-Sáez, J. F. 1970 Organization of argyrophilic nucleolar material throughout the division cycle of meristematic cells. Protoplasma 69, 265–278.

9. Can induced autopolyploidy replace naturally occurring endopolyploidization in roots?*

MARIA J. OLSZEWSKA and ANDRZEJ K. KONONOWICZ

Dept. of Plant Cytology and Cytochemistry and Laboratory of Plant Growth Substances, Institute of Physiology and Cytology, University of Łódź, Banacha 12/16, 90–237 Łódź, Poland

The purpose of this work was to find out whether: 1) colchicine-induced autopolyploidy can replace the spontaneous endopolyploidization; 2) the presence of autopolyploid nuclei in the zones of root growth and differentiation changes the pattern of ^{3}H uridine incorporation into nuclei and nucleoli.

Experiments were carried out on the roots of seedlings of *Zea mays, Helianthus annuus* and *Pisum sativum*. Seedlings were incubated in colchicine (0.25 mg/ml) for 8 h and then transferred into water for 12 h. After this postincubation time, the thickened parts of roots arising as a result of the colchicine treatment, translocate to the suprameristematic zones, i.e., they are localized at a distance of 5–7 mm from the root tip (Fig. 1AB, 2AB, 3AB). Control seedlings

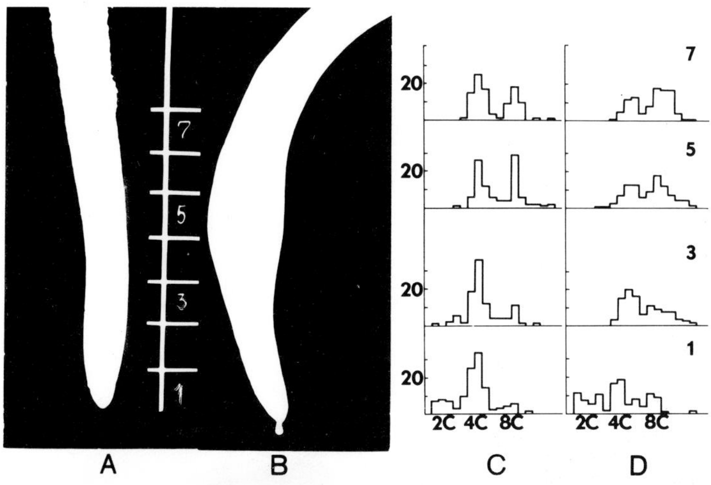

Fig. 1. *Zea mays* roots: A – control; B – after 8 h in colchicine and 12 h postincubation in water; C and D – cytophotometric measurements of relative DNA content in cortex parenchyma nuclei; abscissa: DNA content/nucleus (logarithmic scale); ordinate: number of nuclei; 1, 3, 5, 7 – successive segments from root tip; C – control; D – after colchicine treatment.

* This work was supported by the Polish Academy of Sciences within the project 09.7.1.5.4.

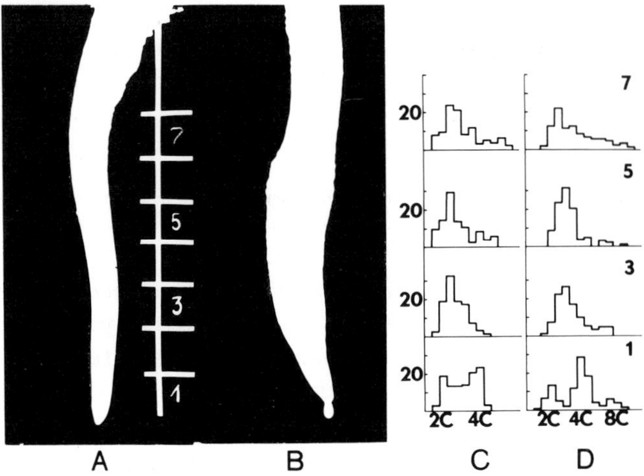

Fig. 2. *Helianthus annuus* roots: A – control; B – after 8 h in colchicine and 12 h postincubation in water; C and D – cytophotometric measurements of relative DNA content in cortex parenchyma nuclei; abscissa: DNA content/nucleus (logarithmic scale); ordinate: number of nuclei; C – control; D – after colchicine treatment.

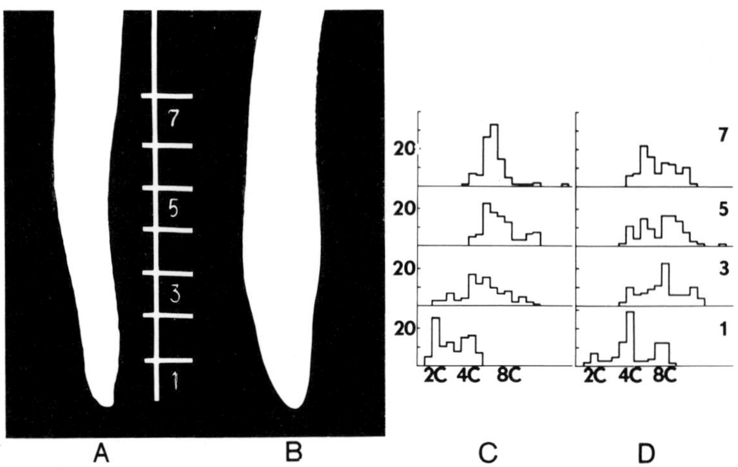

Fig. 3. *Pisum sativum* roots: A – control; B – after 8 h in colchicine and 12 h postincubation in water; C and D – cytophotometric measurements of relative DNA content in cortex parenchyma nuclei; abscissa: DNA content/nucleus (logarithmic scale); ordinate: number of nuclei: C – control; D – after colchicine treatment.

and those postincubated for 8 h in water following colchicine treatment were placed for 4 h in ^{3}H thymidine (2 μCi/ml). Incubation with ^{3}H uridine (1 h, 50 μCi/ml) took place during the last hour of postincubation. Fixed roots were cut into 1 mm segments, squashed and covered with Ilford K2 emulsion.

Cytophotometric measurements of the DNA content (Feulgen's procedure) have shown that cells in all developmental zones (1st, 3rd, 5th and 7th mm segments from root tip) in colchicine-treated roots contain the nuclei at higher level of ploidy than the control roots (Fig. 1CD, 2CD, 3CD). Analysis of autoradiograms have revealed similar or even higher ^{3}H thymidine labelling index (Table 1) and the higher radioactivity of nuclei in the colchicine-treated roots than in the control ones. Except for the meristematic zone of *Helianthus annuus*, colchicine does not modify the intensity of ^{3}H uridine incorporation into nuclei and nucleoli (Table 1).

It may be concluded that: 1) colchicine-induced autopolyploidy cannot replace the spontaneous DNA endoreplication in growing and differentiating cells of roots; 2) the presence of autopolyploid nuclei in growing and differentiating root cells does not modify the pattern of ^{3}H uridine incorporation into nuclei and nucleoli.

Table 1. Incorporation of ^{3}H thymidine and ^{3}H uridine into nuclei of cortex parenchyma of control and colchicine-treated roots

Species	Segment No.	^{3}H thymidine per cent of labelled nuclei		^{3}H uridine, number of grains			
				nucleus		nucleolus	
		control	colch.	control	colch.	control	colch.
Zea mays	1	40.3	45.1	16.0±0.7	18.4±0.7	11.5±0.7	13.8±0.7
	3	22.1	30.8	10.7±0.6	9.8±0.6	4.4±0.3	3.4±0.4
	5	15.1	24.3	8.8±0.5	9.0±0.6	1.9±0.1	1.9±0.1
	7	25.5	22.0	9.7±0.6	7.2±0.6	1.5±0.1	1.9±0.1
Helianthus annuus	1	36.7	36.5	10.6±0.5	19.9±0.3	6.9±0.4	13.5±1.0
	3	6.3	6.2	10.6±0.4	8.5±0.4	3.7±0.3	3.2±0.2
	5	6.2	5.9	8.0±0.3	8.7±0.4	2.1±0.2	2.1±0.2
	7	7.7	7.8	5.8±0.3	7.8±0.4	2.1±0.2	2.6±0.2
Pisum sativum	1	16.9	24.3	7.8±0.5	8.6±0.5	2.8±0.3	4.1±0.3
	3	27.4	23.7	9.3±0.6	8.9±0.7	2.3±0.2	3.0±0.3
	5	10.2	11.5	8.5±0.6	8.5±0.6	2.8±0.3	2.8±0.2
	7	5.0	6.0	8.4±0.5	8.1±0.6	2.7±0.3	2.5±0.3

10. In situ ^{3}H rRNA/DNA hybridization and silver staining of NORs during growth and differentiation of root cortex cells in the presence or absence of DNA endoreplication*

MARIA J. OLSZEWSKA, ANDRZEJ K. KONONOWICZ, HALINA KONONOWICZ and KRYSTYNA JANAS

Dept. of Plant Cytology and Cytochemistry and Laboratory of Plant Growth Substances, Institute of Physiology and Cytology, University of Łódź, Banacha 12/16, 90–237 Łódź, Poland

The purpose of this study was to investigate changes in the number of NoRs in successive zones (1st, 3rd, 5th and 7th mm segments from the root tip) of growth and differentiation of roots in 4 plant species characterized by the presence or absence of endopolyploid nuclei. Cytophotometric measurements of the DNA content (after Feulgen's procedure, hydrolysis for 1 h in 4N HCl at 20 °C) have shown that in supra-meristematic zones (3rd, 5th and 7th segments) in *Haemanthus katharinae* and *Tulipa kaufmanniana* all nuclei have 2C DNA con-

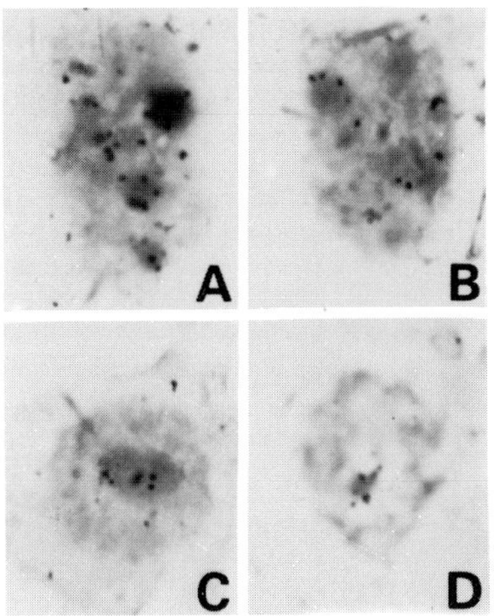

Fig. 1. ^{3}H rRNA/DNA in situ hybridization: A, B – *Tulipa kaufmanniana*, C, D – *Zea mays*; A, C – 1st segment; B, D – 7th segment.

* This work was supported by the Polish Academy of Sciences within the project 09.7.1.5.4.

tent (Figs. 3A and 4A), in *Zea mays* 4C and 8C nuclei are present, while in *Helianthus annuus* in the 5th and 7th segments some nuclei reveal 4C and 8C DNA content (comp. [8]).

^{3}H rRNA was obtained from sterile cultures of *Spirodela oligorrhiza* after 24h incubation with ^{3}H uridine. As the rRNAs in Angiospermae show similarities of nucleotide sequences [4], this rRNA may be considered as homologous for the species under study. After an in situ hybridization (procedure according to [1]), in autoradiograms (Ilford K2 emulsion, Mayer's hematein staining) the grains were localized over nucleoli (Figs. 1 A–D and 2AB). The number of grains corresponds to the DNA content in *Haemanthus katharinae* and *Tulipa kaufmanniana,* i.e. the greatest number was found in the meristematic zones (DNA content 2C–4C, Figs. 3 AB, 4 AB). In spite of endopolyploidization, in *Zea mays* the number of grains diminishes in the supra-meristematic zones (Fig. 5A). In *Helianthus annuus,* after diminishing in the 3rd mm, the number of grains increases slightly in the 5th and 7th mm, i.e., in these segments where endopolyploid nuclei appear (Fig. 6A).

Silver staining, identifying the number of NoRs and the fibrillar component of nucleoli [2, 3, 9] was performed according to [9]. In meristematic nucleoli, active in rRNA synthesis [5–8], deeply stained structures against the light backround are visible. In suprameristematic zones the nucleoli are almost homogenously deeply stained (Fig. 2 CD). Cytophotometric evaluation of silver staining (extinction value × surface area of nucleoli) have indicated that the quantity of the

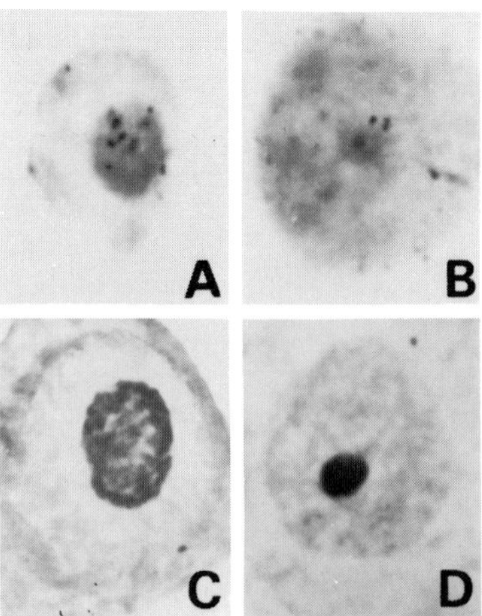

Fig. 2. *Helianthus annuus:* A, B – ^{3}H rRNA/DNA in situ hybridization; C, D – Ag-staining of nucleoli; A, C – 1st segment; B, D – 7th segment.

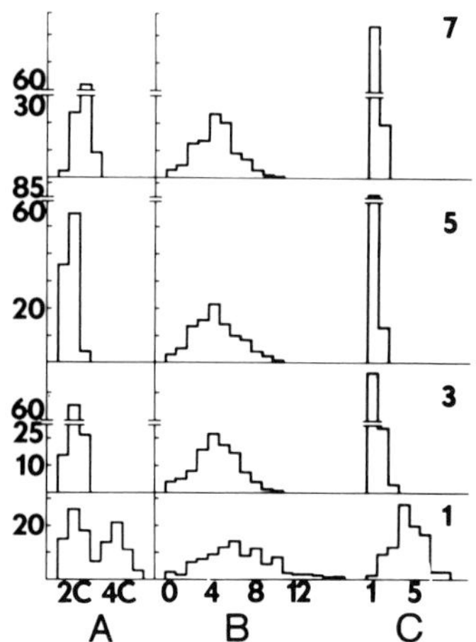

Fig. 3. *Haemanthus katharinae*: A – cytophotometric measurements of relative DNA content in cortex parenchyma nuclei; abscissa: DNA content/nucleus (logarithmic scale); ordinate: number of nuclei; B – frequency distribution of grains over nucleoli after in situ ^{3}H rRNA/DNA hybridization; abscissa: number of grains; ordinate: per cent of nuclei; C – cytophotometric evaluation of Ag-staining; abscissa: quantity of product in arbitrary units; ordinate: per cent of nuclei; 1, 3, 5, 7 – successive segments from root tip.

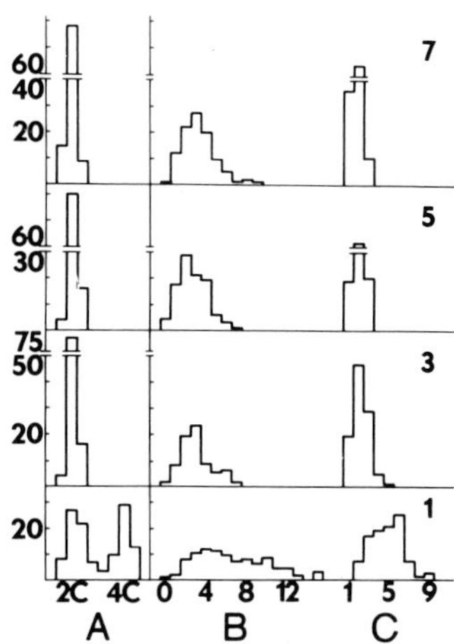

Fig. 4. *Tulipa kaufmanniana*: A – cytophotometric measurements of relative DNA content in cortex parenchyma nuclei; abscissa: DNA content/nucleus (logarithmic scale); ordinate: number of nuclei; B – frequency distribution of grains over nucleoli after in situ ^{3}H rRNA/DNA hybridization; abscissa: number of grains; ordinate: per cent of nuclei; C – cytophotometric evaluation of Ag-staining; abscissa: quantity of product in arbitrary units; ordinate: per cent of nuclei; 1, 3, 5, 7 – successive segments from root tip.

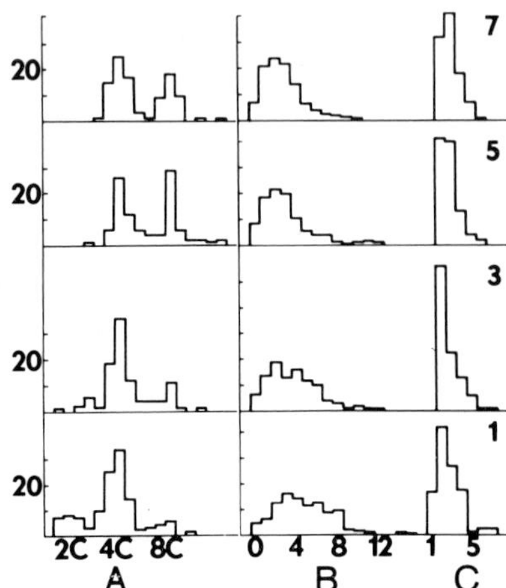

Fig. 5. *Zea mays*: A – cytophotometric measurements of relative DNA content in cortex parenchyma nuclei; abscissa: DNA content/nucleus (logarithmic scale); ordinate: number of nuclei; B – frequency distribution of grains over nucleoli after in situ ^{3}H rRNA/DNA hybridization; abscissa: number of grains; ordinate: per cent of nuclei; C – cytophotometric evaluation of Ag-staining; abscissa: quantity of product in arbitrary units; ordinate: per cent of nuclei; 1, 3, 5, 7 – successive segments from root tip.

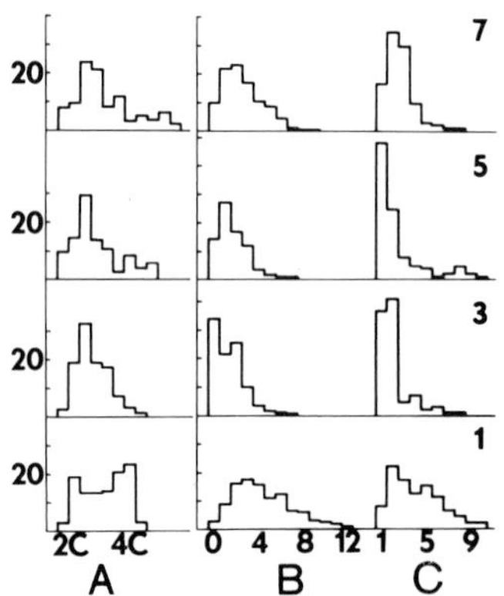

Fig. 6. *Helianthus annuus*: A – cytophotometric measurements of relative DNA content in cortex parenchyma nuclei; abscissa: DNA content/nucleus (logarithmic scale); ordinate: number of nuclei; B – frequency distribution of grains over nucleoli after in situ ^{3}H rRNA/DNA hybridization; abscissa: number of grains; ordinate: per cent of nuclei; C – cytophotometric evaluation of Ag-staining; abscissa: quantity of product in arbitrary units; ordinate: per cent of nuclei; 1, 3, 5, 7 – successive segments from root tip.

brown product was the greatest in all species in the meristematic zones and decreased in the successive developmental zones in *Haemanthus katharinae* and *Tulipa kaufmanniana,* and also – in spite of endopolyploidization – in *Zea mays* (Fig. 3C, 4C, 5B). In *Helianthus annuus* these values slightly increased in the 5th and 7th mm (Fig. 6B).

The results obtained with the aid of ^{3}H rRNA/DNA in situ hybridization and those obtained with silver staining of NoRs are consistent with our data concerning ^{3}H uridine incorporation into nucleoli in successive zones of growth and differentiation of root cortex cells: in the same experimental conditions, in *Zea mays* radioactivity of nucleoli declines between the 1st and 7th segments about 10 times, while in *Helianthus annuus* it diminishes 2.5 times only [8]. These results indicate a possibility of the under-replication of rDNA during DNA endoreplication in short-lived seminal roots of *Zea mays,* as has been found in polytene chromosomes of *Drosophila melanogaster* [10] and might explain similar reduction of rRNA synthesis in differentiating root cortex cells in the absence [5, 7] or presence [7, 8] of endomitosis. Our further investigations concerning rDNA replication during endopolyploidization will include the roots of other species of mono- and dicotyledonous plants.

REFERENCES

1. Avanzi, S., Durante, M., Cionini, G. and D'Amato, F. 1972 Cytological localization of ribosomal cistrons in polytene chromosomes of Phaseolus coccineus. Chromosoma (Berl.) 39, 191–203.
2. Giménez-Martin, G., de la Torre, C., Lopez-Saez, J. F. and Esponda, P. 1977 Plant nucleolus: structure and physiology. Cytobiologie 14, 421–462.
3. Hernandez-Verdun, D., Hubert, J., Bourgeois, C. A. and Bouteille, M. 1980 Ultrastructural localization of Ag-NoR stained proteins in the nucleolus during the cell cycle and in other nucleolar structures. Chromosoma (Berl.) 79, 349–362.
4. Maggini, F., De Dominicis, R. I. and Salvi, G. 1976 Similarities among ribosomal RNA's of Angiospermae and Gymnospermae. J. Mol. Evol. 8, 329–335.
5. Kononowicz, A. K., Kuran, H. and Olszewska, M. J. 1976 Cytochemical and ultrastructural studies of changes in the nucleus and nucleolus during cell differentiation in the root cortex of Haemanthus katharinae. Folia Histochem. Cytochem. 14, 151–164.
6. Olszewska, M. J. 1976 Autoradiographic and ultrastructural study of Cucurbita pepo root cells during their growth and differentiation. Histochemistry 49, 157–175.
7. Olszewska, M. J. and Kononowicz, A. K. 1979 Activities of DNA polymerases and RNA polymerases detected in situ in growing and differentiating cells of root cortex. Histochemistry 59, 311–323.
8. Olszewska, M. J. and Kononowicz, A. K. 1981 Can induced autopolyploidy replace naturally occurring endopolyploidization in roots? Proc. 2nd Symp. on Structure and Function of Roots, this volume, pp. 49–51.
9. Schubert, I., Anastassova-Kristeva, M. and Rieger, R. 1979 Specificity of NoR staining in Vicia faba. Exp. Cell Res. 120, 433–435.
10. Spear, B. 1974 The genes for ribosomal RNA in diploid and polytene chromosomes of Drosophila melanogaster. Chromosoma (Berl.) 48, 159–179.

11. Abnormalities in the ultrastructure of root cap cells in *Raphanus sativus* L. induced by actinomycin D, hydroxyurea, α-naphthylacetic acid and streptomycin

FRANCISZEK KADEJ and AGNIESZKA KADEJ*

Central Laboratory and Institute of Biology of the M. Curie-Sklodowska University, 30–031 Lublin, Plac M.C.S. 3, 30–031 Lublin, Poland, Dept. of Anatomy and Cytology of the M. Curie-Sklodowska University, Lublin, Poland*

The influence of such chemicals as actinomycin (AD), hydroxyurea (HU) and streptomycin (S) is well documented in the medical and biochemical literature. These chemicals are known to be inhibitors of plant growth [1, 2, 4] and α-naphthylacetic acid (NAA) simultaneously influences morphogenesis [e.g. 5]. However, there have been few reports of the changes in ultrastructure of plant cells following treatment with these chemicals.

In normal root caps, one or two peripheral layers are rich in hypertrophied dictyosomes, large secretory vesicles and a well developed system of ER (Figs. 2, 3). Furthermore, there are mitochondria, amyloplasts and normal nuclei. Cells of the middle layers contain amyloplasts filled with statolithic starch, mitochondria and particularly large ER. Dictyosomes seem to be less active than those of the peripheral layers. The root cap initials are poor in all organelles [3].

Young seedlings of *Raphanus sativus* were grown in water culture with antibiotics or other chemicals for about 5, 24, 48 and 96 h. The roots which retained the ability of growth in pure water were used for purposes of evaluation. No irreversible inhibition of root growth was caused by: 0.05 μg/ml HU, 0.1 μg/ml AD, 0.25 μg/ml streptomycin and $5\text{–}10^{-6}$ M NAA.

In order to obtain additional data on the redifferentiation of the ultrastructure, the root cap was removed by cutting it just beneath the constructional centre of the root meristem and cultivating the regenerates in water with HU.

By routine methods of fixation and preparation, the cells were observed in EM Tesla 613. In the root cap, the cells of the outer and middle layers and the root cap initials were analyzed.

The root cap of all treated roots preserves its typical cellular pattern composed of a relatively constant number of cells arranged in 8–10 horizontal layers (Fig. 1).

The effects of treating the roots with AD and HU seem to be identical and they cause the same deformation of the ultrastructure. The cells of the outer layers are

R. Brouwer et al. (eds.), Structure and Function of Plant Roots, 59–65.

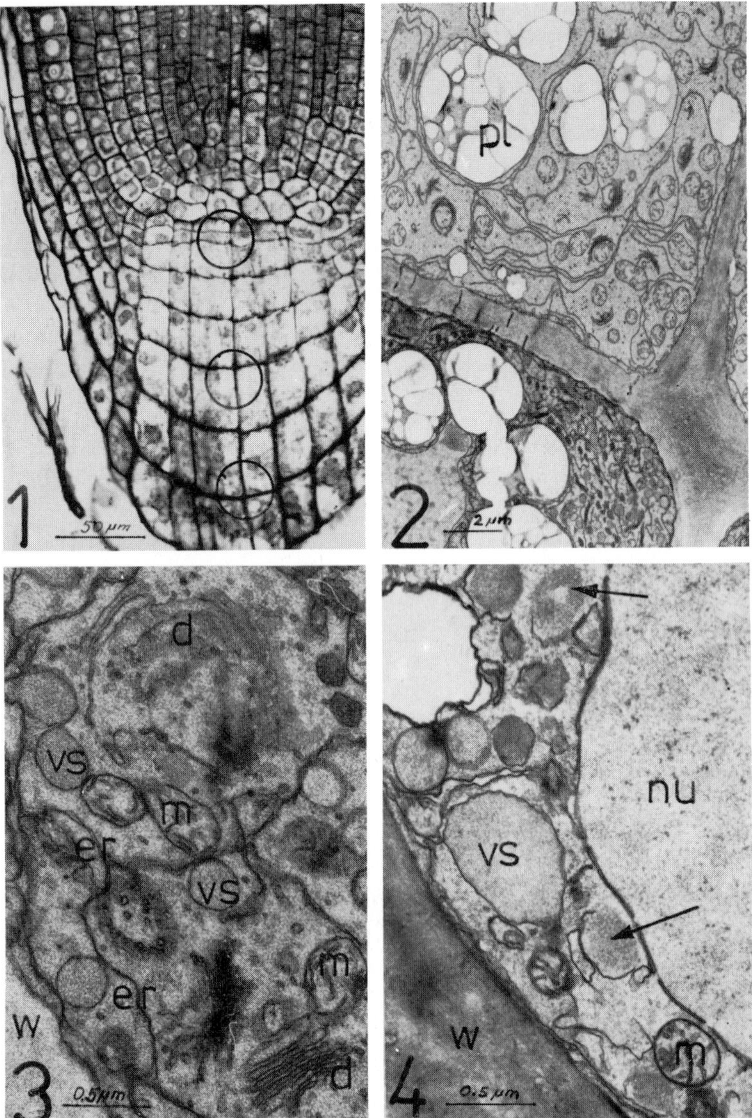

Fig. 1. Root apical meristem of *Raphanus sativus*. Circles mark analyzed parts. ×240.

Figs. 2–3. Cells of the outer layers from the normal root cap. er = endoplasmic reticulum, m = mitochondria, d = hypertrophied dictyosomes, vs = secretory dictyosomal vesicles. ×4000 a. 24 000.

Fig. 4. Cells of the outer layers treated with HU for 5 h. Note the early stage of the disintegration of the secretory vesicles (arrows). nu = nucleus. ×24 000.

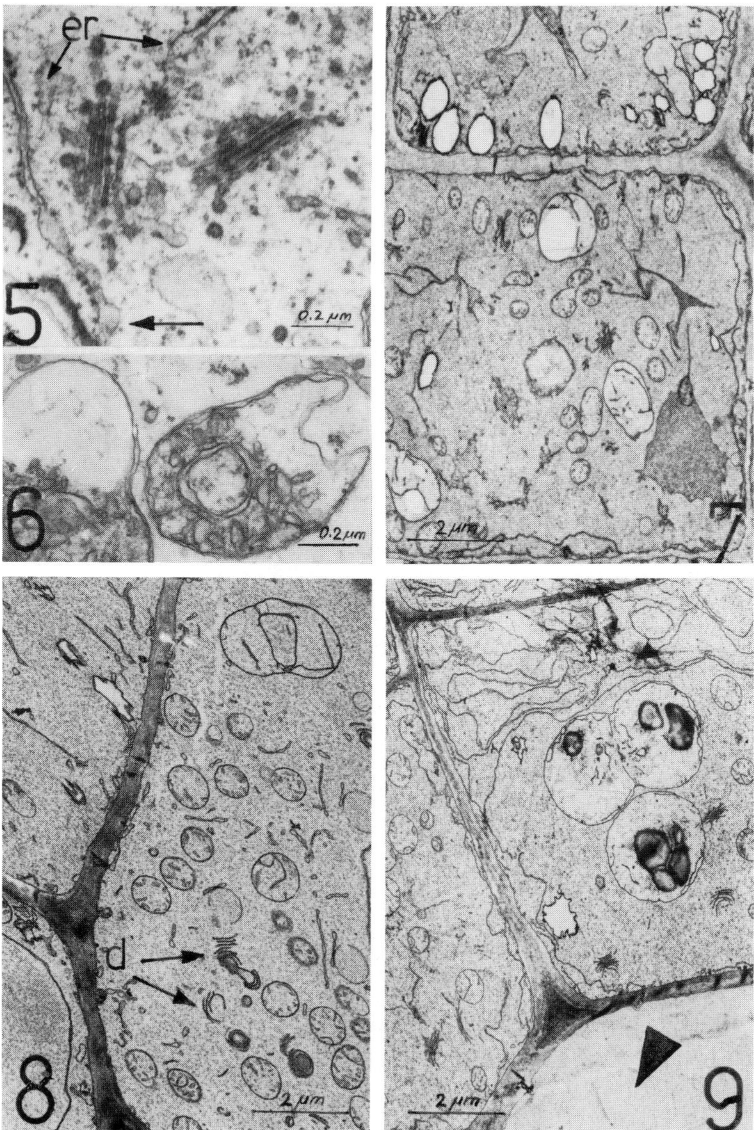

Fig. 5. Two dictyosomes after treatment with HU for 24 h. Note their disintegration and the swelling of the ER which probable precedes its disintegration (arrows). ×40 000.

Fig. 6. The swollen mitochondria from the cell showed in Fig. 5.

Fig. 7. Root cap initials treated with actinomycin D for 48 h. The ultrastructure of these initials without visible changes. ×6400.

Fig. 8. Cells from the middle layers of the root cap showed in Fig. 7. Note the more advanced disintegration of ER and dictyosomes (arrows). ×6400.

Fig. 9. Cells of the middle layers treated with HU for 48 h and after putting into fresh water for further 48 h. Note both the system of the rebuilt ER constituted of long cisternae and statolithic starch grains and dictyosomes. The protoplast of the more outer layer cells is completely disintegrated (arrow). ×6400.

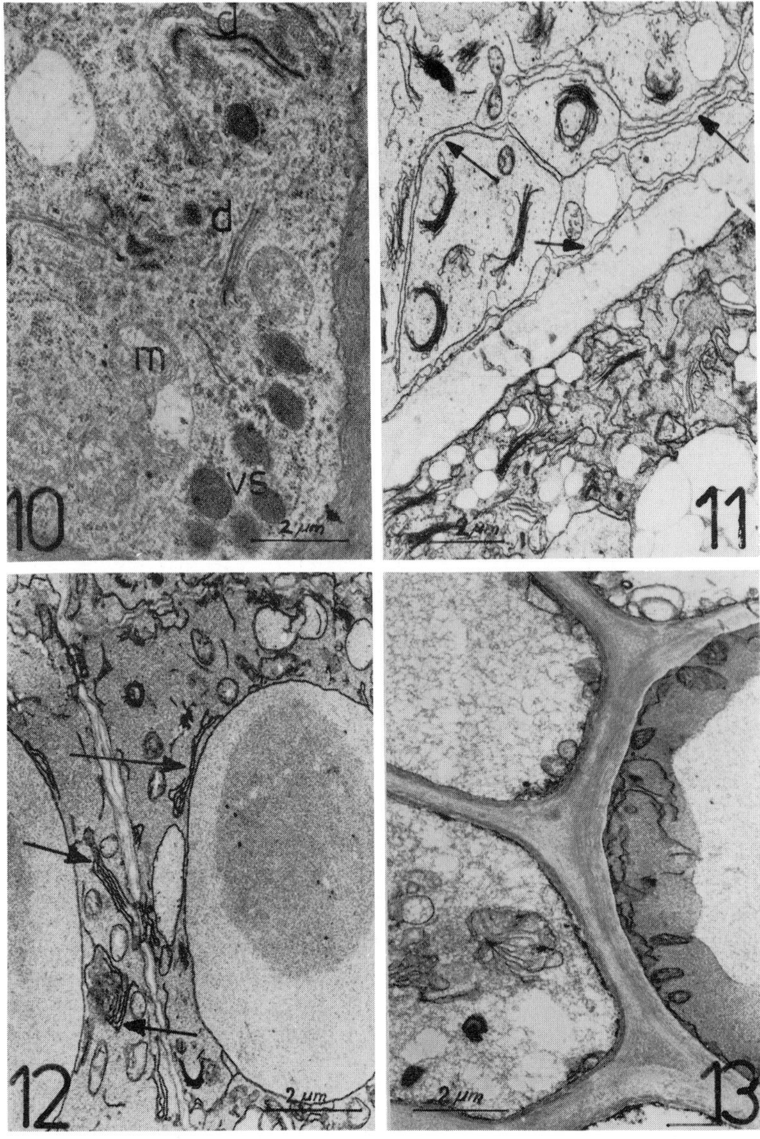

Figs. 10–11. Cells of the outer layers treated with streptomycin for 24 h (in Fig. 10 – fixed in glutaraldehyde and OsO_4, in Fig. 11 – in $KMnO_4$). ER (arrows) remains well developed, dictyosomes (d) and secretory vesicles (vs) preserve their individuality. ×6400.

Figs. 12–13. Cells treated with NAA for 48 h. Fig. 12 – from the middle layer. Note modified forms of ER (arrows). Fig. 13 – from the outer layer. The protoplasm is partly disintegrated. ×6400.

attacked earliest and most drastically. After 5 h bursting of the secretory vesicles and dissolution of their contents are observed (Fig. 4). ER at this time also undergoes disintegration (Fig. 5). The cisternae become shorter, less numerous and practically disappear and mitochondria and plastids gradually disintegrate. The majority of them show remarkable local blebbing (Fig. 6).

In the middle part of the root cap, the response to the chemicals is seen mainly in the reduction of the ER and dictyosomes (Fig. 8). No visible changes of the ultrastructure have been observed in the root cap initials (Fig. 7).

A common effect in the cells of all developmental zones of the root cap seems to be disturbance of the characteristic plasmalemma which breaks off from the cell wall and is visible following fixation with glutaraldehyde (Figs. 14, 16).

Slightly different changes take place in the ultrastructure of the root cap cells when treated with streptomycin. Although in the secretory system this antibiotic causes transformation of the vesicles of dictyosomal origin, in the dictyosomes themselves and in the ER system, these organelles preserve their individuality for a long time. ER is dissolved and develops into a more complex system (Fig. 11). In the presence of NAA morphological transformation is caused mainly by ER which occurs profusely and most frequently in greater densities (Fig. 12). All the chemicals cause irreversible changes in cells of the outer root cap leyers, resulting in death of the cells.

The ability to continue growth, after removing the chemical is preserved only by the cells of the middle layers and by the root cap initials and the ultrastructure characterizing the normal root cap cells is also restored (Fig. 9). The organelles form according to the principle of regeneration processes specific for the regeneration of a new root cap (Fig. 7). The cells with preserved ability to continue their growth, in the vicinity of the dead cells, are subsequently differentiated into cells of the outer root cap layers with a highly developed ER system, with dictyosomes and plastids filled with statolithic starch (Fig. 9).

After surgical excision of the root cap and cultivation of the roots in the presence of HU, the cells of the previous initial layer undergo insignificant redifferentiation and the ultrastructure of these traumatic cells does not show any changes. ER is still poorly developed and cisternae are generally short while dictyosomes remain inactive without hypertrophied vesicles around. Mitochondria are well formed and plastids preserve the form of proplastids (Fig. 14, 15) and only the outer part of the cytoplasm seems to be disturbed. Signs of its destruction and detachment from the cell wall, resembling plasmolysis, can also be seen (Fig. 14, 16).

Because HU and AD are known as DNA inhibitors their influence in the mature outer root cap cells seems to be interesting. For the same reason it is not

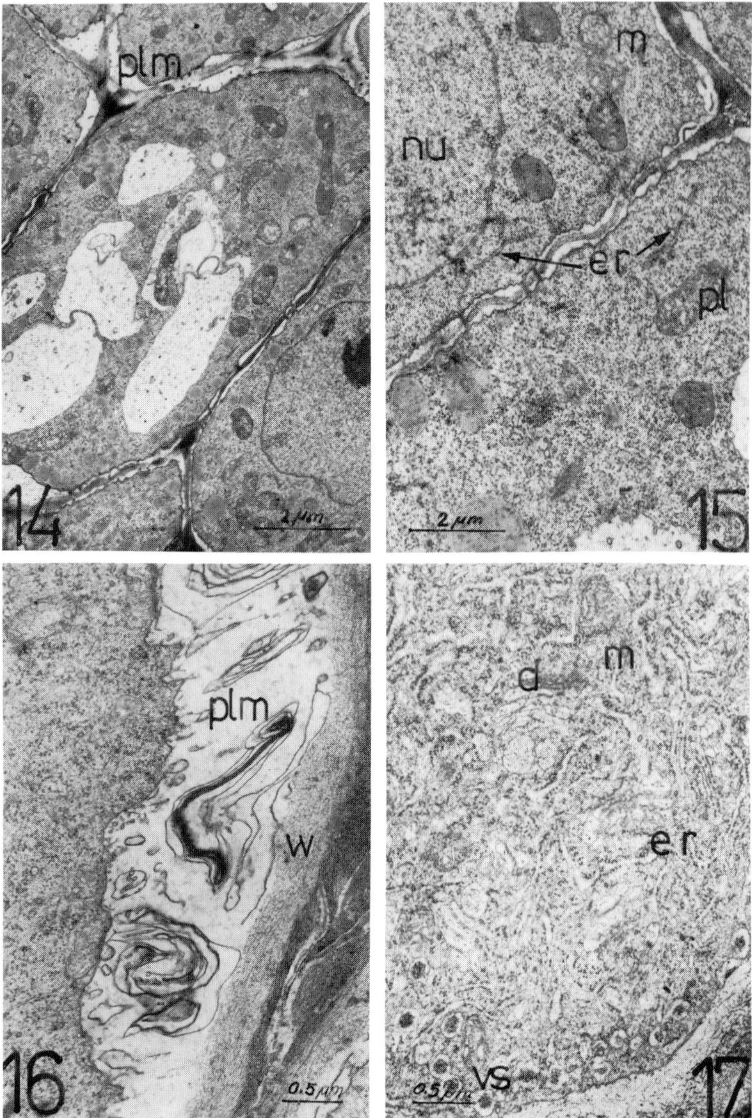

Figs. 14–16. The previous root cap initials after excising of the root cap and growing in the presence of HU for 48 h. No noticeable transformation of the organelles is seen. plm = the detachment from the cell wall resembling plasmolysis. × 6400 and 16 000.

Fig. 17. The previous root cap initials after excising the root cap and regenerating in normal conditions. Note the well developed system of ER (er), hypertrophied dictyosomes (d) and secretory vesicles (vs). × 16 000.

easy to understand that the ultrastructure of cells in the meristematic zone remains without visible changes in their organelles.

REFERENCES

1. Filho, S. A., Pereira de Almeida, E. R. and Gander, E. S. 1978 The influence of hydroxyurea and colchicine on growth and morphology of Trypanosoma cruzi. Acta Trop. 35, 3, 229–238.
2. Frank, V. 1977 Restoration of mitotic activity in novobiocin and actinomycin D treated root apices of Allium cepa L. Biologia (Bratisl.) 32, 12, 967–972.
3. Kadej, F. and Kadej, A. 1970 Ultrastructure of the root cap in *Raphanus sativus*. Acta Soc. Bot. Pol. XXXIX, 4, 733–737
4. Rosenkranz, H. S., Garro, A. J., Levy, J. A. and Carr, H. S. 1966 Studies with hydroxyurea. I. The reversible inhibition of bacterial synthesis and the effect of hydroxyurea of the bactercidal action of streptomycin. Biochim. Biophys. Acta 114, 501–515.
5. Torrey, J. G. 1957 Auxin control of vascular pattern in regenerating pea root meristems grown in vitro. Am. J. Bot. 44, 859–870.

12. DNA content in metaxylem of barley roots

ŠTEFAN KUBICA

Institute of Experimental Biology and Ecology, C.B.E.V. S.A.V., Dúbravská cesta 14, 885 34 Bratislava, C.S.S.R.

Differentiation of tracheary elements is a unique phenomenon among plant cells. The xylem cells grow quickly, form a secondary cell wall, autolyze their contents and perforate their transversal wall, forming coherent tube-like vessels. The changes of the DNA content in the nuclei are best studied during the first steps of differentiating the cells of the root. On the other hand, little information concerned with the whole process of forming vessels exists.

For studying the dependence of the DNA content on the degree of differentiation cells of the root metaxylem of barley have been chosen. The cells of the central metaxylem can be observed in the initials, differing by their size from the other cells of the central cylinder. The DNA content was determined cytophotometrically in the cell nuclei of the central metaxylem stained with Schiff's reagent, beginning with the cap junction up to the perforation of the transversal walls. An increase in the DNA content was observed in the nuclei from 2C to 32C.

In the first 1.0 mm segment of the metaxylem, the number of nuclei with a DNA content of 2C was low, being between 2C–4C at the prevailing number of cells. Cells with nuclei containing 2C of DNA were found up to a distance of 0.17 mm; however the cells divided up to 0.24 mm from the cap junction. This means a shortening of the G_1 phase, as, in the opposite case, cells had to be found with 2C DNA content in the whole area of cell division.

The cells of the central metaxylem do not stop DNA synthesis at values of 2C or 4C, but they continue and become polyploidic. As early as 32 mm from the cap junction, cells were noticed with DNA of 4C–8C and at a distance 0.57 mm with a DNA content of 8C–16C in their nuclei. The increase of the DNA content in the nuclei of the central metaxylem was of exponential character up to a value of 16C.

Further from the tip, increase of the DNA content was slower, up to a value of 32C, which was reached by the cells at a distance of 7–9 cm from the cap junction. During the linear increase of the DNA content in the nuclei, beginning with a distance of 6–6.5 cm, secondary cell wall formation was observed at the cells of the central metaxylem. After forming the secondary cell wall, some increase of

R. Brouwer et al. (eds.), Structure and Function of Plant Roots, 67–69.

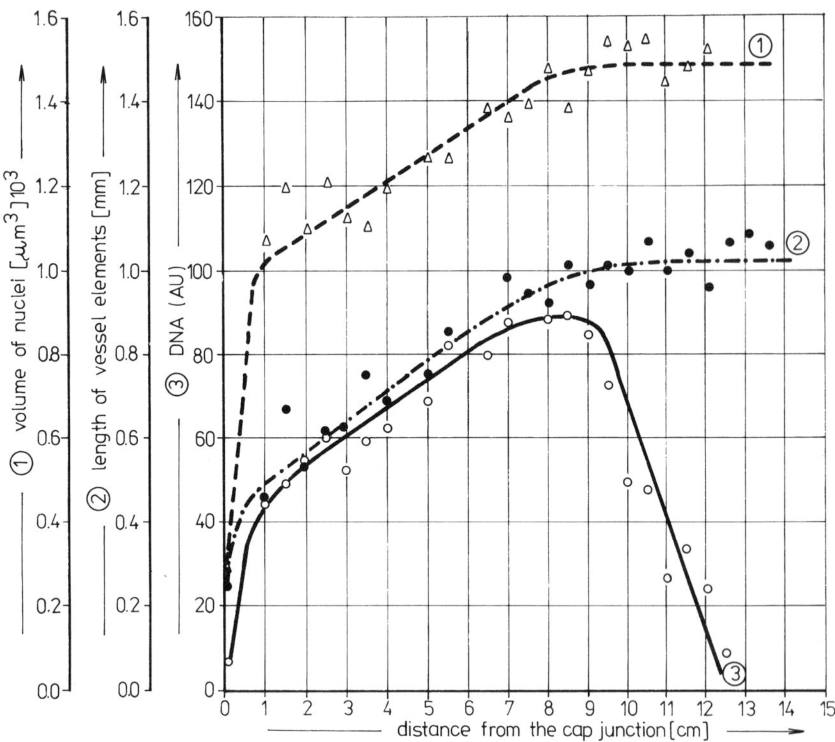

Fig. 1. Arrangement of the nuclei in central metaxylem with a different DNA content [3], different volumes of the nuclei [1] and different length of the cells [2].

the DNA content was observed in the nuclei. When a value of 32C was reached, no further DNA synthesis was observed in the nuclei, but it decreased in the nuclei of the central metaxylem, until reaching values of 8C–4C at a distance of 11–13 cm from the cap junction. Nuclei with a DNA content lower then 8C could not readily be determined, as DNA decrease lowered staining intensity, without changing the volume of the nuclei. The volumes of the nuclei and lengths of the cells increased similarly with the DNA content until the value of 32C was reached. A DNA content of 2C has been observed in embryos of *Pinus pinea* and *Lactuca sativa* [2] in the provascular cells. Similarly it can be seen from work dealing with the influence of cytokinin and auxin on xylem cell formation in tissue cultures [4], [5] that development of the secondary cell wall proceeds also in cells with a low DNA content. However, formation of the secondary cell wall in the central metaxylem of barley was observed before the end of the last endocycle. A small increase of the DNA was observed in polyploid xylem cells of maize leaves, which have already formed their secondary cell walls [3]. It is possible that a suitable ratio of auxin to cytokinin is reached during the last endocycle. According to the opinion of Barlow and Mc Donald [1] the relation of auxin and cytokinin levels determines only the process of DNA synthesis, from the viewpoint of the quality of the cycle and of its duration. Simultaneously the level of growth substances influences ultrastructure and synthetic ability of the cells, either directly, or indirectly through the cell cycle.

According to Roberts [6] a sufficient amount of specific functional RNA is necessary for initiating cytodifferention, in order to produce initiating protein and this is assumed to be only formed at a particular step of the cell cycle. It is, therefore, necessary for forming cytodifferentiating protein that the cells remain at the required period of the cell cycle for a certain time, either during mitosis, or during the endomitotic cycle. In the central metaxylem of the barley root formation of the secondary cell wall was observed during the long duration last endocycle. Prolonging the last endocycle seems to be important for initiating processes, directly related to the formation of the secondary cell wall.

REFERENCES

1. Barlow, P. W. and MacDonald, P. D. M. 1973 An analysis of the mitotic cycle in the root meristem of *Zea mays*. Proc. R. Soc. London B 183, 385–398.
2. Brunori, A. and D'Amato, F. 1967 The DNA content of nuclei in the embryo of dry seeds of Pinus pinea and Lactuca sativa. Caryologia 20, 153–161.
3. Lai, V. and Srivastava, L. M. 1976 Nuclear changes during differentiation of xylem vessel elements. Cytobiologie 12, 220–243.
4. Libbenga, K. R. and Torrey, J. G. 1973 Hormone-induced endoreduplication prior to mitosis in cultured pea cortex cells. Amer. J. Bot. 60, 293–299.
5. Matthyse, G. A. and Torrey, J. G. 1967 Nutritional requirements for polyploid mitoses in cultured pea root segments. Physiol. Plant 20, 661–672.
6. Roberts, L. W. 1976 Cytodifferentiation in plants. Cambridge, London, New York, Melbourne: Cambridge University Press.

13. A light microscopic study of the central metaxylem ontogenesis in the root of barley (*Hordeum distichum* L.)

ALEXANDER LUX JR.

Dept. of Plant Physiology, Comenius University, Odborárske nám. 5, 886 04 Bratislava, Czechoslovakia

The cells of the central metaxylem in the barley root differentiate from isodiametrical meristematic cells into mature cylindrical vessel members of maximal length up to 1,3 mm [4]. The prospective cells of the central metaxylem divide only briefly and the mitosis soon passes into endomitosis with a continuing DNA synthesis, thus giving rise to endopolyploid nuclei with the DNA content of as much as 32C [4]. During the differentiation of the central metaxylem in the barley root, changes have been followed of the size of vessel members, their activity and organization of their protoplasts with the light microscope. Seedlings of barley (*Hordeum distichum* L., cv. Slovenský dunajský trh) were grown on sheets of moist filter paper at 25 °C. Segments of primary roots were fixed by glutaraldehyde and osmium tetroxide and embedded into Durcupan ACM Fluka. Approximately 1 μm thick sections were stained with toluidine blue and basic fuchsine (A. Lux, in press).

RESULTS

Cells of the central metaxylem are easily recognizable by their position and size (Fig. 1A). Prospective vessel members form a longitudinal cell column, which can be observed from the third or fourth cell from the root cap junction (Fig. 1B, C). The first 5–7 cells in the longitudinal column are approximately isodiametrical of an average size of 11 μm. The cells have a large spherical nucleus with one nucleolus. Basipetally, from 100 μm to 320–350 μm from the root cap junction a mitotically active group of 30–35 disc-like cells can be seen (Fig. 1B, 2A). The height of these cells varies from 3 to 10 μm, in average 7.3 μm. Their nuclei are flattened, their shape conforms with the shape of the cells and 1–3 nucleoli were observed in them. This group of cells is characterized by the presence of small vacuoles. The disc-like cells are followed by cells which start to elongate (Fig. 2B). Their nuclei regain a spherical shape and each contains a single nucleolus. The vacuoles of these cells are still small and numerous, and can best be seen in the sections of the peripheral parts of cells (Fig. 2C). From 550 μm

R. Brouwer et al. (eds.), Structure and Function of Plant Roots, 71–76.

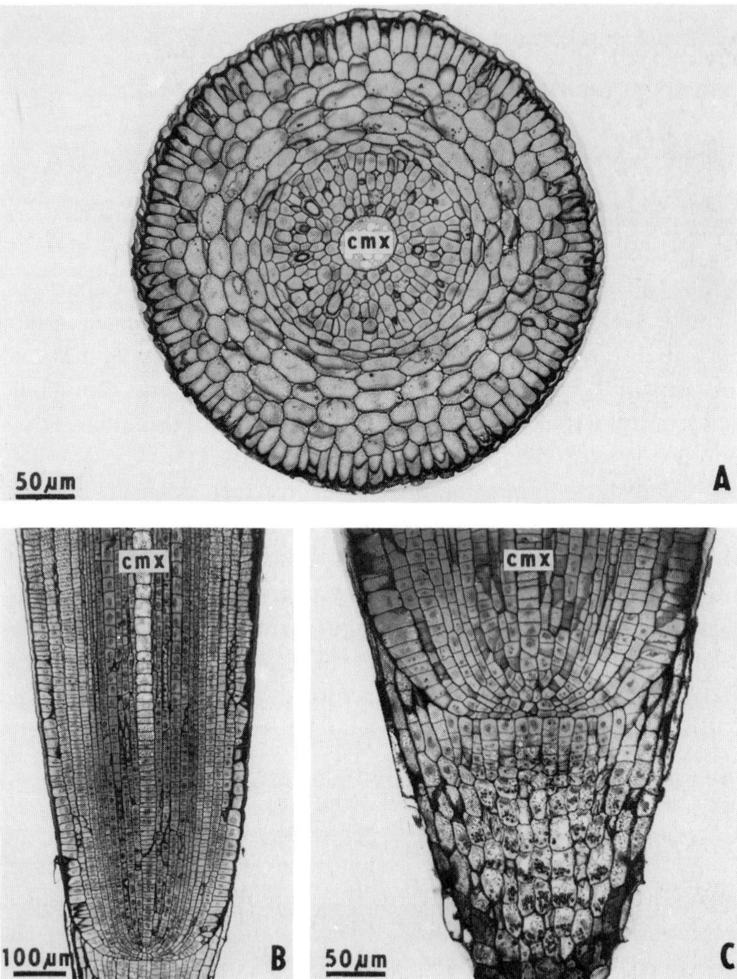

Fig. 1. Primary root of barley in transverse (A) and median longitudinal (B, C) sections. cmx – central metaxylem.

the elongation of cells is intensified and in these cells the fusion of small vacuoles and formation of larger ones was observed (Fig. 2D). In the second millimetre, the metaxylem vessel members continue to elongate, becoming cylindrical with a longer longitudinal axis (Fig. 3A). In this region the primary thickening of the end walls is observable and enlarging vacuoles gradually press the cytoplasm and the nucleus to the cell walls. The centre of the cell contains a central vacuole with

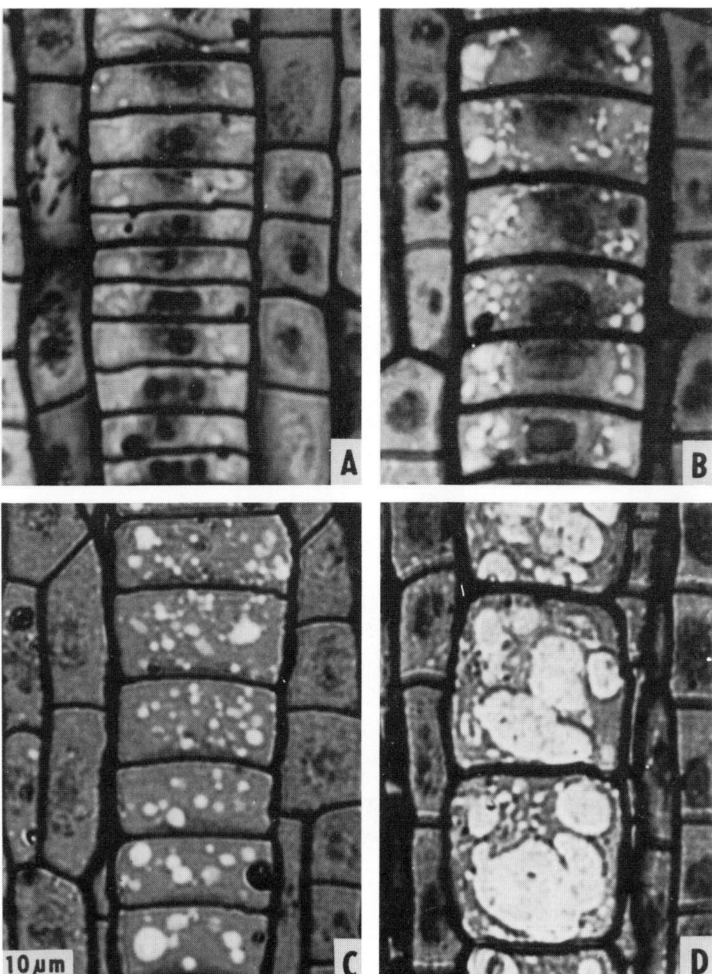

Fig. 2. Prospective vessel members of central metaxylem of barley root in longitudinal sections. A. Mitotically active cells. B. Cells at the beginning of elongation growth. C. The same cells as in B, the section of the peripheral parts of cells. D. The cells at a distance of approximately 560 µm from the root cap junction.

thin cytoplasmic bridges. In the third millimetre the vessel members have the form of elongated cylinders (Fig. 3B). The nucleus is situated beside the longitudinal cell wall approximately in its centre. After marking the root surface in one millimetre regions the most intensive elongation was found in the region of the 1st and 2nd mm. Elongation stops at a distance of 4–5 mm from the root cap

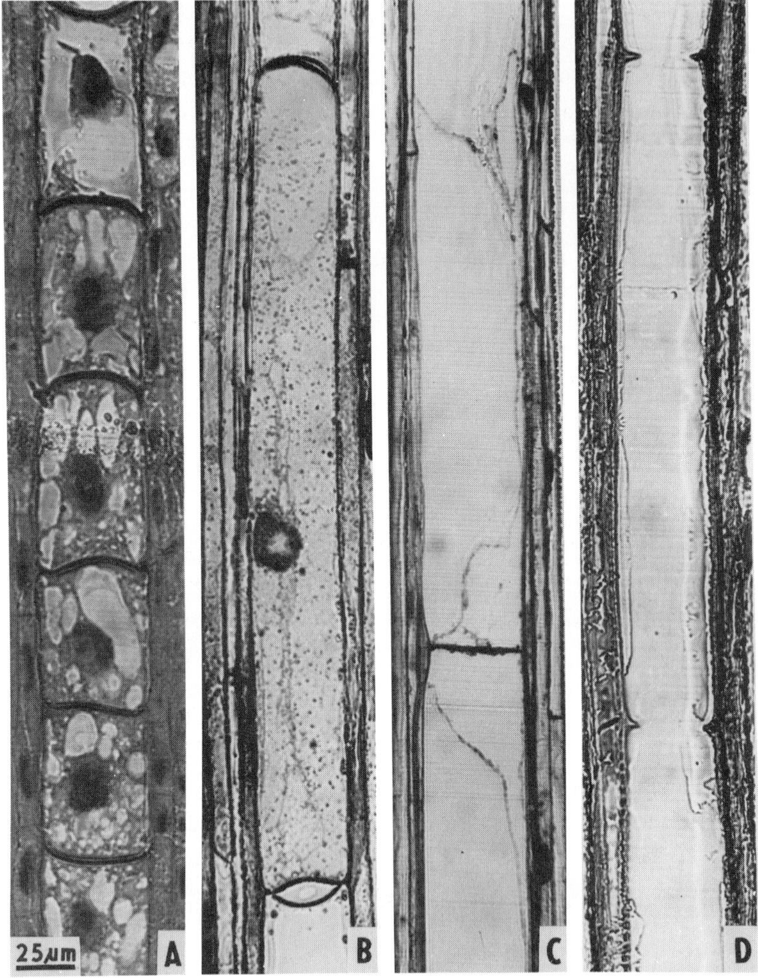

Fig. 3. Vessel members of central metaxylem of barley root in longitudinal sections at a distance A – 2 mm, B – 3 mm, C – 10 mm from the root cap junction. D. Detail of mature vessel at a distance of 145 mm.

junction and the course of differentiation of the vessel members slows up after attaining their final size. Fig. 3C shows a detail of vessel members from the tenth millimetre with cytoplasmic bridges and with thick end walls. The first signs of longitudinal wall thickening and formation of secondary walls were observed with the electron microscope at a distance of approximately 50 mm (A. Lux, unpublished). The perforation of end walls occurs at a distance of 120–140 mm. The fragments of the end walls of the individual vessel members are inside the mature vessel in the shape of narrow rims. In sections they appear as triangular protrusions from the longitudinal walls (Fig. 3D). Vessel members of the central metaxylem are fused, after perforation of the end walls, into continuous tubular vessels.

DISCUSSION

The large central metaxylem vessel is one of the characteristic anatomical features of the primary root of barley [3, 2]. Prospective vessel members of this solitary vessel divide almost exclusively in a transverse manner. Longitudinal cell division and the consequent formation of two central cell columns has been observed only rarely [2].

Cell division in the central metaxylem of the barley root cultivated at 25 °C stops at a distance of 300–350 μm from the root cap junction [5]. The value referred to, which is also in agreement with our finding, was confirmed also by the establishment of the DNA content in the nuclei of these cells [4]. In the region up to 320 μm from the root cap junction, the cells in the central metaxylem exhibited a DNA content 2C, 2C–4C and 4C. Within the range from approximately 320 μm to 620 μm, the DNA content in the nuclei was 4C–8C and 8C. More proximally the DNA content 8C–16C occurred up to the distance of approximately 1000 μm. The increase of the DNA connected with the first endocycle, was obviously realized in the cells at an approximately equal stage of differentiation in the region of 350–550 μm from the root cap junction. Intensification of elongation and vacuolation in the cells situated basipetally from the distance of 550 μm appears to be a specific feature of the second endocycle.

Early vacuolation has been reported, at a distance of 100 μm [3] in the cells of the central metaxylem. In the electron microscope, however, provacuolar bodies and small vacuoles are already visible from the beginning of the central cell column (A. Lux, unpublished).

The large cells of the central metaxylem in the barley root, although determined as the first ones, exhibit a slow course of differentiation. The mitotically active section of the cell column is short and is limited to 35–40 cells. Within approximately 4 mm the cells complete their elongation and remain without secondary walls up to the distance of 50 mm. The late perforation of end walls, in the investigated case at a distance of 120–140 mm, is generally characteristic of the metaxylem [1]. It is presumed that only after perforation the central metaxylem fully is able to carry out the function of transport of aqueous solutions.

REFERENCES

1. Esau, K. 1965 Vascular differentiation in plants, New York: Holt, Rinehart and Winston.
2. Hagemann, R. 1957 Anatomische Untersuchungen an Gerstenwurzeln. Die Kulturpflanze 5, 75–107. Berlin: Akademie Verlag.
3. Heimsch, Ch. 1951 Development of vascular tissues in barley root. Amer. J. Bot. 38, 523–537.
4. Kubica, Š. 1976 DNA content in nuclei of metaxylem in the root of barley during ontogenesis (in Slovak). Thesis. Bratislava: Ú.E.B.E. SAV.
5. Luxová, M. 1975 Some aspects of the differentiation of primary root tissues. In: Torrey, J. G. and Clarkson, D. T. (eds.), The development and function of roots, pp. 73–90. London: Academic Press.

14. Correlative morphogenesis of endodermis and xylem elements in the developing barley root

M. F. DANILOVA

Komarov Botanical Institute of the Academy of Sciences of the U.S.S.R., Leningrad 197022, U.S.S.R.

Some data on the ultrastructure of barley root tissues has been published earlier [3, 4]. This paper presents the results of an electron microscopic study of endodermis and xylem elements at successive levels in barley root. Based on the external features of the differentiation 6 zones can be recognized in the root. Ultrastructural characteristics of endodermis and xylem elements along the zones are as follows.

(1) Zone preceding that of the root hairs (3–5 mm from root apex). Casparian bands in the endodermis are not yet formed (Fig. 1a); protoxylem tracheary elements are at the stage of differentiation; metaxylem elements are at the stage of growth and vacuolation; cytoplasm in all xylem elements appears very active (Fig. 1b, c).

(2) Zone of growing root hairs (6–9 mm). The formation of the Casparian bands in the endodermis is completed (Fig. 2a) and is followed by maturation of protoxylem tracheary elements (Fig. 2b). Central vacuoles are formed and secondary cell wall deposition in the peripheral vessels can be observed (Fig. 2b) in both peripheral and central metaxylem elements; cytoplasm in the central metaxylem cells appears active (Fig. 2c).

(3) Zone of mature root hairs (15–18 mm). The endodermis is in the state I. The peripheral metaxylem vessels are mature and xylem parenchyma and central vessels have active cytoplasm.

(4) Zone of lateral root initiation (60–70 mm). Most of the endodermal cells are in the primary state, but in some of them incipient suberization is observed; peripheral metaxylem vessels are mature; secondary cell wall deposition occurs both in the central metaxylem vessel, and in the peripheral metaxylem parenchyma.

(5) Zone of unbranched lateral roots (100–110 mm). The endodermis is in the state II (Fig. 3a) and partly in the initial state III. Secondary cell wall formation occurs in the central vessel and sclerification of xylem parenchyma cells is in progress (Fig. 3b).

(6) Zone of branched lateral roots (150–180 mm). The endodermis is in state III

R. Brouwer et al. (eds.), Structure and Function of Plant Roots, 77–83.

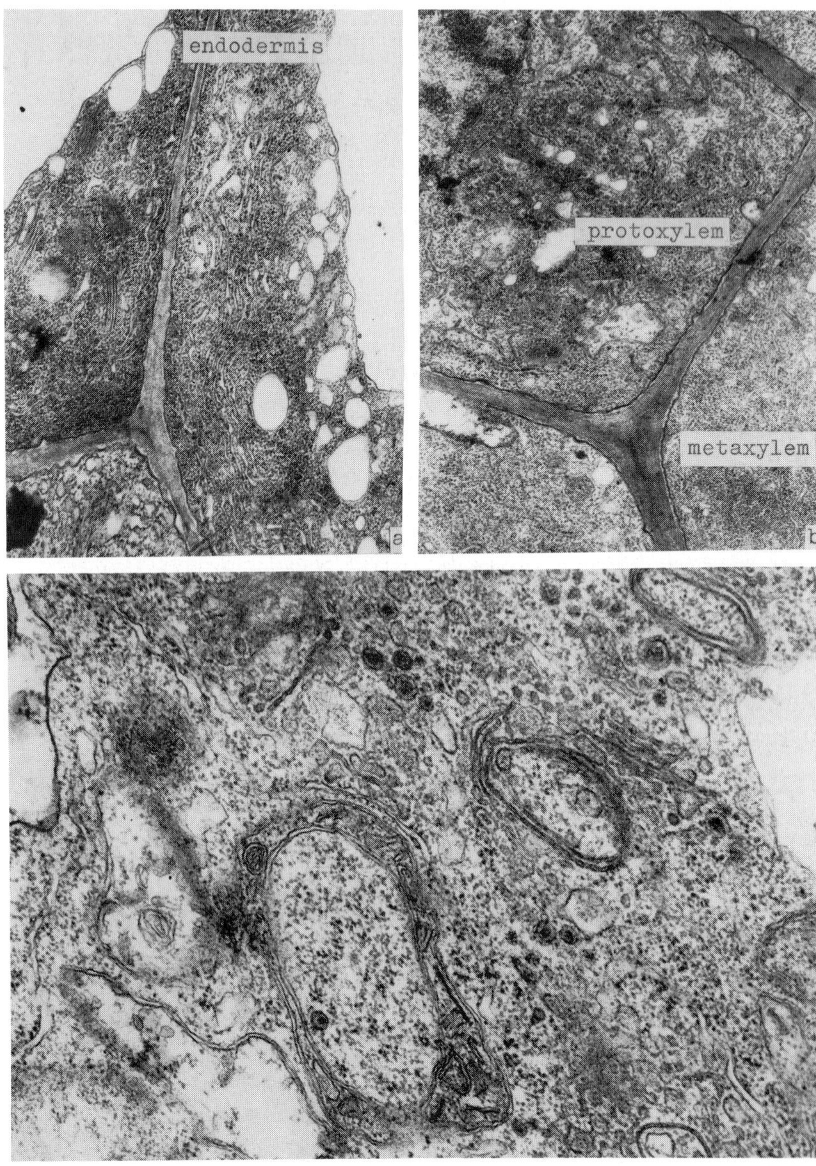

Fig. 1. Ultrastructure of endodermis and tracheary elements in the zone preceding the zone of root hairs: cell wall between the two endodermal cells (a), proto- and metaxylem tracheary elements (b), fragment of the central vessel (c).

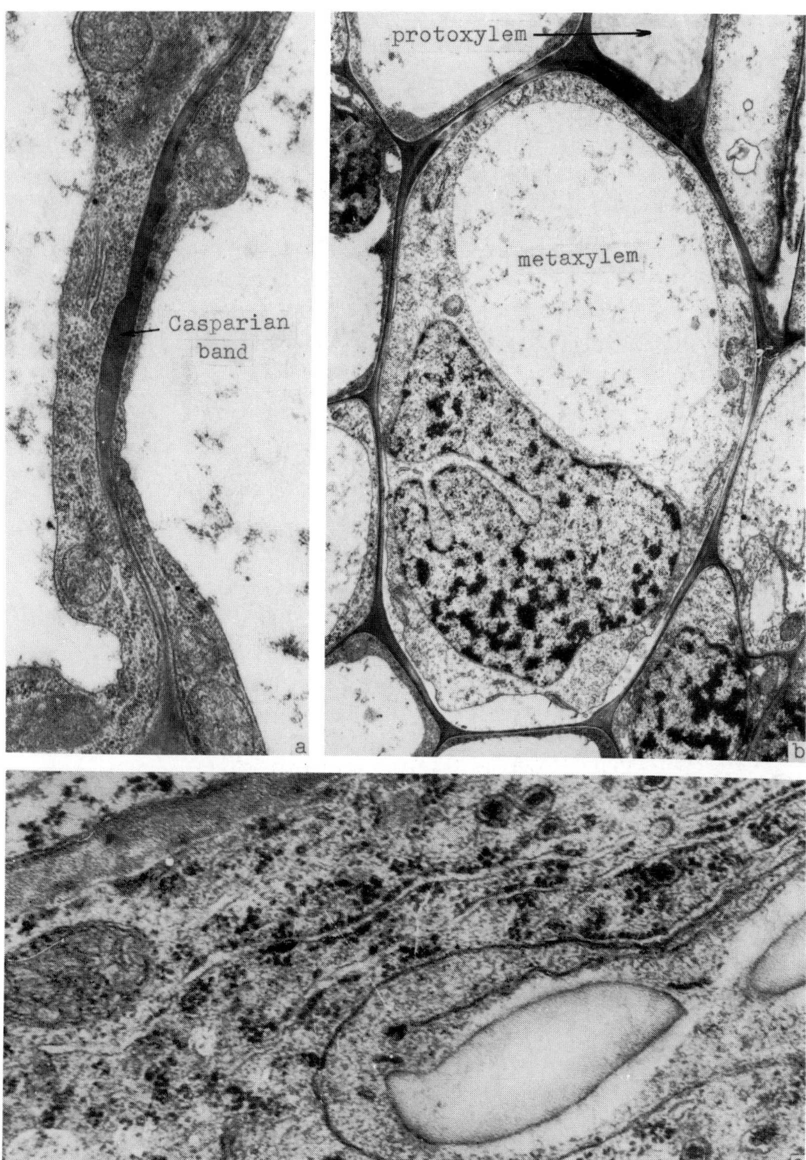

Fig. 2. Ultrastructure of endodermis and tracheary elements in the zone of growing root hairs: cell wall with Casparian band between two endodermal cells (a), proto- and metaxylem elements (b), fragment of differentiating central vessel (c).

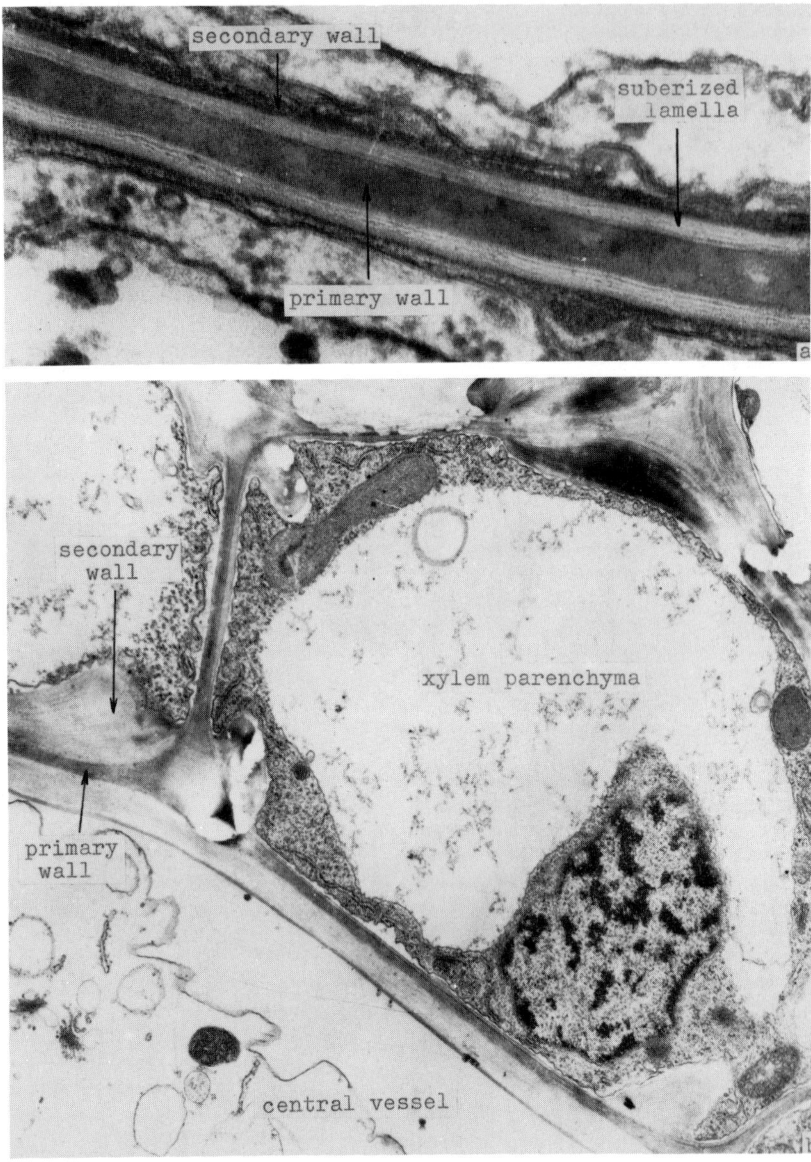

Fig. 3. Suberization of endodermal cell walls (a) and secondary depositions on the walls in xylem parenchyma cells (b) in the zone of unbranched lateral roots.

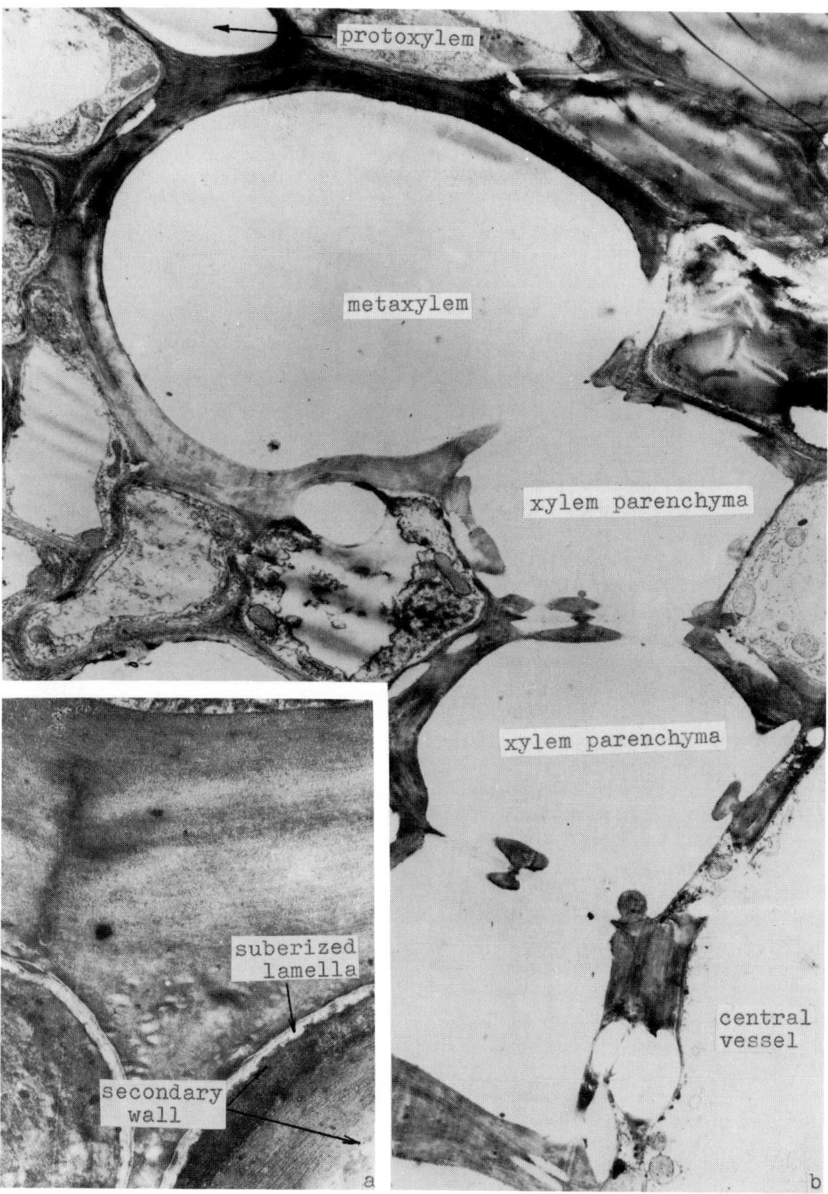

Fig. 4. Endodermis (a) and xylem cells (b) in the zone of branched lateral roots.

(Fig. 4a). Some of the xylem parenchyma cells have lost their cytoplasm and the central vessel has a nucleus and a thin layer of cytoplasm (Fig. 4b).

These observations show that differentiation of specialized elements of the root tissues terminates at different levels of the developing root. This apparent asynchrony is under genetic control and is part of the essential pattern of the root development, which determines the strong correlation between tissue structure and function. In the primary state of the root growth there is no level at which all stelar cells are completely mature. At any level beginning from the zone of growing root hairs immature xylem parenchyma cells and differentiating vascular elements occur side by side with mature elements.

The ultrastructure of every given cell type is different at different stages of the development. At the early stage of differentiation all types of xylem cells are very rich in cytoplasm. This is the feature which was used as a structural base for an assumption of the secretory function of xylem parenchyma cells [3]. However, these cells cannot be regarded as specialized for ion secretion since they lack specific features of ion secreting cells [5]. In spite of the number of essential structural and functional differences between tracheary elements and xylem parenchyma cells some patterns of their specialization in barley root are rather similar. Both types of cells develop secondary cell walls and undergo sclerification followed by the loss of the cytoplasm. Therefore, the general ultrastructural characteristics of the cytoplasm of all xylem components are rather similar. Differences observed at the same level of the root are the consequence of their asynchronous differentiation.

The data on developmental dynamics and ultrastructure of the stelar cells support the Hylmö [2]–Higinbotham [1] hypothesis of the active role of differentiating tracheary elements in the production of xylem exudate. Differentiating vessels at early developmental stages represent the final link in the symplast of primary root tissues. During the growth period the future vessels are in a state of high metabolic activity and obtain the necessary nutritive elements from the neighbouring cells by the symplastic pathway. These stages of development of peripheral and central metaxylem elements belong to the zone situated between the root apex and zone of growing root hairs. During the following prolonged period of differentiation until the onset of lignification the vessels lose their symplastic connections with neighbouring cells but remain viable and can absorb ions from the stelar free space. Most of the metabolic products and unused ions are accumulated in their vacuoles. The vacuolar sap filling the cavities of tracheary elements after disintegration of their protoplasts is transformed into most of xylem exudate. In barley and other grasses, xylem parenchyma cells while losing their cytoplasm at the final step of differentiation can also add to the xylem exudate some contents of their vacuoles which enters the vessels through lateral pits.

Casparian bands appear on the radial walls of endodermal cells in the region of growing root hairs, prior to the maturation of protoxylem elements. Distal from this region two radial transport pathways, symplastic and apoplastic, operate in the root. Owing to the Casparian bands the free space (the apoplast) of the root tissues becomes divided into two parts, the outer and inner ones. However, no barriers appear between cortex and stele along the symplastic pathway. Ion leakage from the dead tracheary elements into the free space of the cortex is prevented by the presence of Casparian bands in the endodermis and these bands function as barriers for both inward and outward ion movement.

REFERENCES

1. Higinbotham, N., Davis, R. F., Mertz, S. M. and Shumway, L. K. 1973 Some evidence that radial transport in maize roots is into living vessels. In: Ion transport in plants, pp. 493–506. London: Anderson.
2. Hylmö, B. 1953 Transpiration and ion absorption. Physiol. Plantarum 6(2), 333–405.
3. Läuchli, A., Kramer, D., Pitman, M. G. and Lüttge, U. 1974 Ultrastructure of xylem parenchyma cells of barley roots in relation to ion transport to the xylem. Planta 119, 85–99.
4. Robards, A. W., Jackson, S. M., Clarkson, D. T. and Sanderson, J. 1973 The structure of barley roots in relation to the transport of ions into the stele. Protoplasma 77, 291–312.
5. Thomson, W. W. 1975 The structure and function of salt glands. In: Plants and saline environments, pp. 118–146. Springer-Verlag.

15. Division and differentiation during regeneration at the root apex

PETER W. BARLOW

Agricultural Research Council, Letcombe Laboratory, Wantage, Oxon, OX12 9JT, England

If cells are removed from a plant organ the remaining cells respond by forming a seal over the wound and in some cases can even regenerate the lost tissue. The cells involved in regeneration or repair first dedifferentiate their already established state and then redifferentiate along new developmental pathways [2].

In animals, regeneration of lost cells or tissues proceeds either with the accompaniment of cell division (epimorphosis) or in the absence of division (morphallaxis). In the latter situation the new tissue is regenerated by a remodelling of the pre-existing tissue. Regeneration in both plants and animals is normally epimorphic and, perhaps as a consequence of this, it has been postulated that cell division is an event essential for switching a cell from one developmental pathway to another [see 3]. Such a theory would lead us to conclude that epimorphosis is the norm since redifferentiation of the remaining cells could not occur unless they also divided. However, there is little incontrovertible evidence that division is essential for differentiation and exceptions to the interdependence of the two processes have been reported [3]. In view of all this it is of interest to know whether morphallactic regeneration can occur in plant tissues.

EXPERIMENTAL

When the root cap of *Zea mays* is removed the cells remaining at the apex divide and a new cap eventually regenerates [1]. This is normally an epimorphic system, but it is also one which can be used to test whether a morphallactic response can be forced upon it. Markers of cap regeneration include the presence of a particular distribution of cells containing large amyloplasts (statenchyme), the synthesis and secretion of slime by the outer cells and their sloughing from the cap surface, and the loss of mitotic potential in maturing cap cells. Since the root cap is normally the site of graviperception, the recovery of a gravitropic response can serve as a marker of the functional state of the regenerated cap.

Decapped roots were grown in solutions of various inhibitors of cell division for up to 4 days at 22 °C; after this time cap regeneration is complete in water-grown (control) roots. Inhibitors which seemed to interfere least with the cellular

R. Brouwer et al. (eds.), Structure and Function of Plant Roots, 85–87. All rights reserved.

properties used to assess regeneration were 5-aminouracil (AU) and hydroxyurea (HU). Both these agents limit cell division by interfering with DNA synthesis. 10 mM HU and 5 mM AU permitted, respectively, a maximum of 0 and 2 divisions to occur within the decapped apex during the 4 day period. In control decapped apices at least 3–4 divisions occurred during this period. The extent of the statenchyme within the apices of inhibitor-treated roots was similar to that found in control regenerating roots. Slime synthesis occurred in the outer layer of cells at the tip of both control and experimental roots. Moreover, in the latter roots these outermost cells did not divide following decapping, yet they exhibited the same sequence of differentiation (starch build-up, then its breakdown, and slime synthesis) as would normally occur with the accompaniment of mitosis. Such roots were also positively gravitropic suggesting that within the remodelled cap-like apex the hormonal system thought to mediate gravitropism had also regenerated.

Roots grown in 10 mM HU for 4 days were returned to water whereupon the cells resumed division. In the apices of these roots divisions occurred in a position equivalent to that normally occupied by the meristem, but did not occur in the most distal cells which had already differentiated as cap cells. This result suggests that the internal conditions that normally specify the extent of division within the cap tissue had also been regenerated in spite of the fact that no division had actually taken place. The pattern of division in these recovering roots reflects the pattern of the internal division-limiting conditions.

CONCLUSIONS

Following decapping, a group of cells with the properties of cap cells can develop at the apex even though cell division may be either totally inhibited or severely impaired. The various cell-types within this cap also occupy the correct spatial arrangement. Therefore, the apex can remodel itself to form a cap-like tissue; that is, it shows a morphallactic response.

A conclusion from these observations is that differentiation of cap cells is not dependent on mitosis. Differentiation may be viewed as the consequence of a domain of morphogens that exists within the tissue; cells are important because they are the units within this domain that give expression to differentiation in ways specified by the morphogens. In the absence of division (or even in its presence) regeneration of the cap occurs because capdetermining gradients of morphogens are regenerated within the remaining apex; cells recognizable as cap-cells differentiate in response to these gradients. If this view of differentiation is acceptable, then controversy over whether cell division and differentiation are mutually exclusive events [see 3] loses meaning since these two programmes that operate in developmental processes may be viewed as being executed independently and in parallel, rather than interdependently and in series.

REFERENCES

1. Barlow, P. W. 1974 Regeneration of the cap of primary roots of Zea mays. New Phytol. 73, 937–954.
2. Gautheret, R. J. 1966 Factors affecting differentiation of plant tissues grown *in vitro*. In: Cell differentiation and morphogenesis, pp. 55–95. Amsterdam: North-Holland Publishing Co.
3. Roberts, L. W. 1976 Cytodifferentiation in plants. Xylogenesis as a model system. Cambridge University Press.

16. Root growth and gravireaction: endogenous hormone balance

PAUL EMILE PILET

Institute of Plant Biology and Physiology of the University, CH-1005 Lausanne, Switzerland

The role originally ascribed to indolyl-3-acetic acid (IAA) in the classical hypothesis first proposed by Cholodny [1] and Went [17] has been filled – for root growth and gravireaction – by abscisic acid (ABA) which could be considered [4, 18] as one of the endogenous growth inhibitors [16] produced (at least for maize roots) by the cap cells and moving in the basipetal direction [4–8; 19, 20].

IAA was found – using the GC-MS technique – to be present in the maize root tip [14] as showed in Table 1. But ABA was also detected by a similar method and for the same material [15] as indicated in Table 2. These two series of data, previously discussed [7, 8, 13, 37] in terms of hormonal balance, clearly indicate

Table 1. IAA content (GC-MS determinations) of different regions of maize (cv. Kelvedon 33) roots

Root parts (in mm from the tip)	µg of IAA* per kg F.W.	µg of IAA (± standard error) per g. D.W.
0.0–0.5[1]	356.6 ± 16.2	2.674 ± 0.121
0.5–1.0[2]	179.9 ± 23.4	1.158 ± 0.151
1.0–4.0[3]	76.5 ± 28.3	0.562 ± 0.208

* F.W.: fresh weight D.W.: dry weight
[1] cap cells
[2] apex (quiescent centre, meristem)
[3] elongation zone

Table 2. ABA content (GC-MS determinations) of different regions of maize (cv. Kelvedon 33) roots (see Table 1)

Roots parts (in mm from the tip)	µg of ABA per kg F.W.	µg of ABA (± standard error) per g. D.W.
0.0–0.5	36.1 ± 7.6	0.271 ± 0.057
0.6–1.0	66.5 ± 18.4	0.428 ± 0.118
1.0–4.0	33.3 ± 0.9	0.245 ± 0.006

R. Brouwer et al. (eds.), Structure and Function of Plant Roots, 89–93.

that in the cells of the root cap, IAA and other auxins could be accumulated and ABA and other growth inhibitors could be formed or released.

Endogenous IAA plays an essential role in the regulation of root growth [2, 3, 11] and it has been observed [11] that elongation of maize root apical segments which had had their endogenous IAA concentration reduced by an exodiffusion technique, was stimulated by IAA applied to their basal cut ends, whereas growth of untreated root segments was inhibited by all tested concentrations of IAA. In the former assays, the endogenous IAA content – monitored by a highly specific and sensitive spectrophoto-fluorimetric method – of detipped apical root segments was strongly reduced by allowing the segment to transport IAA acropetally into buffered agar blocks.

Data related to the effects of exogenous auxin on root gravireaction have tended to be contradictory. In some reports, it has been noticed that high concentrations of IAA may reduce the gravicurvature, while low concentration may enhance the gravireactivity. In some others, it has been claimed that IAA may abolish or reverse the graviresponses. This confusing situation might be explicable on the basis of differences in endogenous auxin content before an IAA treatment [Elliott M. C. and Pilet P. E., unpublished; Pilet P. E. and Rebeaud J. E., in preparation] and also by the fact that auxin level, in growing roots, may change with age [3].

In a recent paper [10] it was reported that the gravireaction of apical root segments of maize (cv. Orla) was abolished by detipping the segments and was then restored by replacing the tips upon the apical cut section. After exodiffusion of endogenous IAA the retipped segment showed a significantly lower gravireaction. Application of low IAA concentration to the basal cut end of the root segments increased the graviresponse of retipped segments. These results clearly indicate that endogenous IAA may play an essential role in the regulation and the control of the root graviresponses.

For most root species, ABA was found to significantly reduce their root elongation [9]. But little is known about ABA effects on root gravitropism. It has been reported [18] that an ABA pretreatment may induce the gravireaction of maize roots kept in the dark, these roots being only gravireactive in light. However, ABA cannot induce graviresponses in darkexposed decapped roots. Quite similar data were obtained with apical segments of maize (cv. LG 11) treated or not with ABA and gravireacting both in the dark or in light [12]. As can be seen in Table 3, ABA enhanced the gravireactivity and this much more strongly for root segments kept in darkness than for those illuminated.

Table 3. Gravireaction* (in degrees ± standard error) of maize (cv. LG 11) root segments pretreated (1 h of total immersion) with a buffered (pH 6.1) solution containing (+) or not (−) ABA at 10^{-5} mol dm^{-3}

Pretreatment	Darkness	Light
− ABA (C)	19.3 ± 2.6	42.4 ± 6.3
+ ABA (TR)	35.2 ± 5.3	55.3 ± 4.8
per cent of stimulation[1]	82	30

* Measurement after 6 ± 0.5 h
[1] per cent = 10^2.(TR–C)/C

Table 4. ABA content (in ng per 100 segments and in ng per mg of D.W.) in 5 ± 0.1 mm root tips, after 2 h in horizontal position, of 10 ± 0.2 mm maize (cv. LG 11) root segments maintained in darkness or in light

	Darkness (D)	Light (L)
ng*/100 segments	22.8 ± 2.4	31.1 ± 4.0
per cent[1]	0	36
ng*/mg D.W.	0.57 ± 0.01	0.69 ± 0.03
per cent[1]	0	21

* Measurement after 6 ± 0.5h
[1] per cent = 10^2.(L–D)/D

On the other hand, it has been also observed [12] that the ABA level for these segments maintained horizontally was lower when kept in the dark than in light as presented in Table 4. More recently [Chanson A. and Pilet P. E., unpublished] it was observed that ABA – applied on the tip of roots, using a buffered droplet – significantly enhanced the gravireaction of maize root.

In a paper in press [Pilet P. E. and Rivier L.], the redistribution of endogenous ABA, under gravity action (transversal gradient) was analysed for the elongating – and gravicurving – part of the gravireacting (positive, negative and ageotropic responses) maize (cv. LG 11) apical root segments kept in humid air. The significant increase in the ABA content in the lower half (and the concomitant decrease in the upper part) of the horizontally placed root showing a positive graviresponse strongly indicates that ABA essentially acts on the regulation and the control of the growth and the georeaction of maize roots under the action of gravity.

In conclusion, it seems quite clear now that the control of the axial growth of the primary roots as well as the gravireaction of these roots involves the interaction – in the elongation zone – of several endogenous hormones [13]. Two of these regulators (IAA and ABA) may at least play an essential role in these processes. IAA (and other auxins) moves preferentially in the acropetal direction and ABA (and some growth inhibiting substances) is transported preferentially in the acropetal direction.

Consequently, the importance of the root cap is now quite evident. It not only acts as a part of the root which protects the root meristematic cells and which facilitates the penetration of the root into the soil (slime, produced by the cap cells may lubricate the tip). It is the site of the perception of gravity and, for some roots, the perception of light. The root cap has to be considered as the site of auxin accumulation (flow from the base to the apical part) and as the site of formation and the release of the growth inhibitors (flow from the tip to the base).

REFERENCES

1. Cholodny, N. 1926 Beitrage zur Analyse der geotropischen Reaktion. Jahrb. wiss. Bot. 65, 447–459.
2. Elliott, M. C. 1977 Auxins and the regulation of root growth. In: Pilet P. E. (ed.), Plant Growth Regulation, pp. 100–108. Berlin, Heidelberg, New York: Springer.
3. Pilet, P. E. 1961 Auxins and the process of aging in root cells. In: Klein R. M. (ed.), Plant Growth Regulation, pp. 167–178. Iowa State University Press, Ames.
4. Pilet, P. E. 1975 Abscisic acid as a root growth inhibitor: physiological analyses. Planta 122, 299–302.
5. Pilet, P. E. 1976 The light effect on the growth inhibitors produced by the root cap. Planta 130, 245–249.
6. Pilet, P. E. 1976 Effects of gravity on the growth inhibitors of geostimulated roots of Zea mays L. Planta 131, 91–93.
7. Pilet, P. E. 1977 Growth inhibitors in growing and geostimulated maize roots. In: Pilet P. E. (ed.), Plant Growth Regulation, pp. 115–128. Berlin, Heidelberg, New York: Springer.
8. Pilet, P. E. 1979 Kinetics of the light-induced georeactivity of maize roots. Planta 145, 403–404.
9. Pilet, P. E. and Chanson, A. 1980 Effect of absicis acid on maize root growth. A critical examination. Plant Sci. Lett. (in press).
10. Pilet, P. E. and Elliott, M. C. 1981 Some aspects of the control of root growth and georeaction: the involvement of IAA and ABA. Plant Physiol. (in press).
11. Pilet, P. E., Elliott, M. C. and Moloney, M. M. 1979 Endogenous and exogenous auxin in the control of root growth. Planta 146, 405–408.
12. Pilet, P. E. and Rivier, L. 1980 Light and dark georeaction of maize roots: effect and endogenous level of abscisic acid. Plant Sci. Lett. 18, 201–206.
13. Reinhold, L. 1978 Phytohormones and the orientation of growth. In: Letham, D. S., Goodwin, P. B. and Higgins, T. J. V. (eds.), Phytohormones and related compounds: a comprehensive treatise, Vol. II, pp. 251–289. Amsterdam: Elsevier, North-Holland.
14. Rivier, L. and Pilet, P. E. 1974 Indolyl-3-acetic acid in cap and apex of maize roots: identification and quantification by mass fragmentography. Planta 120, 107–112.
15. Rivier, L., Milon, H. and Pilet, P. E. 1977 Gas chromatography-mass spectrometric determinations of abscisic acid levels in the cap and the apex of maize roots. Planta 134, 23–27.
16. Wain, R. L. 1977 Root growth inhibitors. In: Pilet P. E. (ed.), Plant Growth Regulation, pp. 109–114. Berlin, Heidelberg, New York: Springer.

17. Went, F. W. 1928 Wuchsstoff und Wachstum. Rec. Trav. bot. néerl. 25, 1–116.
18. Wilkins, H. and Wain, R. L. 1975 Abscisic acid and the response of the roots of Zea mays L. seedlings to gravity. Planta, 126, 19–23.
19. Wilkins, M. B. 1977 Geotropic response mechanisms in roots and shoots. In: Pilet, P. E. (ed.), Plant Growth Regulation, pp. 199–207. Berlin, Heidelberg, New York: Springer.
20. Wilkins, M. B. 1979 Growth control mechanisms in gravitropism. In: Haupt, W. and Feinleib, M. E. (eds.), Encyclopedia of Plant Physiology NS Vol. 7, Physiology of Movements, pp. 601–626.

17. ABA effects on root growth and gravireaction of *Zea mays* L.

ALAIN CHANSON and PAUL EMILE PILET

Institute of Plant Biology and Physiology of the University, CH-1005 Lausanne, Switzerland

The curvature of roots seems to be associated with an overall inhibition of growth [1]. These reactions appear to be controlled by some growth inhibiting substances produced (or released) by the cap of maize roots in response to gravity [2, 3] and to light [4]. Moving in the basipetal direction, they accumulate in the lower part of the roots, causing a downward bending (positive georeaction [5]. Abscisic acid (ABA) is present in maize roots and more especially in their tips [10]. If exogenous ABA is asymmetrically applied on the lower part of the apical cut surface of a decapitated root segment, a strong positive curvature is obtained [6, 7]. Furthermore, it has been reported that an ABA pretreatment of intact

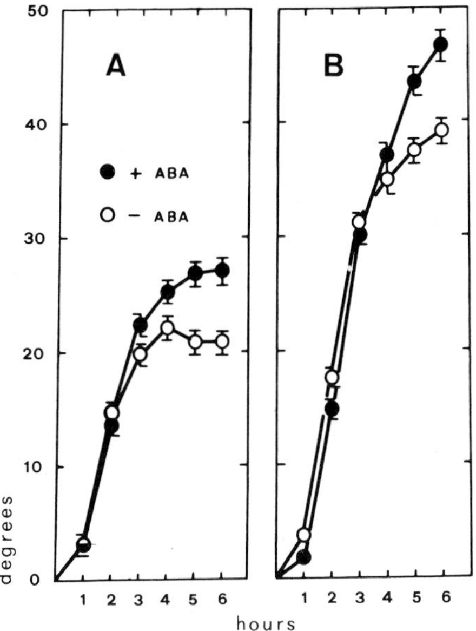

Fig. 1. Gravireaction (curvature in degrees ± standard error) of the apical segments of primary roots of maize (cv. LG 11) pretreated for 1 h with ABA applied to the tip and maintained horizontally in darkness (A) and in light (B).

R. Brouwer et al. (eds.), Structure and Function of Plant Roots, 95–97.

roots (5 apical mm) induced a growth inhibition and a gravireaction of horizontal roots in darkness [11]. Similar data were obtained with apical segments of the same cultivar: an ABA treatment (roots being immersed 1 h in a buffered ABA solution) enhanced significantly the georeactivity and this much more strongly for the segments kept in darkness than those illuminated [9].

The aim of this paper was to analyse the effect of exogenous ABA on the growth [see Pilet P. E. and Chanson A.] and gravireaction of maize (cv. LG 11) apical root segments geostimulated both in the dark and in light.

MATERIAL AND METHODS

Intact seedlings of *Zea mays* L. cv. LG 11 were pretreated vertically (1 h in the dark) with a droplet of buffered (pH 6) solution (5 μl) applied on the root tip, with or without ABA (5×10^{-5} mol.dm^{-3}). The apical root segments (10 ± 0.1 mm) were then placed horizontally for 6 h in both light and darkness, and their growth and downward curvature were measured, using 110 + 6 segments for each set of experiments.

RESULTS AND DISCUSSION

An ABA pretreatment increased significantly the gravireaction of the apical root segments both in the dark (Fig. 1A) and in light (Fig. 1B). The positive curvature was larger in the light than in darkness. This could be related to a higher level of endogenous ABA found in roots when they are exposed to light [9]. Both ABA and light produce – as previously reported [8] – an inhibitory effect on the elongation of the apical root segments (Table 1). Therefore, the effect of ABA was less pronounced in the light. Consequently, it is difficult not to consider ABA

Table 1. Elongation (mm ± standard error) of the apical segments of primary roots of maize (cv. LG 11) pretreated for 1 h with ABA applied to the tip and maintained horizontally for 3 and 6 h

Time (h)		Root elongation (mm)		Inhibition (per cent of control)
		−ABA	+ABA	
Dark	1–3	0.762 ± 0.047	0.584 ± 0.058	23.3
	1–6	1.079 ± 0.053	0.995 ± 0.099	9.3
Light	1–3	0.592 ± 0.059	0.515 ± 0.057	13.0
	1–6	0.993 ± 0.067	0.932 ± 0.065	6.1

as one of the inhibiting substances which have been shown to be formed or released in the cap cells of growing maize roots in response to light and gravity.

REFERENCES

1. Audus, L. J. and Brownbridge, M. E. 1957 Studies on the geotropism of roots. I. Growth rate distribution during response and the effects of applied auxins. J. Exp. Bot. 8, 105–124
2. Gibbons, G. S. B. and Wilkins, M. B. 1970 Growth inhibitor production by root caps in relation to geotropic responses. Nature 226, 558–559.
3. Pilet, P. E. 1971 Root cap and georeaction. Nature 233, 115–116.
4. Pilet, P. E. 1973 Inhibiteur de croissance et énergie lumineuse. C.R. Acad. Sci. Paris 276, 2529–2531.
5. Pilet, P. E. 1977 Growth inhibitors in growing and geostimulated maize roots. In: Pilet, P. E. (ed.), Plant Growth Regulation, pp. 115–128. Berlin, Heidelberg, New York: Springer.
6. Pilet, P. E. 1975 Abscisic acid as a root growth inhibitor: physiological analyses. Planta 122, 299–302.
7. Pilet, P. E. 1976 The light effect on the growth inhibitors produced by the cap. Planta 130, 245–249.
8. Pilet, P. E. and Chanson, A. 1981 Effect of abscisic acid on maize root growth. A critical examination. Plant Sci. Lett. (in press).
9. Pilet, P. E. and Rivier, L. 1980 Light and dark georeaction of maize roots: effect and endogenous level of abscisic acid. Plant Sci. Lett. 18, 201–206.
10. Rivier, L., Milon, H. and Pilet, P. E. 1977 Gas chromatography-mass spectrometric determinations of abscisic acid levels in the cap and the apex of maize roots. Planta 134, 23–27.
11. Wilkins, H. and Wain, R. L. 1975 Abscisic acid and the response of the roots of Zea mays L. seedlings to gravity. Planta 126, 19–23.

18. Critical study of the elongation and gravireaction of maize roots: light effect

ROLAND BEFFA and PAUL EMILE PILET

Institute of Plant Biology and Physiology of the University, CH-1005 Lausanne, Switzerland

When maize roots were placed in a horizontal position, a positive gravitropic effect appeared [1, 10, 15]. The root bent downwards and its 6 h elongation was shorter than that of roots kept vertically [3]. The georesponses of these roots, at least for a period of 5 h, were for some varieties (cv. Kelvedon 33, . . .) strongly light dependent [6, 8]. In contrast, maize roots of cv. Anjou 210 similarly georeacted both in light and in the dark [11]. The roots of cv. LG 11 showed a greater curvature in light than that observed in darkness [14]. In roots exposed to gravity, some inhibitors – produced or released by the root cap [7, 10, 11, 15] – accumulate in the lower part of the elongating roots, causing the downward bending (positive gravitropism) and the inhibition of elongation [9, 15].

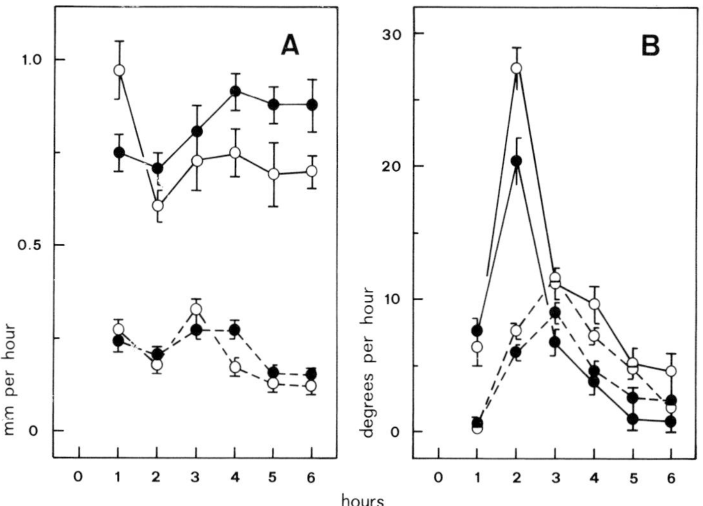

Fig. 1. Rates of elongation (mm per h.) (A) and curvature (degrees per h.) (B) of apical root segments (broken lines) or intact roots (continuous lines) of *Zea mays* L. cv. LG 11. Roots are maintained horizontally for 6 h. in darkness (filled circles) or in white light (open circles). The points and vertical lines represent the mean value ± standard error respectively.

R. Brouwer et al. (eds.), Structure and Function of Plant Roots, 99–101.

Moreover, an acropetal flow of other growth regulators from the basal regions of the root, from the seed [2, 5, 12, 13] and the shoot [4] could suggest that for the growing and georeacting root, a balance between the different growth regulators [7] and the trophic factors [16] which reached the elongating zone, could play an important part in the amplitude of elongation and curvature. Using a macro-photographic technique (R. Beffa and P. E. Pilet, in preparation), it has been possible, by a kinetic study, to show the correlations between the rates of growth and curvature (two systems were used, i.e. intact roots and apical root segments).

With maize (LG 11) roots, it was established that the growth rate (Fig. 1A) was depressed in the 2nd hour after the beginning of the gravistimulus much more in light than in darkness. This occurs both with apical segments and intact roots. At the same time, the rate of gravicurvature (Fig. 1B) was optimal for the intact roots, but not for the apical segments for which it shifted to one hour later. Consequently, the *maximum* rate of gravicurvature corresponds to a growth rate which is *minimum* in intact roots and *maximum* in apical segments. Moreover, the inhibiting effects of light on growth rate and their promoting effects on rate of curvature were more important on intact roots. Thus, some factors coming from the caryopsis [5] – trophic factors, growth regulators and some precursors of growth inhibitors – and their interaction with the growth inhibitors produced by the root cap, are not only important for the regulation of the total elongation and the gravireaction, but also seem to play an essential part in the correlations between growth and graviresponses.

REFERENCES

1. Audus, L. J. 1975 Geotropism in roots. In: Torrey, J. G. and Clarkson, D. T. (eds.), The development and function of roots, pp. 327–363. New York: Academic Press.
2. Burström, H. G. 1969 Influence of the tonic effect of gravitation and auxin on cell elongation and polarity in roots. Am. J. Bot. 56, 679–684.
3. Larsen, P. 1953 Influence of gravity on rate of elongation and autotropic reactions in roots. Physiol. plant 6, 735–774.
4. Martin, H. V., Elliott, M. G., Wangermann, E. and Pilet, P. E. 1978 Auxin gradient along the root of maize seedling. Planta 141, 179–181.
5. Ney, D. and Pilet, P. E. 1980 Importance of the caryopsis in root growth and georeaction. Physiol. plant 50, 166–168.
6. Pilet, P. E. 1975 Effects of light on the georeaction and growth inhibitor content of roots. Physiol. plant 33, 94–97.
7. Pilet, P. E. 1975 Abscisic acid as a root inhibitor: physiological analyses. Planta 122, 299–302.
8. Pilet, P. E. 1976 The light effect on the growth inhibitors produced by the root cap. Planta 130, 245–249.
9. Pilet, P. E. 1976 Effect of gravity on the growth inhibitors of geostimulated roots of Zea mays L. Planta 130, 91–93.
10. Pilet, P. E. 1977 Growth inhibitors in growing and geostimulated maize roots. In: Pilet, P. E. (ed.), Plant Growth Regulation, pp. 115–128. Berlin, Heidelberg, New York: Springer.

11. Pilet, P. E. 1978 The role of the cap in the geotropism of roots exposed to light. Z. pflanzenphysiol. 89, 411–426.
12. Pilet, P. E., Elliott, M. G. and Moloney, M. M. 1979 Endogenous and exogenous auxin in the control of root growth. Planta 146, 405–408.
13. Rivier, L. and Pilet, P. E. 1974 Indolyl-3-acetic acid in cap and apex of maize roots: identification and quantification by mass fragmentography. Planta 120, 107–112.
14. Wilkins, H. and Wain, R. L. 1975 The role of the root cap in the response of the primary roots of Zea mays L. seedlings to white light and to gravity. Planta 123, 217–222.
15. Wilkins, M. B. 1979 Growth-control mechanism in gravitropism. In: Haupt, W. and Feinleib, M. E. (eds.), Encyclopedia of plant physiology, new series, 7, pp. 601–626. Berlin, Heidelberg, New York: Springer.
16. Zippel, R. and Ehwald, R. 1980 Accumulation of 2-deoxy-D-glucose in the maize radicle after import via the phloem. Biochem. physiol. pflanzen 175, 676–680.

19. The differentiation of rhizodermal cells in grasses

VICTOR N. FILIPPENKO

Institute of General and Inorganic Chemistry of the Academy of Sciences of the U.S.S.R., Moscow 117071, U.S.S.R.

It is the purpose of the present communication to demonstrate the cell lineages in rhizodermis of festucoid (wheat) and panicoid (maize) grasses. The developmental potentials of individual rhizodermal cells were investigated taking into account that cells in longitudinal files represent the progenies of the same initials. The cytological mechanisms of cell diversification were also studied. The investigations were conducted with the aid of cytomorphological techniques and X-irradiation (10 kR) which selectively inhibited the mitoses, but did not prevent cell elongation and differentiation.

RHIZODERMAL CELL DIFFERENTIATION AFTER COMPLETE INHIBITION OF MITOSES BY X-IRRADIATION IN WHEAT AND MAIZE SEEDLINGS GROWING UNDER AIR-HUMID OR WATER ENVIRONMENTS

The rhizodermal cells from the same parts of the meristem in X-irradiated wheat roots grown in humid air or in water differentiated with similar patterns. Cells from the basal part of the meristem differentiated to hair and hairless cells but all cells from the apical part of meristem differentiated only to hair cells. The pattern of rhizodermal cell differentiation in maize was conditioned by environment. Cells from the basal part of meristem in maize roots grown in humid air differentiated to hair and hairless cells, and cells from the apical part of meristem differentiated only to hair ones. All meristematic cells in roots grown in water differentiated to hairless ones.

Thus we can conclude that meristematic rhizodermal cells in a festucoid grass, wheat, are determined initially as potentially hair cells. In contrast, in a panicoid grass, maize, meristematic cells are initially undetermined.

From the literature [2, 4] we know that in festucoid grasses differentiation of rhizodermal cells takes place in the basal part of the meristem during asymmetrical differentiative mitoses (ADM) but the data about differentiation in panicoid grasses are contradictory [1–4]. It is not certain whether the determination of rhizodermal cell differentiation occurs during the mitotic divisions or at the stage of cell elongation.

R. Brouwer et al. (eds.), Structure and Function of Plant Roots, 103–105.

INVESTIGATION OF DETERMINATION OF RHIZODERMAL CELL DIFFERENTIATION IN MAIZE ROOTS

We have studied the pattern of rhizodermal cell differentiation in intact and X-irradiated maize seedlings after transfers of them from air-humid environment to water and vice versa.

The direction of rhizodermal cell differentiation can be changed only for cells located in the meristem at the moment of transfer, but not for elongating cells. Thus the final determination in maize occurs in accordance with the external conditions during the beginning of cell elongation.

SYMMETRY OF LAST AND/OR LAST BUT ONE MITOSES IN THE RHIZODERMIS OF MAIZE ROOTS GROWING IN HUMID AIR OR WATER ENVIRONMENTS

Symmetry of last and/or last but one mitoses was measured as the ratio between the lengths of daughter cells from the basal part of meristem and the zone of morphological differentiation. The degree of symmetry of differentiative mitoses was calculated from the cell length ratio in heterocomplexes, i.e., the pairs and triplets with heterotypical cells. The symmetry degree of proliferative mitoses was established in homocomplexes.

It was established that the last and/or last but one mitoses in maize rhizodermis may have various degrees of symmetry. Among these mitoses only asymme-

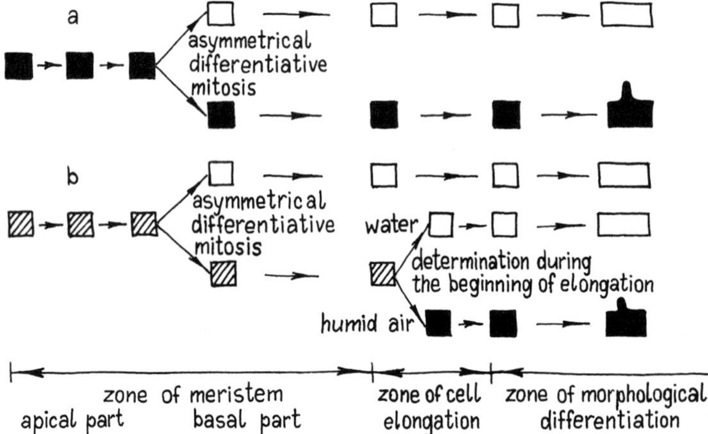

Fig. 1. Cell lineages in rhizodermis of festucoid (a) and panicoid (b) grasses. The dark shaded squares represent the hair cell initials, the hatched squares – the undetermined cell initials and the light squares – the hairless cell initials.

trical divisions are differentiative. The divisions usually resulted in the appearance of short apical and large basal cells. Thus ADM also occur in rhizodermis of a panicoid grass, maize. Hence cell lineages in rhizodermis of grasses may be represented as follows (Fig. 1).

In festucoid grasses the final determination of rhizodermal cells to hair and hairless initials occurs during ADM in the basal part of the meristem. The role of ADM in festucoid and panicoid grasses is similar, it operates by 'prohibition' of root hair development for larger daughter cells. However, in panicoid grasses in contrast to festucoid ones the determination takes place in two stages. At the first stage the larger daughter cells originating from ADM in the basal part of the meristem determine to hairless initials. At the second stage the undetermined cells determine during the beginning of elongation in accordance with external conditions. In humid air they determine to hair cell initials and in water to hairless ones. Thus during the final determination the rhizodermal cells in panicoid grasses determine in humid air to hair and hairless cell initials but in water only to hairless ones.

REFERENCES

1. Abrew, S. L., Rothwell, N. V. and Lewis, R. F. 1973 An autoradiographic analysis of the root epidermis of swith grass (Panicum virgatum). Amer. J. Bot. 60, 496–504.
2. Cormack, R. G. H. 1962 Development of root hairs in angiosperms. II. Bot. Rev. 28, 448–464.
3. Rothwell, N. V. 1966 Evidence for diverse cell types in the apical region of the root epidermis of Panicum virgatum. Amer. J. Bot. 53, 7–11.
4. Sinnott, E. W. and Bloch, R. 1939 Cell polarity and the differentiation of root hairs. Proc. Nat. Acad. Sci. (U.S.A.) 25, 248–252.

20. Distribution of plasmodesmata in root epidermis

ELENA B. KURKOVA

K. A. Timiriazev Institute of Plant Physiology, Academy of Sciences of the U.S.S.R., Moscow 127 296, U.S.S.R.

It was shown earlier by electrophysiological techniques that root hairs of the aquatic plant *Trianea bogotensis* Karst. (Hydroharitaceae) were characterized by intracellular K^+ activity approximately twice that of hairless cells, therefore it was assumed that there are more plasmodesmata on the pathway of K from the root hair to xylem vessels than from hairless cells [2].

We counted plasmodesmata on transverse root sections of about 70 nm thickness in the root hair zone. In cell walls of root hairs plasmodesmata are frequent especially in tangential walls between root hairs and cortical cells, where they form groups (Fig. 1). Frequency of plasmodesmata per 1 μm^2 and per total

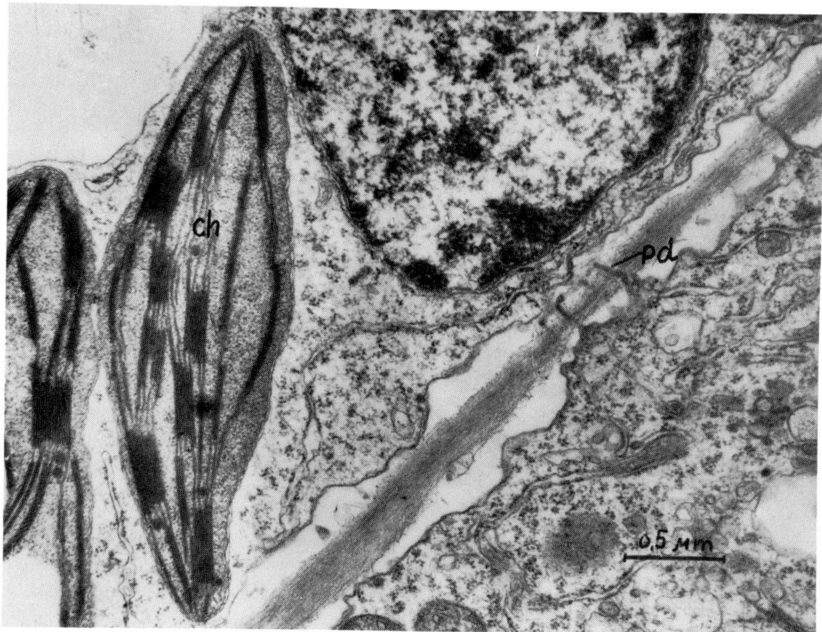

Fig. 1. Plasmodesmata in root hair tangential wall of *Trianea bogotensis*. Cortical cell with chloroplasts. Abbreviations: pd – plasmodesmata, ch – chloroplast.

R. Brouwer et al. (eds.), Structure and Function of Plant Roots, 107–109.

Table 1. Distribution of plasmodesmata in root epidermis of *Trianea bogotensis*

Cell wall type	Number of plasmodesmata	
	per 1 μm²	per total cell wall
	Tangential walls	
Root hair-cortical cell	2.06	10412
First hairless-cortical cell	0.10	630
Second hairless-cortical cell	0.12	756
	Radial walls	
Root hair-first hairless cell	0.60	1764
First hairless cell – Second hairless cell	0.21	529
Second hairless cell – Third hairless cell	0.14	353

surface of root hair tangential walls was 17–20 times higher than that in tangential walls of hairless cell (Table 1). Plasmodesmatal frequency in radial walls gradually declined with increasing the distance from the root hairs. The results indicate that root hairs are mainly responsible for the initial uptake and transport of K^+ across the root of *Tr. bogotensis*.

Table 2. Distribution of plasmodesmata in root epidermis of *Raphanus sativus*

Cell wall type	Surface area of the cell wall, μm²		Number of plasmodesmata		
	zone 1	zone 2	per μm²		per total cell wall
			zone 1	zone 2	
	Tangential walls				
Root hair–cortical cell	240.8	1651.2	1.13	0.16	272.8
Hairless–cortical cell	150.0	2132.8	1.00	0.07	150.0
	Radial walls				
Root hair–cortical cell	180.6	949.4	0.09	0.02	16.2
Hairless cell–hairless cell	285.0	1860.0	0.06	0.01	16.3

Zone 1 – beginning of elongation
Zone 2 – root hairs

For the terrestrial plant *Raphanus sativus* L. (Cruciferae) it was shown that intracellular K^+ activity in root hairs and hairless cells is identical [3].

In epidermal and cortical root cells of terrestial plants plasmodesmata are extremely rare, therefore their count is very difficult [1] particularly in the zone of root hairs, where frequency of plasmodesmata is very low and the size of cells is immense. Therefore, we counted plasmodesmata at the beginning of the elongation zone and made a recalculation for the root hair zone taking into account the changes of cell wall surface during the growth of the root. The frequency of plasmodesmata per 1 μm^2 and per the total surface of the root hair tangential wall was about double that in the tangential wall of the hairless cell (Table 2). However, the values of plasmodesmatal frequencies in radish root are an order of magnitude lower than those in *Tr. bogotensis*. Thus, it appears that the symplastic pathway for ions in radish root epidermal cells is minimal and that the root hairs have no particular role in ion uptake.

REFERENCES

1. Robards, A. W. and Clarkson, D. T. 1976 The role of plasmodesmata in the transport of water and nutrients across roots. In: Gunning, B. E. S. and Robards, A. W. (eds.), Intercellular Communication in Plants: Studies on Plasmodesmata, pp. 181–201.
2. Vakhmistrov, D. B. and Kurkova, E. B. 1979 Symplastic connection in rhizodermis of Trianea bogotensis Karst. (in Russian), Fiziol. rast. 26, 943–952.
3. Vakhmistrov, D. B., Kurkova, E. B. and Zlotnikova, I. Ph. 1981 Symplastic connection and intracellular K^+ activity in rhizodermis of Raphanus sativus L. (in Russian). Fiziol. rast. 28 (in print).

21. Suberization and browning of grapevine roots

DENNIS RICHARDS and JOHN A. CONSIDINE

Horticultural Research Institute, Dept. of Agriculture, Burwood Highway, Knoxfield, Victoria 3180, Australia

As roots of woody dicolytedonous plants develop and age they undergo changes both internally, in the organization and differentiation of their tissues, and externally, in their physical size and appearance. Terms such as suberization, browning and secondary growth used to describe these developmental changes are, however, often used loosely and even interchangeably. This has led to some confusion with regard to the internal changes in root structure that may accompany alterations in the physical appearance of the root. In an attempt to clarify the use of these terms portions of grapevine (*Vitis vinifera* L.) roots were sectioned at various positions along the root axis.

RESULTS AND DISCUSSION

Roots were collected in mid-summer from soil cores taken from a field grown planting of grapevines. The samples yielded an assortment of roots of various diameters, age and colour. Zones of interest were sectioned, stained if necessary, and viewed under bright field or ultra-violet epi-fluorescence (Olympus microscope, exciter filter UG–1, barrier filter L420). Sections of the root tissue are represented diagramatically in Fig. 1 (A–E).

At a distance of about 20 mm from the tip of young creamy-white roots the endodermis was suberized. Histochemical tests (e.g., oil red 0 stain) and autofluorescence under UV light indicated that suberin were deposited around the endodermal cells (Figs. 1A and 2). Suberization of the endodermis was not associated with any alteration in the external appearance of the root.

Not all primary roots were white. Some roots had developed a brown appearance which extended to and, in some cases, encompassed the root tip (Fig. 5). Sections of this tissue (Figs. 1B and 3) showed that suberin lamellae were deposited around the hypodermis (intercutis [7], exodermis [2] which fluoresced under UV light. The brown colour was associated with the death and collapse of the outer epidermal cells and was presumably due to oxidation of phenols normally enclosed within the vacuole. The onset of this peripheral suberization was not accompanied by any degeneration of cortical tissue. It appears that the

R. Brouwer et al. (eds.), Structure and Function of Plant Roots, 111–115.

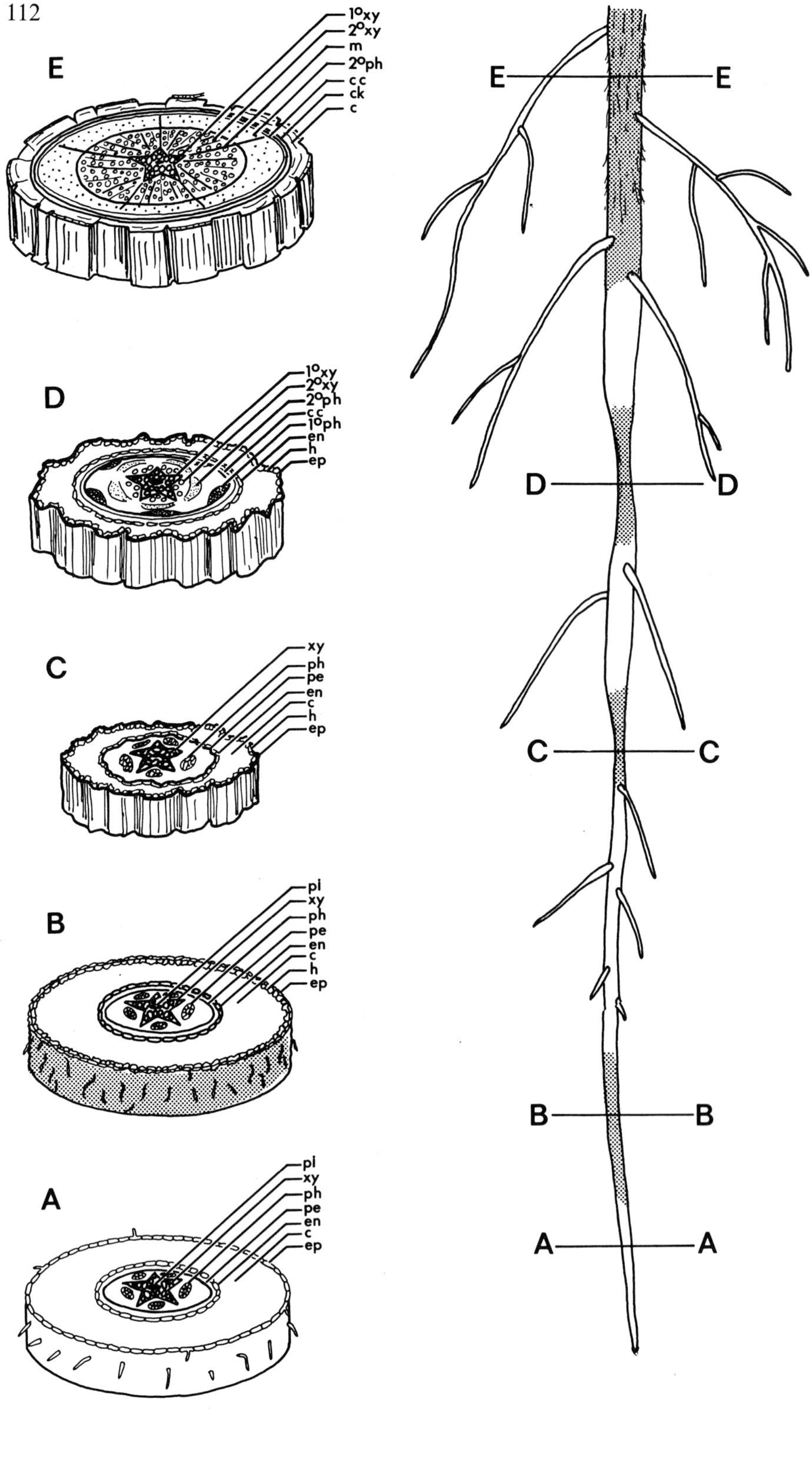

E

1°xy
2°xy
m
2°ph
cc
ck
c

E — E

D

1°xy
2°xy
2°ph
cc
1°ph
en
h
ep

D — D

C

xy
ph
pe
en
c
h
ep

C — C

B

pi
xy
ph
pe
en
c
h
ep

B — B

A

pi
xy
ph
pe
en
c
ep

A — A

development of the hypodermis as an external barrier is a means by which primary roots can tolerate unfavourable soil conditions without desiccation [8]. Preliminary observations on overwintering grapevine roots (M. Marson, unpublished) indicated that suberized hypodermal cells also contain high levels of phenols which may impart some resistance to infection if cells are damaged.

Sometimes browning occurs in patches along an otherwise white and fleshy primary root and is associated with a marked decrease in root diameter (Fig. 6). Examination of root sections shows that this results from a massive breakdown of cortical cells. This phenomenon has been noted in other deciduous perennial plants such as *Malus* [4] and *Prunus* [1]. The term 'cortical browning' has been used to describe this process [4] but many authors have used the unqualified term 'suberization' [1, 5]. Although death and browning of the cortex may be a natural process it is unknown whether it results from internal changes in the root associated with the onset of secondary development, changes due to soil temperatures and desiccation, or to the feeding of soil flora and fauna [4]. In fact in some of the sampled grapevine roots cortical and endodermal breakdown preceded any secondary development and in such instances the stele degenerated (Fig. 1C) ultimately resulting in the death of the distal portion of the root. In cases where secondary cambial development had already begun (Fig. 1D and 4) the cortex and endodermis were damaged and shed without the death of the central cylinder.

As secondary development continues the cork cambium (phellogen) produces layers of heavily suberized cells externally, while the vascular cambium produces secondary xylem and phloem elements internally (Fig. 1E). Thus the root assumes a brown, woody appearance, the diameter steadily increases, and the cortical and endodermal tissue slough off.

Many publications making reference to root appearance, structure and function do so without adequate description or definition of terms. Thus reference to 'suberization' may be specific, referring to, say, endodermal suberization [3, 6] or general, referring to any root that is brown in colour [1, 5]. As shown here, a brown appearance in primary roots can be due to either death of the epidermis,

---

Fig. 1. Diagramatic representation of a grapevine root showing anatomical and structural changes along its length. A – White primary root, epidermis intact with root hairs. B – Epidermis collapsed giving a brown appearance, hypodermis and endodermis suberized. C – Cortical collapse (browning) and degeneration of all tissues. D – Cortical collapse with expansion of the stele due to development of secondary growth. E – Woody, secondary growth with remnants of the cortex still attached.

Abbreviations: ep, epidermis; h, hypodermis; c, cortex; en, endodermis; pe, pericycle; ph and 1° ph, primary phloem; xy and 1° xy, primary xylem; pi, pith; cc, cork cambium; ck, cork; m, medullary ray; 2° ph, secondary phloem; 2° xy, secondary xylem.

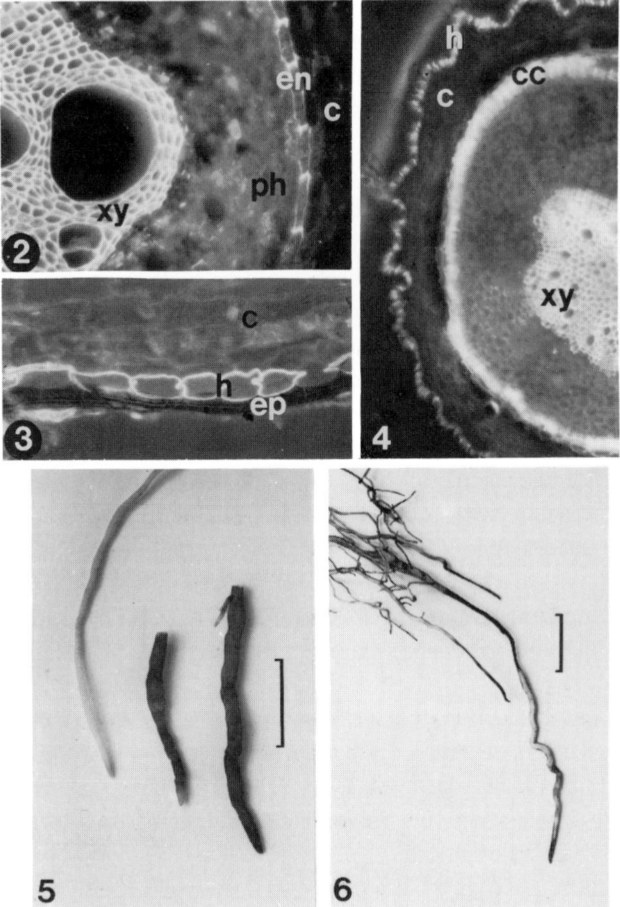

Figs. 2–6.

Fig. 2. TS of white root as position shown in Fig. 1A showing fluorescence of suberin lamellae around endodermis under UV illumination (× 400). (For abbreviations see caption to Fig. 1.)

Fig. 3. LS of brown root at position shown in Fig. 1B showing fluorescence of suberin lamellae around hypodermis and outer dead epidermis (× 400). (For abbreviations see caption to Fig. 1.)

Fig. 4. TS of brown root at position shown in Fig. 1D showing degenerated cortex and development of secondary growth within the stele. Hypodermis and newly formed cork cambium are fluorescing under UV illumination (× 200). (For abbreviations see caption to Fig. 1.)

Fig. 5. Brown and white primary roots of field-grown grapevines corresponding to Fig. 1A, B. Bar = 10 mm.

Fig. 6. Portion of grapevine root system showing browning and constriction corresponding to either Fig. 1C or Fig. 1D. Bar = 10 mm.

to cortical breakdown or even to degeneration of all the tissues of the root axis. Further, in secondary roots, the development of a cork cambium also results in a brown appearance.

If the functional consequences of changes in root anatomy and structure are to be elucidated, clarification is needed in the use and definition of terms.

REFERENCES

1. Bhar, D. S., Mason, G. F. and Hilton, R. J. 1970 *In situ* observations on plum root growth. J. Amer. Soc. Hort. Sci. 95, 237–239.
2. Esau, K. 1977 Anatomy of seed plants, pp. 217–221, 2nd Edn. New York: John Wiley and Sons.
3. Harrison-Murray, R. S. and Clarkson, D. T. 1973 Relationships between structural development and the absorption of ions by the root system of Cucurbita pepo. Planta 114, 1–16.
4. Head, G. C. 1973 Shedding of roots. In: Kozlowski, T. T. (ed.), Shedding of plant parts, pp. 237–293. New York and London: Academic Press.
5. Kramer, P. J. and Bullock, H. C. 1966 Seasonal variations in the proportions of suberized and unsuberized roots of trees in relation to the absorption of water. Amer. J. Bot. 53, 200–204.
6. MacKenzie, K. A. D. 1979 The development of the endodermis and phi layer of apple roots. Protoplasma 100, 21–32.
7. Perold, A. 1927 A treatise on viticulture. London: Macmillan and Co. Ltd.
8. Robards, A. W., Clarkson, D. T. and Sanderson, J. 1979 Structure and permeability of the epidermal/hypodermal layers of the sand sedge (Carex arenaria L.). Protoplasma 101, 331–347.

II. METABOLISM OF ROOTS

22. Regulation of enzymes involved in inorganic N metabolism in pea roots

JOSEF SAHULKA

Institute of Experimental Botany of the Czechoslovak Academy of Sciences, 166 30 Praha 6 Vokovice, Czechoslovakia

In the past decade I have studied the regulation of nitrate reductase (NR), nitrite reductase (NIR), glutamine synthetase (GS), and glutamate dehydrogenase (GDH) levels in isolated pea roots [3, 4, 5, 6]. This paper deals with results obtained since October 1979.

THE EFFECTS OF SUGARS

The levels of the enzymes can be influenced by sugar supply [6]. Exogenous sucrose enhances NIR induction in roots exposed to both nitrate and nitrite but does not significantly influence NIR levels in roots cultivated without them. Sugar starvation brought about by cultivating isolated roots in nitrate containing medium lacking sugars effects NIR induction to a much lesser extent than NR induction. The NIR level is increased to a larger extent by low sucrose, glucose and fructose concentrations (1 and 2 gl^{-1}) in roots exposed to nitrate than in those exposed to nitrite. Glucose and fructose are as efficient as sucrose in enhancing NIR induction at all concentrations tested. Mannose enhances NIR induction in roots exposed to nitrate at concentrations up to 10 $g\,l^{-1}$, but not at 20 $g\,l^{-1}$. In roots exposed to nitrite, mannose is more efficient at 2 $g\,l^{-1}$ than sucrose, glucose and fructose, but tends to inhibit NIR induction at 20 $g\,l^{-1}$. Galactose does not support NIR induction in roots exposed to nitrate at any concentration tested (up to 20 $g\,l^{-1}$) and inhibits it in roots exposed to nitrite. Xylose increases NIR level in roots exposed to both nitrate and nitrite, while L-sorbose is only effective in roots exposed to nitrate.

Apparently NIR induction is also dependent on the supply of metabolizable sugars and the response of NIR level to sugars differs from the responses of the other enzymes studied [6], which confirms our earlier conclusion that the level of each of the four enzymes is regulated by sugars via different mechanisms.

Exogenous sucrose not only maintains high levels of GS in isolated pea roots but can also modulate the effect of the end-product, i.e. of L-glutamine, on GS level: GS level is not affected by exogenous L-glutamine in roots cultured with

R. Brouwer et al. (eds.), Structure and Function of Plant Roots, 119–121.

saturating concentrations of sucrose, but the decrease caused by sugar starvation in roots cultured without sucrose is enhanced by it. A similar response, though not so clear-cut, can be observed with casamino acids.

Although the decrease in GS level and the increase in GDH level due to sugar starvation have also been observed in leaf tissue and in maize roots [1, 2, 7] and seem to be a general rule, the rates, extents and time courses of these changes can differ according to the material used, which has also been observed with pea roots where cultivar and perhaps seed vigour differences may play an important role.

THE EFFECT OF H^+ CONCENTRATION

Another very important factor in the regulation of the levels of the enzymes is the concentration of H^+ ions. Increased H^+ concentration increases GDH level [3] and decreases GS level [4, 5], and in the presence of nitrate can increase NR level in pea roots. The author has recently demonstrated the effect of H^+ ions on GS and NR levels by applying L-glutamic acid and Na glutamate to nutrient media containing sucrose in which isolated pea roots were cultured. Glutamate can be considered a substrate for GS and one of the end-products of NR. According to the generally accepted postulates on the effect of substrates and end-products on enzymes, glutamate should either increase or not influence GS level and decrease NR level. However, when applied as free acid, glutamic acid decreases GS level and increases NR level, and when applied as sodium glutamate, it does not significantly affect GS level, but decreases NR level, this decrease being reversed by L-glutamic acid. Nitric acid and α-ketoglutaric acid also increase NR level in pea roots when added to solutions containing nitrate salts at saturating concentrations and when pH values of the solutions fall to 4. The effect of L-glutamic acid on the GS level is less pronounced in roots cultured in nitrate containing media than in those lacking nitrate. This can be at least partially explained by OH^- production in the course of nitrate assimilation. L-glutamic acid can reverse not only the decrease in NR level caused by Na glutamate, but also by L-glutamine and by casamino acids and can depress GS level in roots cultured with sucrose and these substances. Sodium glutamate does not affect GS level even in roots cultured without sucrose. These results and the results reported earlier suggest that H^+ concentration in the cells plays an important role in the regulation of the levels of these enzymes. However, the mechanism of action of increased H^+ concentration is not clear and may differ with each enzyme.

THE EFFECT OF AUXINS ON GS

The effects of indole-3-acetic and of naphthaleneacetic acids on GS were examined because (a) auxins are known to induce H^+ extrusion from plant tissues and acidification in the cells, and (b) increased H^+ concentration results in increased GDH and decreased GS levels, and (c) because auxins increase GDH level in pea roots. If auxins acted on the enzymes via increased H^+ concentration, they would have to decrease GS level sharply. The results obtained have shown however, that in roots cultured with sucrose, naphthaleneacetic acid either does not influence or increases GS level, with these effects being dependent on NAA concentration, and that indole-3-acetic acid tends to decrease GS level only slightly. In roots cultured without sucrose, both auxins slow down the decrease in GS level caused by sugar starvation. This suggests that the effect of auxins on GDH (established previously) may represent a specific effect of auxins which is not mediated by increased H^+ concentration. The slow-down effect observed in roots cultured without sucrose may partly be the result of slower consumption of endogenous sugars due to the inhibition of apical growth. However, auxins could have several other effects which may be indirectly reflected in GS level.

REFERENCES

1. Nauen, W. and Hartman, T. 1980 Glutamate dehydrogenase from Pisum sativum L. Localization of the multiple forms and of glutamate formation in isolated mitochondria. Planta 148, 7–16.
2. Oaks, A., Stulen, I., Jones, K., Winspear, M. J., Misra, S. and Boesel, I. L. 1980 Enzymes of nitrogen assimilation in maize roots. Planta 148, 477–484.
3. Sahulka, J. and Gaudinová, A. 1976 Enhancement of glutamate dehydrogenase activity in excised pea roots by exogenously supplied acids. Z. Pflanzenphys. 78, 13–23.
4. Sahulka, J. and Lisá, L. 1979 Regulation of glutamine synthetase in isolated pea roots. I. Differential effects of ammonium salts in sugar-supplied roots. Biochem. Physiol. Pflanzen. 174, 646–652.
5. Sahulka, J. and Lisá, L. 1979 Regulation of glutamine synthetase level in isolated pea roots. II. The effect of exogenously supplied acids and bases in sugar-supplied and sugar-starved roots. Biochem. Physiol. Pflanzen. 174, 653–659.
6. Sahulka, J. and Lisá, L. 1980 Effect of some disaccharides, hexoses and pentoses on nitrate reductase, glutamine synthetase and glutamate dehydrogenase in excised pea roots. Physiol. Plant 50, 32–36.
7. Thomas, H. 1978 Enzymes of nitrogen mobilization in detached leaves of Lolium temulentum during senescence. Planta 142, 161–169.

23. The effect of manganese nutrition on nitrogen assimilation in roots

EBERHARD PRZEMECK and BEREND SCHRADER

Institute of Agricultural Chemistry of the Georg-August-University Göttingen, 3400 Göttingen, F.R.G.

It has been pointed out that there are genetical differences in the nitrogen assimilation capacity of roots, and that they are effected by a typical nitrate intensity in relation to that of the shoot [4]. Nitrogen assimilation from nitrate as the only nitrogen source requires, apart from the substrate and the enzyme nitrate reductase, carbon structures, energy rich compounds and reduction equivalents. For this the root has to be supplied with carbohydrates by the shoot. It is well known that carbon metabolism is strongly influenced by the manganese supply of the plant.

In the experiments described in this paper nitrogen assimilation was investigated in the roots of plants which had been grown at different levels of manganese supply, from Mn deficiency to Mn toxicity. The object was to quantify the turnover of the N assimilation processes in the roots, and to determine the transport rates of inorganic and organic nitrogen compounds with the xylem sap of the roots in relation to Mn nutrition conditions.

RESULTS

In the experiments inorganically and organically bound nitrogen was determined in the roots and in the bleeding sap of the roots as well as the nitrate reductase activity in the root tissue. The trials were performed with 27 day old pumpkin plants, for it is not too difficult to get root xylem sap from detached roots of this plant in a volume which is suitable for the required chemical analysis. Root xylem sap was needed to study the transport phenomena. The plants were grown with a spray culture system [2] under controlled temperature, light and air moisture conditions. Details of the methods including analytical methods have been published recently [3].

Differences in the manganese concentration of the nutrient solution (0.00 – 0.025–0.050–0.100–0.300–1200 ppm) had effects on shoot growth. Absence of Mn brought about the typical Mn deficiency symptoms. 0.050 ppm Mn formed the highest yield. Increasing Mn supply beyond that effected declining shoot

R. Brouwer et al. (eds.), Structure and Function of Plant Roots, 123–127.

yields, and 1200 ppm Mn produced the first characteristic Mn toxicity symptoms on the leaves, but at this stage plants looked physiologically uninjured. The weight of the roots showed little differences with an optimum again in the 0.050 ppm variant and with declining root yields towards Mn deficiency and Mn surplus or toxicity.

The amount of nitrate in the root was highest in the high yielding variant (0.050 ppm Mn), and this was similar with respect to the amount of organically bound nitrogen (Table 1). In the bleeding sap the transport rates of nitrate were low in the deficiency and toxicity situations and in the high yield variant it was also low. The transport rate of organically bound nitrogen was highest in this variant, while deficiency and toxicity caused a significantly smaller transport of reduced nitrogen. These results point at differing turnover rates of the nitrate in the roots, depending on Mn supply.

The regular nitrate in the nutrient solution was exchanged for ^{15}N nitrate immediately before cutting off the root from the shoot to collect the xylem sap, thus enabling the estimation of the nitrate reduction turnover in the 60 min of bleeding sap collection. Total nitrate and organically bound nitrogen and the percentage of labeled nitrogen (as atom per cent excess) in these fractions of the roots and of the xylem fluid was analysed.

It can be seen in Fig. 1 that in the roots the percentage of label in the nitrate was lowest in the high yield variant, highest labeling was found in the roots of the deficient and toxic treatments. In contrast, in the bleeding sap the labeling of nitrate was lowest in Mn stress conditions, which demonstrates that the excretion of the very recently absorbed ^{15}N nitrate into the xylem and its transport was reduced in both Mn deficiency and toxicity. The fraction of reduced nitrogen in the bleeding sap of the roots was lowest in Mn deficient plants. The Mn toxicity plants had a significantly higher percentage of ^{15}N in the organically

Table 1. Amounts of nitrate N and organically bound N in the roots and in the bleeding sap

N fractions	ppm Mn without	0.025	0.050	0.100	0.300	1200
			roots (mg/root)			
$N_{(NO_3)}$	0.8	1.2	1.6	1.0	1.1	1.0
$N_{(org)}$	4.1	7.1	7.4	5.5	5.7	5.5
			bleeding sap (µg/h/root)			
$N_{(NO_3)}$	340	520	360	470	520	390
$N_{(org)}$	97	50	177	98	81	63

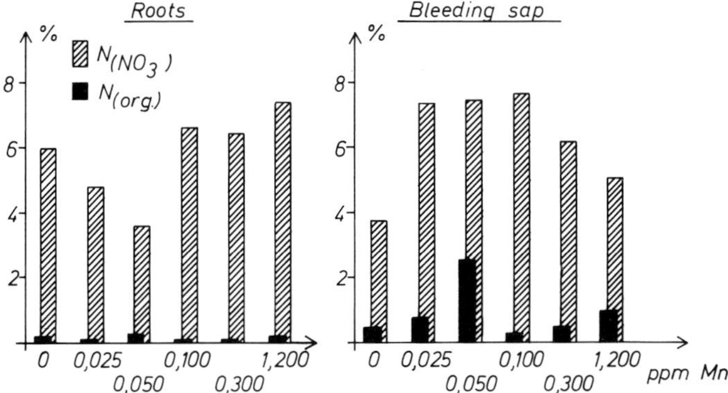

Fig. 1. ^{15}N labeling (atom per cent exc.) of nitrate and organically bound N in the 1st hour after detaching the root from the shoot.

bound nitrogen, but the high yield variant had the maximum of labeling in this fraction of the transport phase. The results demonstrate that there are genuine differences in the physiology of nitrate turnover and transport in the roots, depending on the Mn supply.

The turnover intensity of the nitrate reductase in the root system was investigated and the results calculated as enzyme activity per root system of one

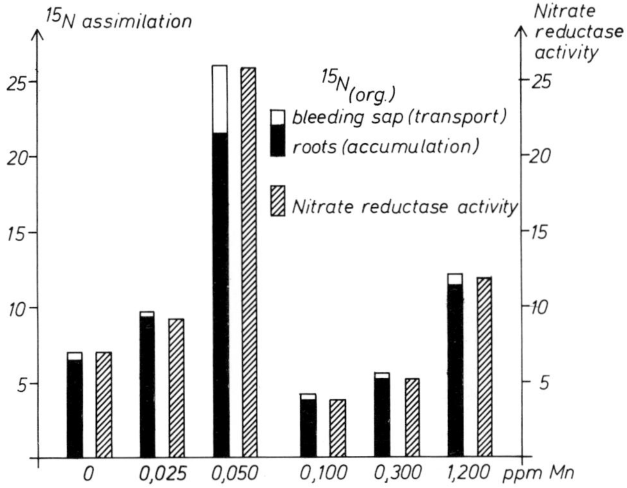

Fig. 2. ^{15}N assimilation (µg/plant/h) and nitrate reductase activity (µg $N_{(NO_2)}$/plant/h) of the roots in the 1st hour after detaching the root from the shoot.

plant and for one hour (Fig. 2). It is evident that the efficiency of the nitrate reductase in the roots is drastically reduced beyond the optimum of Mn supply, i.e. in deficient plants and in latent deficiency as well as in surplus. But roots showing Mn toxicity have considerably higher activities of this enzyme compared to deficient roots.

The intensity of ^{15}N assimilation in the roots during bleeding sap collection can be calculated, too, by considering the absolute amount of nitrogen in the root and the volume of sampled bleeding sap and with respect to the labeling percentage of the used nitrate salt in the nutrient solution. The results show (Fig. 2) that a large amount of organically bound ^{15}N is accumulated in the roots. The high yield variant showed the most intensive ^{15}N assimilation. The Mn toxicity roots had high assimilation rates, too, but most intensely labeled products of organically bound nitrogen remained in the root and only a small proportion could be found in the xylem sap. The results for nitrate reductase activity per root and per hour are very closely related to the results of ^{15}N assimilation intensity.

DISCUSSION

The experiments are concerned with nitrogen metabolism in the root tissue and the transport of nitrogen compounds out of the root. The amount of nitrate and organically bound nitrogen in the roots is primarily correlated with dry matter production of the shoot. But the transport of reduced nitrogen in the bleeding sap only differed a little between the manganese variants, except in the extreme Mn stress situations of deficiency and toxicity. Concentrations and amounts of recently absorbed nitrate are governed by the nitrate turnover, according to the short term experiments with ^{15}N.

With respect to Mn deficiency and Mn toxicity a strong relation exists between N assimilation intensity and the capacity of the nitrate reductase in the root tissue. This was found both in the labeled reduced nitrogen compounds and in nitrate reductase activity. So the results confirm findings [1, 5], where nitrate reductase activity is thought to be the limiting factor of the N metabolism of plants. The results also demonstrate the close relationships between nitrogen assimilation capacity of the root and the mineral nutrition of the plant, which is linked by carbohydrate synthesis in the leaves and the supply to the root of assimilates. The intensity of carbohydrate metabolism is one important factor as a physiological regulation mechanism for metabolical functions of roots.

Moreover, the results point to upsets in the excretion of organic N metabolites into the root xylem under conditions of Mn toxicity but this was not observed in Mn deficient roots. The secretion of compounds into the vessels is an active process, which seems to be affected in the roots by toxic concentrations of manganese in the root medium.

REFERENCES

1. Hornandez, H. H., Walsh, D. E. and Bauer, A. 1974 Nitrate reductase of wheat, its relation to nitrogen fertilization. Cereal Chem. (St. Paul, Minn.) 51, 330–336.
2. Przemeck, E. and Alcalde Blanco, S. 1969 Ein Hydro-Sprühkulturverfahren zum Studium ernährungsphysiologischer Probleme. Angew. Bot. 43, 331–339.
3. Przemeck, E. and Schrader, B. 1980 Stickstoff-Assimilation in der Wurzel und Transport von N-Verbindungen mit dem Blutungssaft der Wurzeln in Abhängigkeit von der Mn-Versorgung. Z. Pflanzenernähr. Bodenkd. 143, 170–181.
4. Radin, J. W. 1977 Contribution of the root system to nitrate assimilation in whole cotton plants. Austral. J. Plant Physiol. 4, 811–819.
5. Witt, H. H. and Jungk, A. 1974 Die nitratinduzierbare Nitratreduktase-Aktivität als Mass für die Stickstoffversorgung von Pflanzen. Landw. Forsch. 30, 1–9.

24. RNA synthesis in the root tip of maize

OTÍLIA GAŠPARÍKOVÁ

Institut of Experimental Biology and Ecology, Slovak Academy of Sciences, 885 34 Bratislava, Czechoslovakia

Root elongation is the result of cell proliferation and cell elongation. Elongation is the most remarkable period in root cell development and in maize it results in a 40–60-fold increase of the cell length [7] and is accompanied by considerable increases in total RNA and protein content [2]. The rate of protein synthesis increases 15–20-fold in these cells [1]. Thus, it may be expected that the activity of the nuclear genetic apparatus is associated with the high rates of protein synthesis. For this purpose, rate and character of ribosomal RNA (rRNA) and transfer RNA (tRNA) synthesis have been studied in consecutive primary root segments of maize seedlings (hybrid CE 380).

RNA synthesis was estimated on the basis of ^{3}H-uridine incorporation (0.37 MBq/ml, 814 GBq/mM) after 2 h incubation of intact seedlings. Total nucleic acid was prepared by the modified phenol procedure [4] and fractionated on polyacrylamide gels [6]. After staining with toluidine blue the gels were scanned at 620 nm and the radioactivity in 2 mm slices was determined.

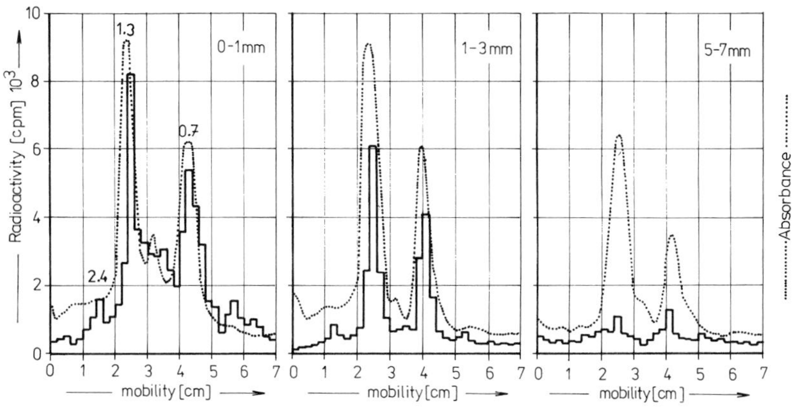

Fig. 1. Polyacrylamide gel fractionation of RNA isolated from different segments of maize root tip. Segments were measured from the cap junction. 35 μg RNA was fractionated on 2.5 per cent polyacrylamide gels for 3.5 h at 50 V. The values refer to the molecular weight × 10^6 daltons of maize rRNA components determined relative to *Escherichia coli* rRNA.

R. Brouwer et al. (eds.), Structure and Function of Plant Roots, 129–132.

Table 1. Comparison of the amount of ³H-uridine incorporated into pre-rRNA and mature rRNA in different segments of root apex.

Segment (mm from cap junction)	Specific activity of rRNA			
	cpm/100 µg RNA			
	2.4*	1.3	0.7	
0–1 mm	5,237	33,094	25,922	64,253
1–3 mm	2,274	19,697	10,854	32,825
5–7 mm	708	4,057	2,794	7,559
	cpm/RNA in 10³ cells			
0–1 mm	7.32	46.33	36.29	89.94
1–3 mm	4.51	39.43	21.70	65.64
5–7 mm	1.34	7.70	5.30	14.34

* 2.4×10^6, 1.3×10^6, 0.7×10^6 molecular weight rRNA.

Maize root cells contain two main RNA components, cytoplasmic rRNAs with molecular weights of 1.3×10^6 and 0.7×10^6 daltons (Fig. 1) and soluble RNAs which represent 5 S rRNA and tRNA (4 S, Fig. 4). The radioactivity profiles presented in Fig. 1 show the highest level of ³H-uridine incorporation into mature rRNAs of meristematic cells, which synthesize a large number of

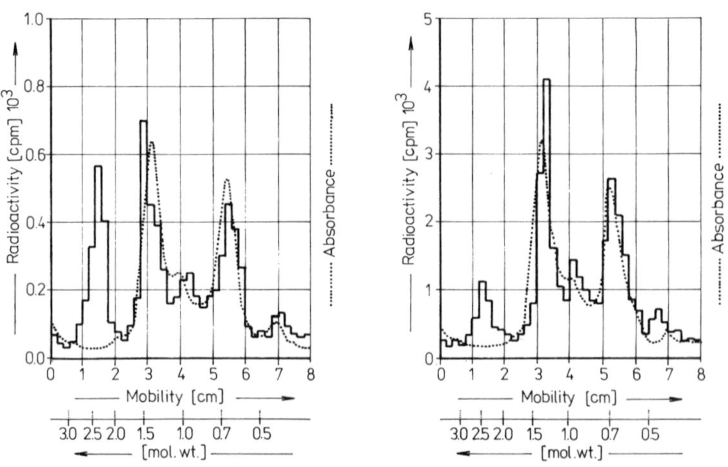

Figs. 2–3. Fractionation of high molecular weight RNA synthetized in maize root cells. Intact roots were labelled with ³H-uridine for 20 min – Fig. 2; and for 2 h – Fig. 3.

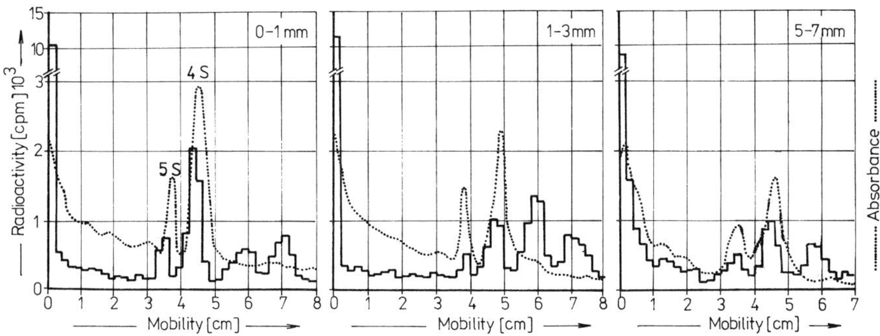

Fig. 4. Fractionation of soluble RNAs isolated from particular maize root growth segments. 35 µg RNA was fractionated on 7 per cent polyacrylamide gels for 3 h at 50 V. The value 5 S refers to 5 S rRNA and that of 4 S to tRNA.

new ribosomes. The rate of synthesis of these RNAs is still relatively high in elongating cells, which like meristematic cells rapidly synthetize labelled larger precursor molecules of rRNA. This fraction shows relatively high rates of specific incorporation but almost no detectable absorbance (Table 1). The experiments with short periods of labelling have proved that actively growing root cells synthetize three rapidly labelled RNAs with molecular weights of 2.4, 1.4 and 0.95×10^6 daltons (Fig. 2). These molecules have only a short half-life and do not accumulate in large amounts. They are macromolecular precursors of rRNA (pre-rRNA) [3]. With increasing period of labelling, the incorporated radioactivity shifts from the high molecular precursors to the mature rRNA species with molecular weight of 1.3 and 0.7×10^6 daltons (Fig. 3). As the cell elongation ceases (7 mm from the cap junction), the incorporation of ^{3}H-uridine into rRNAs diminishes to about 15 per cent and the high molecular pre-rRNA

Table 2. Incorporation of ^{3}H-uridine into tRNA in different segments of root apex.

Segment	Specific activity of tRNA	
	cpm/100 µg RNA	cpm/RNA in 10^3 cells
0–1 mm	13,831	19.36
1–3 mm	7,851	15.75
5–7 mm	7,542	14.24

fraction is hardly labelled at all (Table 1). This is a strong evidence for regulation of rRNA gene activity at the level of transcription. On the contrary, the differences in the specific radioactivity of tRNA are considerably lower in particular root segments (Table 2).

The transcriptional activity of the nucleus is determined by the availability of DNA for transcription and by the activities of RNA polymerases. During the differentiation of maize root cells, the amount of accessible template increases [5] but the activity of RNA polymerases shows little change [8]. Therefore we suppose that the reduction of rRNA and tRNA along the growth region of maize root may be caused by the reduction of the ribonucleoside triphosphates pool or some other factors which regulate the initiation of transcription.

REFERENCES

1. Clowes, F. A. L. 1958 Protein synthesis in root meristems. J. Exp. Bot. 9, 229–233.
2. Gašparíková, O. 1974 The nucleic acid and protein content in growth zones of Zea mays root. In: Kolek, J. (ed.), Structure and function of primary root tissues, pp. 245–250. Bratislava: Veda.
3. Grierson, D. 1976 RNA structure and metabolism. In: Bryant J. A. (ed.), Molecular aspects of gene expression in plants, pp. 53–108. London, New York, San Francisco: Academic Press.
4. Ingle, J. and Burns, R. G. 1968 The loss of ribosomal nucleic acid during the preparation of nucleic acid from certain plant tissues by the detergent-phenol method. Biochem. J. 110, 605–606.
5. Kononowicz, A. K. 1978 RNA synthesis and changes in DNA template activity during the growth and differentiation of parenchyma cells of the cortex of roots of Zea mays and Tulipa kaufmanniana. Folia Histochem. Cytochem. 16, 123–137.
6. Loening, U. E. 1967 The fractionation of high molecular weight ribonucleic acid by polyacrylamide-gel electrophoresis. Biochem. J. 102, 251–257.
7. Luxová, M. 1980 Kinetics of maize root growth (in Ukr.). Ukr. bot. j. 37, 68–72.
8. Olszewska, M. J. and Kononowicz, A. K. 1979 Activities of DNA polymerases and RNA polymerases detected in situ in growing and differentiating cells of root cortex. Histochem. 59, 311–323.

25. Development of protein patterns in the maize root stele

EMIL E. KHAVKIN, INESSA V. ZELENEVA

Siberian Institute of Plant Physiology and Biochemistry, U.S.S.R. Academy of Sciences, Irkutsk 33, 664033, U.S.S.R.
and
EVGENII YU. MARKOV and NATALIA V. OBROUCHEVA

K. A. Timiriazev Institute of Plant Physiology, U.S.S.R. Academy of Sciences, Moscow 127106, U.S.S.R.

In the primary and adventitious roots of maize seedlings, the stele presents an excellent experimental system to study the rearrangement of protein components in relation to differentiation. The fact that the stele and the cortex (plus the epidermis) can be easily separated in the growing part of the root [2] considerably facilitates the biochemical studies of the stele. A survey of the enzyme profiles of the stele and the cortex provided quantitative evidence on the rearrangement of proteins common to all the cell types in the course of differentiation ([5] and I. V. Zeleneva et al., submitted for publication). Two independent approaches of high specificity, e.g., antigen-antibody and lectin-erythrocyte interactions, were employed to expose the stele-specific proteins [2, 3].

The first and the most striking feature of enzyme specialization in the root is compartmentation of catalase (mostly in the stele) and acid glycosidases, especially invertase (mostly in the cortex). The second remarkable feature concerns the specific activities of the enzymes of general (house-keeping) metabolism, e.g., glycolytic and mitochondrial enzymes, whose levels greatly exceed those in the cortex. While contributing less than 15 per cent to the root fresh weight, due to the higher specific activities of most enzymes under study, the stele apparently provides up to 50 per cent of the root metabolic potential. We are inclined to believe that the cells quantitatively predominant in the stele, that is of phloem, xylem and pith parenchyma, must have abundant microbodies and mitochondria that differ in some unknown features from the respective organelles of the cortex cells.

To separate minor stele-specific proteins from the mass of the proteins that were common both to the stele and the cortex, we have successfully employed a series of immunofiltration techniques [2]. Their sensitivity is ca. 1–5 μg/ml specific protein if the latter constitutes over 0.5–1 per cent of the total protein content of the sample. Thus we have found several antigens characteristic of the stele and undetected in the cortex. Most of these antigens are already present in the

R. Brouwer et al. (eds.), Structure and Function of Plant Roots, 133–135.

meristem, and their content increases as cells elongate and differentiate. Yet there seem to be several antigens present only in the vascular cylinder of the grown part of the root.

Recently specific lectins (erythroagglutinins) have been found in the stem phloem exudates [4], and we have suggested that some of the stele-specific antigens may manifest erythroagglutinating activity. Indeed, crude stele extracts, contrary to those of the cortex, agglutinate human erythrocytes. We have purified the stelar lectins ca. 20-fold by $(NH_4)_2SO_4$ precipitation and used this preparation to agglutinate erythrocytes. Proteins desorbed from the latter produced the characteristic stele-specific precipitin band.

We have compared the rates of [^{14}C] leucine incorporation into the specific and common proteins separated by two immunological procedures [1, 2]. Incorporation into the specific antigens exceeded 2–5-fold that of the common antigens in the meristem and in the steles of the primary and adventitious roots, and we have every reason to presume the selective enhancement of the synthesis of the stele-specific proteins in the course of the stele development.

Thus a noticeable rearrangement of protein patterns in the developing vascular cylinder involves (a) different rates of accumulation and turnover of the house-keeping enzymes; (b) the compartmentation of enzymes of a more specialized metabolism, e.g., of lysosomal hydrolases and microbody enzymes; and (c) the selective accumulation of the stele-specific proteins, e.g., the stelar antigens and lectins. By comparison of the seedlings and the embryos of the non-germinated caryopses, we presume that the stele-specific antigens accompany any visible vascular elements, while the stelar lectins appear only in the functioning stele.

We expect that future investigations will supplement our evidence and verify the validity of our presumptions. It seems to be of outmost importance to provide further data on the metabolic profile of several cell types that constitute the vascular cylinder and to pinpoint the stele-specific proteins as products of highly specific differentiation to their respective definite cell types.

REFERENCES

1. Khavkin, E. E., Markov, E. Y. and Mazel', Y. Y. 1980 Synthesis of specific stelar proteins in the primary roots of maize seedlings. Ann. Bot. 45, 127–130.
2. Khavkin, E. E., Markov, E. Yu. and Misharin, S. I. 1980 Evidence for proteins specific for vascular elements in intact and cultured tissues and cells of maize. Planta 148, 116–123.
3. Markov, E. Yu. 1980 Lectins (erythroagglutinins) in the maize caryopsis and seedling. Izv. Sib. Otd. Akad. Nauk S.S.S.R. (in press).
4. Sabnis, D. D. and Hart, J. W. 1978 The isolation and some properties of a lectin (haemagglutinin) from Cucurbita phloem exudate. Planta 142, 97–101.

5. Zeleneva, I. V., Savost'yanova, E. V. and Khavkin, E. E. 1975 Specificity of acid phosphatases and glycosidases in mature cells of the cortex and central cylinder of corn roots. Sov. Plant Physiol. 22, 189–192.

26. Glycosidases in the root tip

KAREL BENEŠ*, VIKTOR B. IVANOV** and VĚRA HADAČOVÁ*

**Institute of Experimental Botany, Prague, Czechoslovakia and **Institute of General and Inorganic Chemistry, Academy of Sciences, Moscow 117071 U.S.S.R.*

The concept of different enzyme phenotypes in particular cells or cell complexes resulting from spatio-temporal regulation of gene activity led to a highly fruitful approach in studies on animal embryogenesis and development [8], whereas its application to plant studies is so far very limited. It is the aim of the present paper to describe a screening of localization of various glycosidases in root tips of several elected species. Attempts will be made also to evaluate the results on the basis of histogenesis.

The objects of the present work were tips of seedling roots of *Cucurbita maxima* Duch. cv. Veltruská obrovská, *Lupinus albus* L. cv. Pflugs ultra, *Pisum sativum* L. cv. Raman, *Vicia faba* L. cv. Chlumecký, *Zea mays* L. cv. Český bílý koňský zub and of adventitious roots of *Allium cepa* L. cv. Hiberna or Vsetatská [4]. Free floating transverse sections from cold calcium–formol fixed material made with a freezing microtome were submitted to simultaneous azocoupling

Table 1. The presence of glycosidases in root tip sections

source \ enzyme	α-fucosidase	β-glucosaminidase	β-glucuronidase	α-mannosidase	β-xylosidase
Allium	−	−	∓	∓	−
Cucurbita	−	∓	∓	∓	+
Lupinus	−	∓	∓	∓	∓
Pisum	−	−	∓	++	−
Vicia	−	∓	∓	++	∓
Zea	−	∓	∓	++	+

++ high, + middle, ∓ low, − no activity

R. Brouwer et al. (eds.), Structure and Function of Plant Roots, 137–139.

reactions at pH 5.5 and 7.5 for 1 to 6 h at lab. temperature [3]. The following substrates were available: 6-bromo-2-naphthyl-α-D-glucuronide, -α-D-mannoside, -β-D-xyloside, 1-naphthyl-N-acetyl-β-D-glucosaminide, 2-naphthyl-α-L-fucoside. Dimethylformamide was used as the substrate solvent [1]. Controls without substrate were run regularly. The sections taken at the level of the beginning of cell elongation in the cortex were compared. Protein content of the particular cells and incubation time were taken into account.

The gross overview of the presence of the particular glycosidases in sections of the root tops at pH 5.5 is presented in Table 1. No activity was detected at pH 7.5.

The localization is mostly of an ubiquitous type with more or less distinct gradients. The results are thus mostly similar to those on β-glucosidase [6] and α- and β-galactosidases [3]. The only case which may be regarded as discontinuous enzyme localization, is that of α-mannosidase in maize where only traces or no activity is found in endodermis and the adjacent cell layer of cortex (see Fig. 1A). However, the described localization pattern appears at definite level of the root tip. In more distal parts, the α-mannosidase activity of the corresponding cell layers does not differ from the remainder of the sections (Fig. 1B).

The localization of α-mannosidase in maize could be understood on the basis

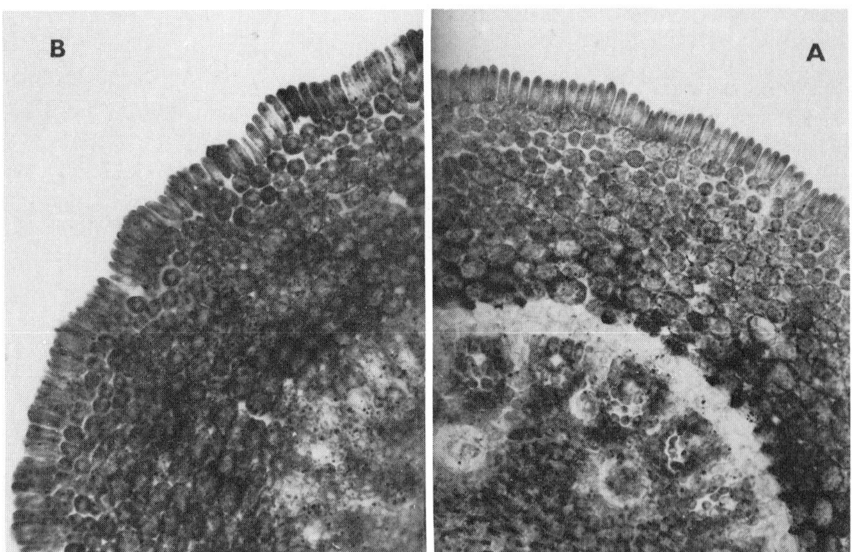

Fig. 1. The localization of α-mannosidase in maize root tip. Prim. magn. 25.6 ×. A. About 1.5 mm from the tip. B. About 1 mm from the tip.

of differential gene activity of a spatio-temporal nature in the 2 cell layers at the root tip). Unfortunately, cases like the present one are hardly accessible for direct and decisive evidence of the regulatory mechanisms involved. From the view-point of histogenesis, 2 cell layers in histogenetically uniform cell complex (cortex) appeared as distinctly different. This could be regarded as another case of differentiation within the given histogen [2]. The localization of α-mannosidase in maize is a clear example of distinct metabolic pattern within the differentiating meristem as shown earlier [5].

Concerning subcellular localization, no striking differences were seen in the resulting colour intensity between the surface and the interior of cells. The technique used does not permit more detailed evaluation under high magnification though methods are available: e.g. the assessment of lysosomal localization of acid phosphatase [7]. However, cases like the localization of α-mannosidase in maize root tip are difficult to explain on the basis of the common concept of lysosomes i.e. homogeneous population of lysosomes within a cell and identical lysosomes in different cells.

REFERENCES

1. Beneš, K. 1974 The use of derivatives of 2-naphthol as substrates for the demonstration of carboxyl esterases in situ and the enhancement of the reaction with DMSO. Histochemistry 42, 193–197.
2. Beneš, K. 1974 Histogenesis and non-specific esterase localization in the root tip. In: Kolek, J. (ed.), Structure and function of primary root tissues, pp. 261–264. Bratislava, Veda.
3. Beneš, K. and Hadačová, V. 1980 The localization and the isoenzymes of α- and β-galactosidases in root tips. Biol. Plant 22, 210–217.
4. Beneš, K. and Kutík, J. 1978 The localization of starch in root tips. Biol. Plant 20, 458–463.
5. Beneš, K. and Opatrná, J. 1964 Localization of acid phosphatase in the differentiating root meristem. Biol. Plant 6, 8–16.
6. Beneš, K., Georgieva Yordanka, D. and Poláčková, D. 1973 The presence and distribution of α- and β-glucosidase in root tip. Biol. Plant 15, 88–94.
7. Beneš, K., Lojda, Z. and Hořavka, B. 1961 A contribution to the histochemical demonstration of some hydrolytic and oxidative enzymes in plants. Histochemie 2, 313–321.
8. Paigen, K. 1979 Acid hydrolases as models of genetic control. Ann. Rev. Genet. 13, 417–466.

27. Glycosidase isoenzymes in root growth zones

VĚRA HADAČOVÁ and KAREL BENEŠ

Institute of Experimental Botany, Czechoslovak Academy of Sciences, Prague, Czechoslovakia

Isoenzyme patterns were studied in order to characterize the proliferative and postproliferative [elongation] cell growth [1] and tissue differentiation phenomena. For this purpose various glycosidases have been examined in particular root growth zones and separately in cortex and central cylinder of root tips.

Extracts from whole tips [12 mm] of seedling roots of *Vicia faba* L., cv. Chlumecký and *Zea mays* L., cv. Český bílý koňský zub and extracts from sections 0–2 mm, 2–4 mm and 8–10 mm from the tip corresponding to zones of division, elongation and maturation and extracts from mechanically separated cortex and central cylinder were submitted to polyacrylamide gel electrophoresis according to our modification of Reisfeld's method [2]. For comparison of isoenzymes in different root tip portions the use of extracts of constant and optimal concentration of proteins is essential. The simultaneous azocoupling reaction [pH 5.5 and 7.4] was used for this purpose.

In whole root tips α-fucosidase and β-glucuronidase were negative in both species tested at both pH 5.5 and 7.4. β-glucosaminidase was weakly active in broad bean and in maize. In both species considerable α-mannosidase and β-xylosidase at activity was found at both pH 5.5 and 7.4. α-mannosidase was more active in broad bean, whereas β-xylosidase predominates in maize.

Characteristic differences in the number and position of isoenzymes were seen in particular root growth zones of both species (Fig. 1). In our previous work [3] results of this kind were obtained only in maize.

Both in broad bean and in maize, differences in the number and position of bands were seen comparing β-xylosidase isoenzymes in cortex and central cylinder, but no such results were found in the case of α-mannosidase (Fig. 1).

Similarly as in our previous work no distinct tendency was apparent when comparing isoenzyme patterns of particular root growth zones. The appearence of new isoenzymes during the postproliferative cell volume growth is not obligatory. The situation is analogous with the isoenzyme patterns of the root cortex and the central cylinder.

R. Brouwer et al. (eds.), Structure and Function of Plant Roots, 141–143.

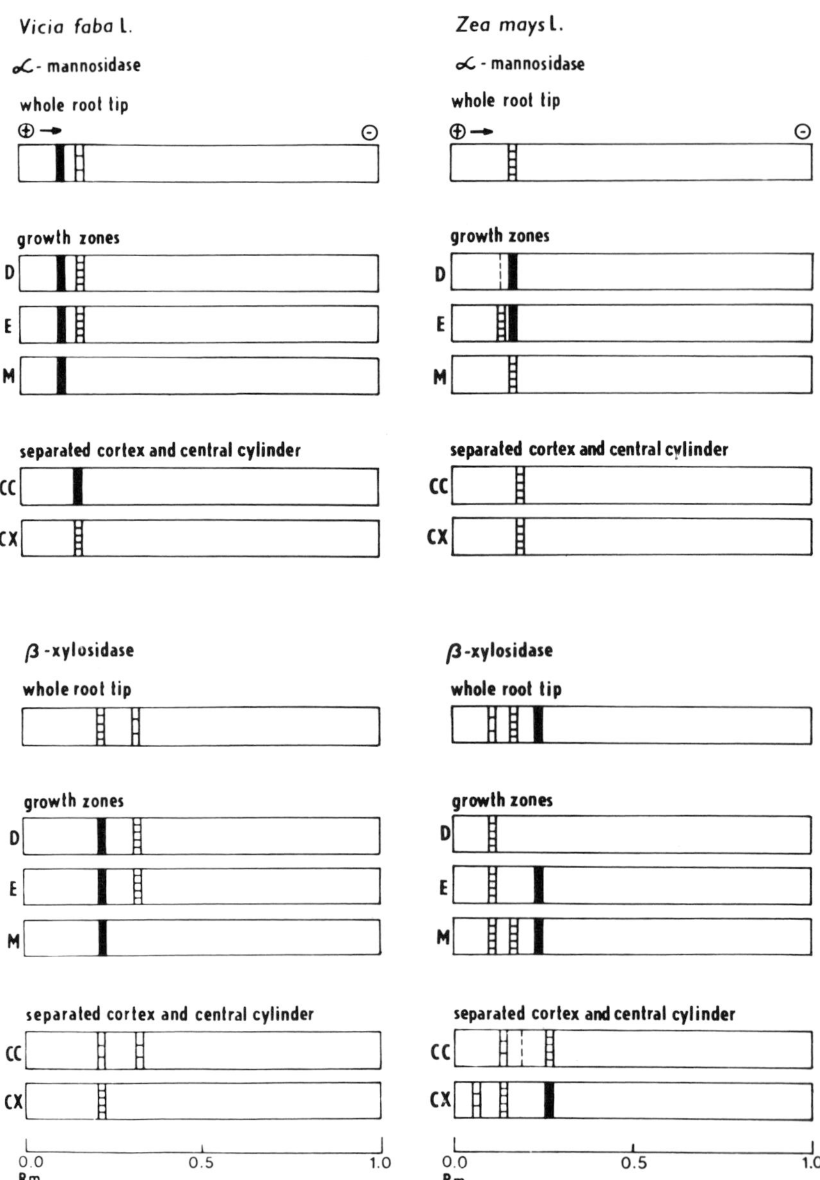

Fig. 1. Isoenzyme patterns of α-mannosidase and β-xylosidase in broad bean and maize root tips. (D – zone of division, E – zone of elongation, M – zone of maturation, CX – cortex, CC – central cylinder, Rm – relative mobility, full bands-high activity, densely hatched bands – middle activity, scarcely hatched bands – low activity, dotted lines – very low and not regularly detectable activity).

REFERENCES

1. Hadačová, V. and Beneš, K. 1977 Investigation of isoenzyme patterns of some oxidoreductases, hydrolases and transferases in different growth zones of broad bean (*Vicia faba* L.). Physiol. Vég 15, 735–745.
2. Hadačová, V. and Beneš, K. 1977 The differentiation of α- and β-glucosidase and α- and β-galactosidase isoenzymes from maize and broad bean root tips using disc electrophoresis in polyacrylamide gels. Biol. Plant 19, 436–441.
3. Beneš, K. and Hadačová, V. 1980 The localization and the isoenzymes of α- and β-galactosidases in root tips. Biol. Plant 22, 210–217.

III. TRANSPORT PHENOMENA

A. WATER TRANSPORT

28. The movement of water across the plant root

JACK DAINTY

Department of Botany, University of Toronto, Toronto, Ontario, Canada M5S 1A1

MARIE KLEINOVÁ and KAREL JANÁČEK

Laboratory for Cell Membrane Transport, Institute of Microbiology, ČSAV, 142 20 PRAHA 4 Krč, Czechoslovakia

This paper arises as part of a review of the processes involved in the movement of water across the plant root, from the soil to the xylem vessels and hence to the shoot. It is really an enquiry as to how far it is possible, or indeed desirable, to analyze the details of the processes; or whether it is more sensible to take only a 'black-box' approach such as is represented by overall equations like those of Fiscus [5, 6, 7]. It is our contention, however, as it is indeed that of most investigators, that we do want to know the details of the process of water movement through the root not only for some kind of intellectual satisfaction but also as a framework to guide new experiments on water transport in the root.

The overall process from some point in the soil to the solution moving up the xylem into the shoot is a flow driven by a force or set of forces which can be described, both as a whole and in each part of the pathway, by the equations of irreversible thermodynamics [2, 3, 5, 13]. These equations more or less correctly include all the forces acting on the water, both those acting directly (the water potential gradient) and those acting more indirectly (frictional drag by moving solutes). The chief parameters involved in an overall approach are a hydraulic conductivity, L_p, a reflection coefficient, σ, an active solute flux, J_s^*, the external osmotic pressure, π_0, and the xylem hydrostatic pressure (negative), P [5, 6, 7]. This approach to the radial movement of water across the root plus the recognition that the longitudinal movement up the xylem has to be described by the so-called standing-gradient osmotic theory [1] gives a workable black-box theory. This theory can be, and has been, used to 'explain' why the resistance of the root apparently varies with the 'driving force', how the xylem osmotic pressure, π_i, varies with flow, and so on.

There is a sense in which the theory must be reasonably correct, but as the criticisms of Newman and others [10, 11, 12] have implied, we need to know more about the inside of the black box; we need to know what the L_p's and σ's are for the individual pathways and barriers, where indeed the solute drag implied by

R. Brouwer et al. (eds.), Structure and Function of Plant Roots, 149–152.

a σ less than one occurs, what the solute concentration and water potential profiles are, where the pumping of solute occurs, and other information.

Some of the earlier worries about the non-linear effects of driving forces on flows were not too sensible. For instance, if the flow is described by the equation: $J_v = L_p(\Delta P - \Delta\pi)$, then we must expect a non-linear relation between J_v and ΔP, because the apparent resistance to pressure, $d(\Delta P)/dJ_v$, is equal to $(1/L_p) + d(\Delta\pi)/dJ_v$, and $\Delta\pi$ varies with the flow rate. The dilemma simply arose because the correct driving force, $(\Delta P - \Delta\pi)$, was not used to calculate the root resistance.

However, there are some experimental data in which even if the apparently correct driving force is used, i.e. $(\Delta P - \Delta\pi) = \Delta\Psi$, the relation between J_v and the water potential difference, $\Delta\Psi$ is non-linear. Fiscus [5, 6, 7] explained away these discrepancies, if they are real, by introducing reflection coefficients, i.e. by saying that the pathway is leaky; he essentially said that if the driving force is modified to $(\Delta P - \sigma\Delta\pi)$ then there would exist a linear relation between flow and force, providing a suitable value of σ is chosen. Thus, there would be no need to think that L_p is not constant.

We make two remarks about this: why should we be so concerned to keep L_p constant? why should it not depend on the flows and driving forces?; and the values of σ used to make the relation between flow and force linear imply quite leaky roots, more leaky to solutes than seems at all reasonable.

If we do not wish to accept values of σ less than one, there are physical processes which make the actual driving force somewhat different from the measured one. These are the sweeping effects of the flow on solute concentrations, both across the root and in the xylem itself and possibly also in the unstirred layers adjacent to the external root surface [2, 11, 13].

At present we have far too little information to estimate what proportion of the water flow takes place in the cortical apoplast, up to the endodermis, and in the stelar apoplast. Methods and theories are being developed [8,9] to investigate and handle these problems, but so far we can only really proceed by guesswork. If most of the solute and water flow up to and beyond the endodermis is in the apoplast, then the water flow will polarise the solute concentration, and therefore the water potential, across the endodermis. It can be shown, for a non-leaky root with $\sigma = 1$, that the apparent driving force $\Delta\Psi$ is reduced by $RT\Delta c$, where Δc is the degree of solute polarisation in the apoplast caused by the sweeping of solute up to and away from the surfaces of the endodermis. Thus if the root resistance is calculated in the normal way by $d(\Delta\Psi)/dJ_v$, the value will not only be incorrect but will be dependent on flow because the solute polarisation is flow-dependent. The effects could be quite large at high flow rates if substantial flow occurs in the apoplast.

The real driving force across the root is also different from the apparent driving force because the solute concentrations in the xylem vessels deep inside the root are not the same as those in the stem, beyond the regions where solutes are 'pumped' into (or absorbed from) the xylem. This has been discussed [1], but much more work, both experimental and theoretical, is required to characterize these standing osmotic gradients in plant roots. It is not even clear in which direction the apparent driving force will be changed. In any case, the effect will be an apparent non-linear resistance, but at present it is impossible to say anything quantitative about this non-linearity.

These polarisation and standing-gradient effects are certainly real and will contribute to the non-linearity of the apparent root resistance, even if L_p is a constant, independent of force or flow. However, there is no real reason to strive too hard to look for explanations which keep L_p constant. We know that in some cells L_p is dependent both on turgor pressure and on solute concentration [4, 14]. Certainly as the driving force across the root changes, usually by changes in the ΔP term, the turgor pressure of, say, the endodermal cells will change. The L_p of these cells, which might well be crucial, could then change and there would thus be non-linearity in the root resistance. Powell [12] has theorized that if the inner membrane of the endodermis had a reflection coefficient less than 1 the solute contents of the endodermal cells would be flushed out by increased flow; this would lead to a drop in turgor and hence a possible change in L_p. But it is surely not necessary to postulate such a drastic mechanism; the turgor would fall without loss of solute as the driving force increased – because of the decrease in the water potential.

We thus reach no conclusions, except that we caution against easy acceptance of low values of σ for any cell membrane for any of the usual solutes. We have tried to indicate where information and theory are really lacking.

REFERENCES

1. Anderson, W. P., Aikman, D. P. and Meiri, A. 1970 Excised root exudation a standing gradient osmotic flow. Proc. Roy. Soc. Lond. B. 174, 445–458.
2. Dainty, J. 1963 Water relations of plant cells. Adv. Bot. Res. 1, 279–326.
3. Dainty, J. 1976 Water relations of plant cells. In: Lüttge, U. and Pitman, M. G. (eds.), Encyclopedia of plant physiology. New Series. Vol. 2A. Transport in Plants, pp. 12–35. Berlin–New York: Springer-Verlag.
4. Dainty, J. and Ginzburg, B. Z. 1964 The measurement of hydraulic conductivity (osmotic permeability to water) of internodal Characean cells by means of transcellular osmosis. Biochim. Biophys. Acta 79, 102–111.
5. Fiscus, E. L. 1975 The interaction between osmotic- and pressure-induced water flow in plant roots. Plant Physiol. 55, 917–922.

6. Fiscus, E. L. and Kramer, P. J. 1975 General model for osmotic and pressure-induced flow in plant roots. Proc. Nat. Acad. Sci. U.S. 72, 3114–3118.
7. Fiscus, E. L. 1977 Determination of hydraulic and osmotic properties of soybean root systems. Plant Physiol. 59, 1013–1020.
8. Green, W. N., Ferrier, J. M. and Dainty, J. 1979 Direct measurement of water capacity of *Beta vulgaris* storage tissue sections using a displacement transducer and resulting values for cell membrane hydraulic conductivity. Can. J. Bot. 57, 981–985.
9. Molz, F. J. and Ikenberry, E. 1974 Water transport through plant cells and cell walls: theoretical development. Soil Sci. Soc. Amer. Proc. 38, 699–704.
10. Newman, E. I. 1976 Interaction between osmotic- and pressure-induced water flow in plant roots. Plant Physiol. 57, 738–739.
11. Nulsen, R. A. and Thurtell, G. W. 1978 Osmotically induced changes in the pressure-flow relationship of maize root systems. Aust. J. Plant Physiol. 5, 469–476.
12. Powell, D. B. B. 1978 Regulation of plant water potential by membranes of the endodermis in young roots. Plant Cell and Environment 1, 69–76.
13. Tyree, M. T. 1970 The symplast concept. J. Theor. Biol. 26, 181–214.
14. Zimmermann, U. and Steudle, E. 1974 Hydraulic conductivity and volumetric elastic modules in giant algal cells: pressure- and volume-dependence. In: Zimmermann, U. and Dainty, J. (eds.), Membrane transport in plants, p. 64. Berlin-Heidelberg-New York: Springer-Verlag.

29. Pressure-flow characteristics of the roots of *Zea mays*

DAVID M. MILLER

Agriculture Canada Research Centre, London, Ont., Canada, N6A 5B7

Root pressure is generally considered to be osmotic in origin. This explanation is not held by all workers, however. An alternative view contends that while an osmotic component certainly exists, the xylem concentrations are inadequate to account for the total pressure and that therefore the remainder must result from some form of active water pumping (see for example [2, 4]). In the ideal method of testing this claim one would allow the root pressure to attain its maximum value, collect a sample of the xylem sap and compare its osmotic pressure to that developed by the root. Unfortunately, this is impossible with present techniques, since to develop maximum pressures requires zero flow, in which case no sap can be collected. One can, however, reduce the flow very nearly to zero and still obtain a small sample for comparison. Such an experiment was performed by collecting 50 μl samples of exudate from each of 22 plants under 2.7 bar pressure and pooling them for analysis. Since the exudation flowrate in this case was reduced to an average of only 7% of the unimpeded rate, the results should represent a close approximation to the ideal case. The osmotic pressure of the exudate collected in this way was found to be 3.2 bar above that of the bathing solution, well in excess of the applied pressure. Thus in the present instance, there is no need to assume that root pressure is generated by any means other than osmotic.

This being so, the volume flow through the root should obey the well-known equation of irreversible thermodynamics

$$J_v = L_p(P_0 - P_x - \sigma RT\Delta S) \tag{1}$$

where J_v is the volume of exudate flowing per second per unit area of root surface, L_p is the hydraulic conductivity of the root, P_0 is the pressure applied to the solution surrounding the root, P_x that at the cut end, R the gas constant, T the absolute temperature and ΔS the sum of the differences in solute osmolality across the membranes. The reflection coefficient, σ, is now generally considered to be 1 for biological membranes [3]. The concentration difference, ΔS, can be obtained by measuring the osmotic pressure difference $\Delta\pi$, which is equal to

R. Brouwer et al. (eds.), Structure and Function of Plant Roots, 153–156.

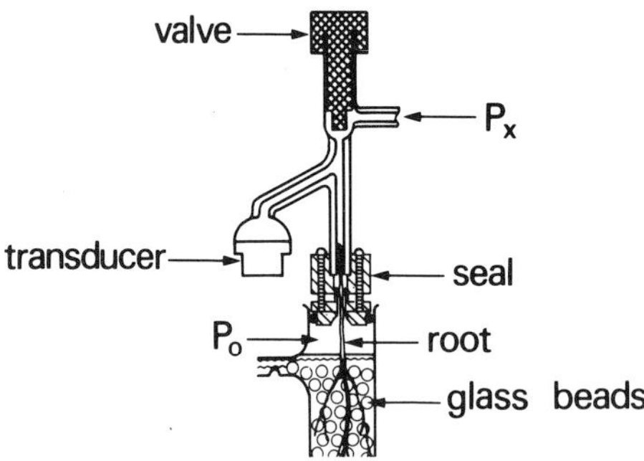

Fig. 1. Apparatus for measuring root pressures.

RTΔC. This measurement must be made under steady state flow, or failing this, so soon after its termination that the concentration gradients established during flow remain unchanged.

A technique for doing this has been reported [1] and will now be described. The apparatus, in a much simplified form, is shown in Fig. 1. It consists essentially of a seal for connecting the cut end of the root to a glass tube to which a pressure transducer and valve are attached. Pressure can be applied to the outside of the root (P_0) or to the cut end (P_x). Closing the valve, by screwing the teflon stem down onto the glass seat, isolates the root, tubing and transducer from the applied pressure P_x so that the transducer records the pressure developed by the root. A recording of this pressure is shown in Fig. 2. Here we see that there is an initial, almost instantaneous jump upon closure of the valve (at time zero) followed by a further gradually slowing increase to a maximum value. Prior to closing the valve a flux of solute, J_s, produced a volume flow, J_v. Reducing J_v to zero, however, does not immediately effect J_s and solutes continue to enter the xylem until their concentration there equals that in the symplasm, at which point the osmotic pressure reaches its maximum value.

The initial pressure jump is partly artifactual since its size is dependent on how tightly the valve is closed. For example, if the stem is forced down onto the seat with a pressure in excess of that necessary to form a seal, the teflon becomes distorted and is forced into the valve throat, compressing the volume below it and raising its pressure. Thus by varying the excess pressure applied, one can obtain a series of recorded pressure jumps of the type shown in Fig. 3.

According to equation (1), when J_v becomes zero, the pressure at the end of the root will assume a new value of $P'_x = P_0 - RT\Delta S$. One might assume then, that

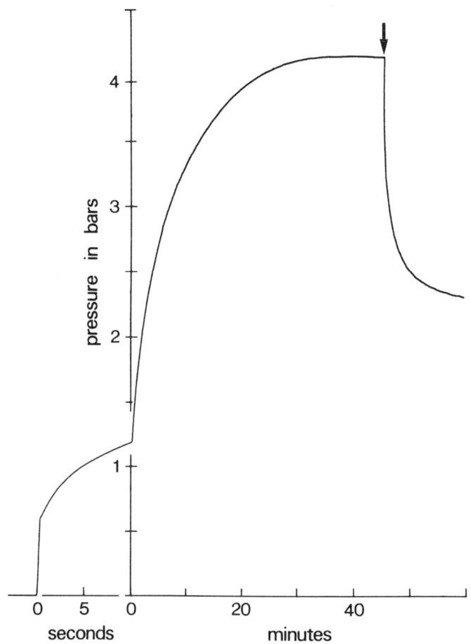

Fig. 2. Transducer pressure recorded as a function of time. The initial pressure applied to the root was atmospheric. At zero time the valve was closed and the resulting pressure, measured in bars above atmospheric, is recorded here. At 10 s the recorder chart speed was reduced, hence the change in time scale. The arrow indicates the point at which a single root of 0.25 mm diameter, branching from near the top of the main root, was severed 1 cm from its origin.

closing the valve should result in an immediate jump to P'_x. Unfortunately this does not occur, as curve A Fig. 3 shows, due to the presence of the rubber seal, the teflon valve stem and the transducer diaphragm, which make the apparatus somewhat distendable. If in addition to closing the valve, however, we were to introduce a pressure equal to P'_x, then J_v would be zero immediately. Conversely, if the pressure is such that J_v equals zero, then we can conclude that it equals the driving force within the root. Our task then, is to determine which of a number of applied pressure jumps of the type shown in Fig. 3, results in an instantaneous initial value of $J_v = 0$.

In Fig. 3 we note that a very high pressure jump (curve H, Fig. 3) is followed by a decline indicating that solution is flowing back through the root and that therefore the pressure initially applied exceeds that in the root. On the other hand, pressures less than that in the root, result in a continuing flow of exudate out of the root producing a pressure which rises rapidly, but at a declining rate, following the pressure jump. This produces a curve which is rounded at the point

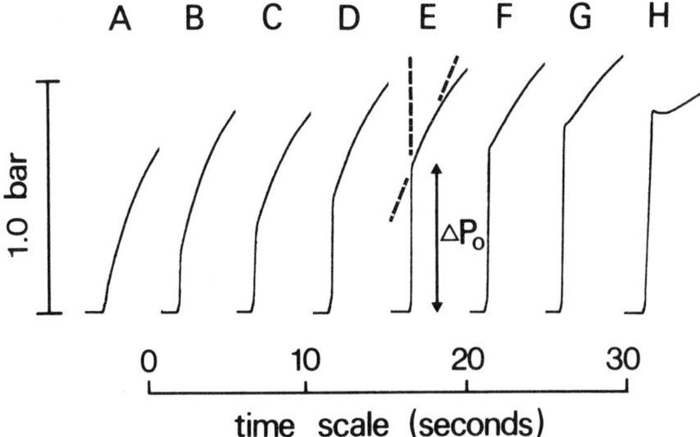

Fig. 3. The effect of increasing the size of the pressure jump, following valve closure, on the pressure recorded by the transducer. The pressure jump was increased progressively from zero at A to its highest value at H.

of transition between the initial jump and the following slower increase in root pressure. Thus the ideal pressure jump is one producing a tracing which breaks sharply away from the initial rise but displays no subsequent hesitation, or inflection as shown by curve E, Fig. 3.

If ΔP_0, the pressure jump, is defined as the difference between the pressure recorded immediately after closure (P'_x) and that immediately before (P_x) (i.e. $\Delta P_0 = P'_x - P_x$) then, $P'_x = \Delta P_0 + P_x = P_0 - RT\Delta C$ or $\Delta P_0 = P_0 - P_x - RT\Delta C$ and expression [1] becomes $J_v = L_p \Delta P_0$. Thus the hydraulic conductivity is simply J_v divided by ΔP_0.

One final note concerns the drastic effects of a small injury to the root. This is illustrated in Fig. 2, where (at the arrow), a small root branch was severed resulting in a rapid decline in root pressure. All roots used in this work were grown in a glass bead supporting medium [1]. These beads were sufficiently large (6 mm diameter) to allow free flow of solution past the root while at the same time providing adequate physical support.

REFERENCES

1. Miller, D. M. 1980 Studies of root function in *Zea mays* I. Apparatus and methods. Canad. J. Bot. 58, 351–360.
2. Mozhaeva, L. V. and Pil'shchikova 1978 Relationship between the value of root pressure components and the rate of water pumping by the root. Dokl. Akad. Nauk SSSR 239, 1005–1008.
3. Stadelmann, E. J. 1977 Passive transport parameters of plant cell membranes. In: Marrè, E. and Ciferri, O. (eds.), Regulation of cell membrane activities in plants, pp. 3–18.
4. Zholkevich, V. N., Sinitsyna, Z. A., Peisakhzon, B. I., Abutalybov, V. F. and D'yachenko, I. V. 1979 On the nature of root pressure. Fiziologiya Rastenii 26, 978–993.

30. On the nature of root pressure

VLADIMIR N. ZHOLKEVICH

K.A. Timiriazev Plant Physiology Institute, Academy of Sciences of the U.S.S.R., Moscow, U.S.S.R.

According to the osmotic conception, the moving force of exudation is determined merely by the difference of the osmotic pressure between the xylem sap and ambient solution; so exudation is impossible in hypertonic solutions. However, it was found by some authors [2, 4, 5], that the osmotic pressure of the ambient solution (OP_e), which stopped exudation, considerably exceeded the osmotic pressure of the exudate (OP_i). Thus there exists a non-osmotic component (NOC) of root pressure.

The results of our experiments [3] with whole root systems and with directly water absorbing root parts of *Helianthus annuus* and *Zea mays* showed that NOC calculated from the difference between OP_e and OP_i amounted averagely to about half of OP_e (Table 1). The errors due to different OP_i values along the root or to an enhancement of OP_i while submerging roots into hypertonic solution

Table 1. Osmotic pressure of ambient solution stopping exudation (OP_e), osmotic pressure of exudate (OP_i) and non-osmotic component of root pressure ($OP_e - OP_i$) in *Helianthus annuus* and *Zea mays*

Object	OP_e	OP_i	$OP_e - OP_i$	
		bar		per cent of OP_e
The whole root system of *Helianthus annuus*	1.59 ± 0.08	0.85 ± 0.05	0.74	46
,,	1.67 ± 0.11	0.74 ± 0.06	0.93	56
,,	1.44 ± 0.04	0.79 ± 0.03	0.65	45
,,	2.38 ± 0.12	0.85 ± 0.07	1.53	64
,,	2.20 ± 0.09	0.61 ± 0.07	1.59	72
The absorbing water root parts of *Zea mays*	3.16 ± 0.05	0.98 ± 0.03	2.18	69
,,	1.41 ± 0.03	0.79 ± 0.01	0.62	44
,,	1.23 ± 0.04	0.73 ± 0.02	0.50	41
,,	1.96 ± 0.02	0.61 ± 0.01	1.35	69
,,	2.26 ± 0.03	0.85 ± 0.02	1.41	62

R. Brouwer et al. (eds.), Structure and Function of Plant Roots, 157–158.

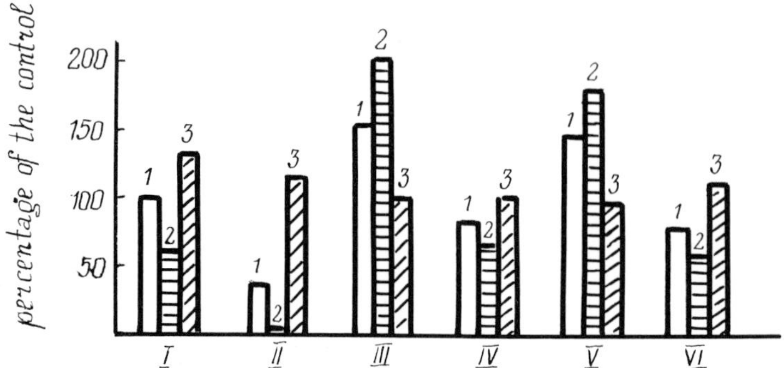

Figure 1. The effect of 2,4-dinitrophenol (5×10^{-4} M, I), pipolphene (4×10^{-4} M, II), $CaCl_2$ (1×10^{-3} M, III), EDTA (7×10^{-3} M, IV), acetylcholine (1×10^{-3} M, V), d-tubocurarine (1.5×10^{-4} M, VI) on the root pressure (1), its NOC (2) and OP_i (3) of *Helianthus annuus*.

have been eliminated. Osmotic pressure was measured microcryoscopically and polyethyleneglycol was used as the indifferent osmotic acting agent.

NOC is energy-dependent and decreases strongly when the cell structure is disturbed. Changes of the NOC value induced by acetylcholine, $CaCl_2$, EDTA, d-tubocurarine (Fig. 1) give evidence for a complicated nature of NOC and support a suggestion that an actinomyosin-like protein participates in root pressure build-up. Such a protein has been isolated from the roots of *Helianthus annuus* in our laboratory [1, 3].

REFERENCES

1. Abutalybov, V. F., Shushanashvili, V. I. and Zholkevich, V. N. 1980 Isolation of actin-like protein from sunflower roots. Doklady Akademii nauk USSR 252, 1023–1024.
2. Broyer, T. C. 1951 Exudation studies on the water relations of plants. Amer. J. Bot. 38, 157–162.
3. Zholkevich, V. N., Sinitsina, Z. A., Peisakhzon, B. I., Abutalybov, V. F. and D'yachenko, I. V. 1979 On the nature of root pressure. Phyziologiya rastenii 26, 978–993.
4. Mozhaeva, L. V. and Pil'shchikova, N. V. 1972 On the nature of pumping water process by plant roots. Izvestiya Timiriazevskoi selskochozaistvennoi akademii 3, 3–15.
5. Overbeek, J. van 1942 Water uptake by excised root systems of the tomato due to non-osmotic forces. Amer. J. Bot. 29, 677–682.

31. Electroosmotic phenomena in longitudinal transport of solutes in roots

ZUZANA HERDOVÁ

Institute of Experimental Biology and Ecology, Slovak Academy of Sciences, 885 34 Bratislava, Czechoslovakia

Electroosmosis is one of a number of electrokinetic phenomena participating in the transport of materials through various barriers. It is supposed that electroosmosis occurs in longitudinal transport through plant roots and the transport of KCl solution through nodal adventitious roots of maize has been examined in the present work.

A mathematical description of electrokinetic phenomena obtained from irreversible thermodynamics was used. It has known form [2]:

$$J = L_{11}\Delta p + L_{12}\Delta\varphi$$
$$I = L_{21}\Delta p + L_{22}\Delta\varphi.$$

Coefficients L_{11} to L_{22} are called phenomenological and they are defined in an earlier paper [1]; Δp is the pressure difference and $\Delta\varphi$ is the potential difference.

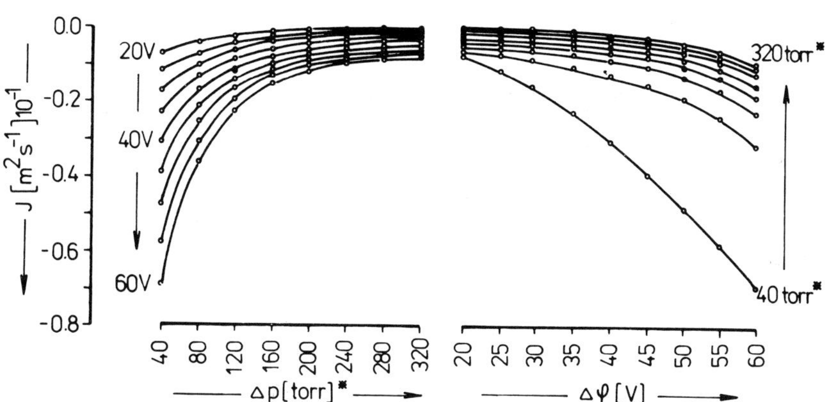

Fig. 1. Dependence of general volume flow J on pressure difference Δp and on potential difference $\Delta\varphi$ for samples of nodal adventitious roots of maize.

R. Brouwer et al. (eds.), Structure and Function of Plant Roots, 159–161.

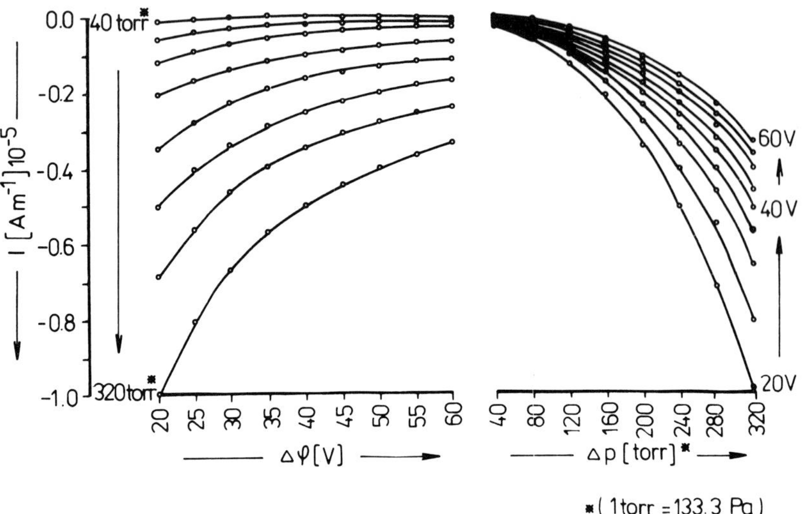

Fig. 2. Dependence of general electric current I on potential difference Δφ and on pressure difference Δp for samples of nodal adventitious roots of maize.

2 cm long segments with a diameter of about 3 mm were prepared and were placed in the electroosmometer consisting of two compartments filled with 8 mM KCl. After application of a potential difference Δφ or pressure difference Δp between the two compartments separated by the sample, volume flow J′ and electric current I′ through the measured sample was noted, alternately.

From the experimental values obtained it was possible to calculate the values of all phenomenological coefficients for nodal adventitious roots of maize. From this base we were able to determine the dependences of general volume flow J and general electric current I passing through the sample under the condition of simultaneous applications of differences in both potential and pressure.

It is apparent that the volume flow J increases with increase of pressure difference (Fig. 1) but only to a limited extent. This extent is determined by the properties of conductive paths in the tissue structure. A similar situation can be seen for the electric current I (Fig. 2) where the consequence of higher electric current is that the tissue dies. The cross dependences are inversely proportional.

The phenomenological coefficient values depend on the age of the plants and measurements over long periods on single samples lead to structural changes so care must be taken when interpreting such data.

REFERENCES

1. Michalov, J. and Palčáková-Herdová, Z. 1979 The electroosmotic model of matter transport through biological barriers. Kybernetika (Praha) 15,6, 482–487.
2. Spanner, D. C. 1964 Introduction to thermodynamics. London-New York: Academic Press.

32. A possible driving force for the transport of water in roots: the temperature gradient

FERENC VETÓ

Biophysical Institute, Med. Univ., H-7643 Pécs, Hungary

The possibility that local temperature gradients play a role in water transport has been raised in recent years [2, 4–10]. The hypothesis is based on the analogy of thermoosmotic model experiments, such as the equation:

$$J_v = L_p\left(\Delta P - \Delta\pi + \frac{Q^*\Delta T}{vT}\right).$$

The direction and the extent of thermoosmosis is shown by the actual sign and value of the heat transfer (Q^*) characteristic of the system in question. In model experiments values of Q^* between $-230\ J\cdot mol^{-1}$ and $+2100\ J\cdot mol^{-1}$ have been found [1, 3]. The aim of the present experiments was to learn if in plant tissues water could be mobilized by controlled temperature gradients and to measure the value of Q^*. In apple tissue a temperature gradient of $1\ K\cdot mm^{-1}$ gave rise to a 4–5 per cent difference in water content over a distance of 10 mm and the value of Q^* amounted to $+450\ J\cdot mol^{-1}$. A ΔT of 3 K between potato tissue and a sucrose solution ($0.6\ mol\cdot kg^{-1}$), the tissue being selectively warmed by microwave irradiation, caused a 15 per cent reduction in exosmosis, in this case the value of Q^* being $-170\ J\cdot mol^{-1}$. Thermoosmosis can be demonstrated in biological tissues but is of slight biological significance according to the available data. Still, a radial temperature gradient may exist in the intensively respiring absorbing zone of the roots and the stele may be warmer than the epiblema. This gradient can be an additional driving force of water absorption, if Q^* has a large negative value.

REFERENCES

1. Dariel, M. S. and Kedem, O. 1975 Thermoosmosis in semi-permeable membranes. J. Phys. Chem. 79, 336–342.
2. Davies, M. 1965 Thermal migration in biological transport? Biophysical J. 5, 651–654.
3. D'Ilario, L. and Canella, M. 1977 Unusual thermal energy transduction by means of a composite membrane system. Polymer 18, 206–207.
4. Ernst, E. 1936 Über die Osmose. Ber. ges. Physiol. exp. Pharm. 94, 658.

R. Brouwer et al. (eds.), Structure and Function of Plant Roots, 163–164.

5. Ernst, E. 1966 Thermoosmosis in biology. Acta Biochim. Biophys. Acad. Sci. Hung. 1, 211–212.
6. House, C. R. 1974 Water transport in cells and tissues. London: Arnold.
7. Katchalsky, A. and Curran, P. F. 1965 Nonequilibrium thermodynamics in biophysics. Cambridge, Mass.: Harvard University Press.
8. Riede, W. 1921 Untersuchungen über Wasserpflanzen. Flora (Jena) 114, 1–118.
9. Spanner, D. C. 1954 The active transport of water under temperature gradients. Symposia of the Society for Experimental Biology 8, 76–93.
10. Vetó, F. 1963 Mobilization of fluids in biological objects by means of temperature gradient. Acta Physiol. Acad. Sci. Hung. 24, 119–128.

33. Conducting efficiency of roots for the longitudinal flow of water

VLADIMÍR KOZINKA

Institute of Experimental Biology and Ecology of CBEV SAV, 885 34 Bratislava, Czechoslovakia

There is an increasing need of quantitative data for modelling water transport in plants. One of the required parameters is the conducting efficiency of roots for the longitudinal flow of water. The complexity of the vascular system of roots is rarely appreciated by investigators of translocation and studies of this kind calls for active cooperation between physiologists and anatomists (Fig. 3; Table 1). There are two kinds of quantitative information available: data of linear pro-

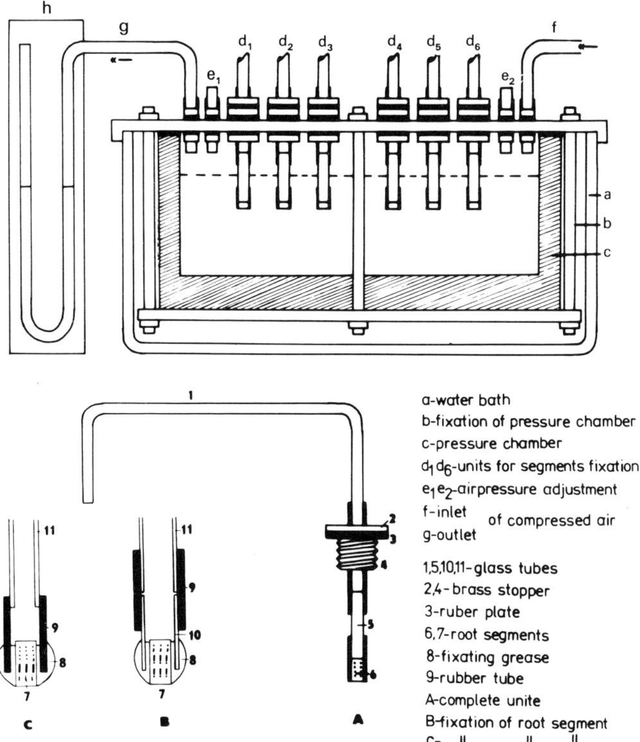

Fig. 1. Apparatus for measuring actual flow volumes of water through detached segment of root.

R. Brouwer et al. (eds.), Structure and Function of Plant Roots, 165–169.

Table 1. Number of protoxylem poles (a) vessels of the early (b) and the late metaxylem (c) and the actual flow volumes of water through detached segments from the basal part of seminal root and adventitious nodal roots of sorghum. (Lenghth of segment 10 mm; applied pressure 0.17 MPa by 20 °C.)

Type of root	Number of conducting elements			Actual flow volumes (mm^3 s^{-1})		
	a	b	c			
Sem. root	7.9	10.3	2.0	1.8	1.2	0.7
1st node	23.2	24.7	7.0	4.0	1.2	1.5
2nd node	34.0	36.9	9.2	6.5	4.7	3.5
3rd node	39.5	47.3	14.6	14.5	11.2	7.8
4th node	59.6	67.9	21.3	8.0	13.8	15.2
5th node	93.4	102.8	32.7	2.0	13.3	20.2
6th node	111.0	115.8	31.5	1.7	13.8	22.7
7th node	112.9	117.1	40.6	—	8.0	10.3
8th node	116.1	117.6	43.0	—	2.5	4.0
1st node	36.9	44.1	13.9	5.7	7.0	6.0
2nd node	70.3	77.5	23.3	3.7	5.5	8.8
3rd node	85.0	86.9	30.2	2.2	9.0	9.3
4th node	85.6	90.0	29.5	—	10.2	5.8
5th node	89.1	93.3	35.0	—	12.2	5.3
6th node	96.1	100.2	44.8	—	2.0	1.0
1st node	48.6	54.8	13.6	2.2	6.3	6.5
2nd node	66.1	70.4	17.5	2.2	5.8	8.5
3rd node	79.9	87.3	23.8	1.7	7.0	8.5
4th node	88.2	91.6	25.8	—	6.0	8.7
5th node	89.7	95.1	32.8	—	2.7	5.5
6th node	88.8	95.0	35.0	—	1.3	2.4
1st node	32.9	38.1	12.7	1.5	2.8	4.5
2nd node	41.2	44.5	13.8	0.5	4.4	10.8
3rd node	53.0	60.1	23.3	—	4.8	8.0
4th node	66.9	69.3	28.0	—	1.2	2.5
5th node	81.2	83.6	28.5	—	1.5	2.2
6th node	91.4	95.1	32.4	—	—	0.4
	Age of plants (days)			56	112	140

Main stalk First tiller
Second tiller Third tiller

gressive rate of flow of water columns in tissues and data of experimental flow through detached segments of roots under controlled conditions.

Determination of measured data was carried out from the results of experimental flow of water through segments of entire roots and central cylinder of *Zea mays L.* and *Sorghum saccharatum L.* (Moench. [1, 2, 3, 4]; Fig. 1). It was found

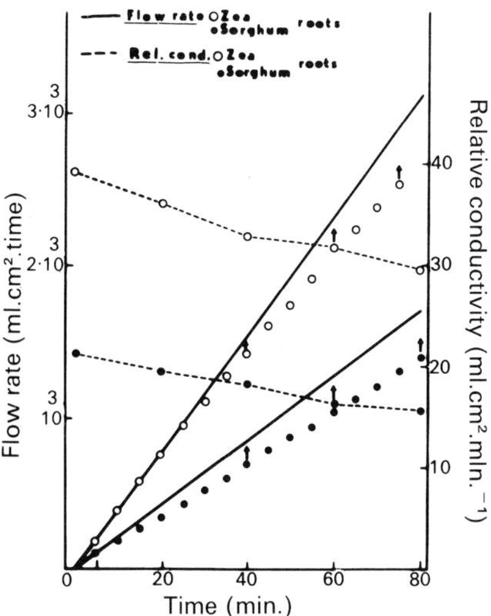

Fig. 2. Changes of actual flow volumes of water and measured values of relative conductivity of detached segments of root with time.

that the relative conductivity decreases with time during an experiment from the time of application of pressure (Fig. 2).

Differences were found between the calculated values of relative conductivity and conducting capacity in which the complete set of vessels is considered as a bundle of capillaries [2, 3]. There can be different causes of such discrepancies. Some of them are: variation in the number of vessels and members of vessels along the segment (Fig. 3; Table 1); variation in the diameter of vessels along the segment; resistance by the residues of the transverse walls of the vessel members and differences in the structure of the internal surface of the vessel. It is not possible to evaluate each of the structural aspects separately. In spite of this fact it is evident from the results that the most significant features are the length of the vessel members together with the area of their perforations.

It has long been recognized that the plant offers resistance to water movement and that a considerable part of it lies in the root. Much attention has been paid to the question as to where within the root lie the main sites of resistance but it is still difficult to answer. It is believed that only when we know the answer we will have a basis for understanding all the features of the root as an organ of water uptake.

The radial difference between the root surface and xylem is usually small, but the tissues do not provide an easy pathway for water movement. Our evidence suggests that there is in fact considerable resistance. Newman has reported

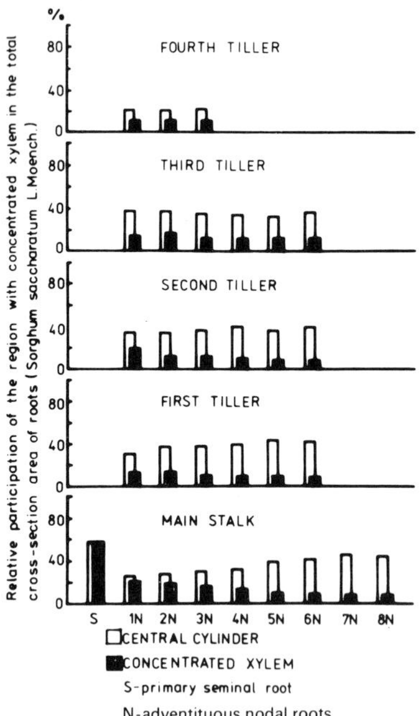

Fig. 3.

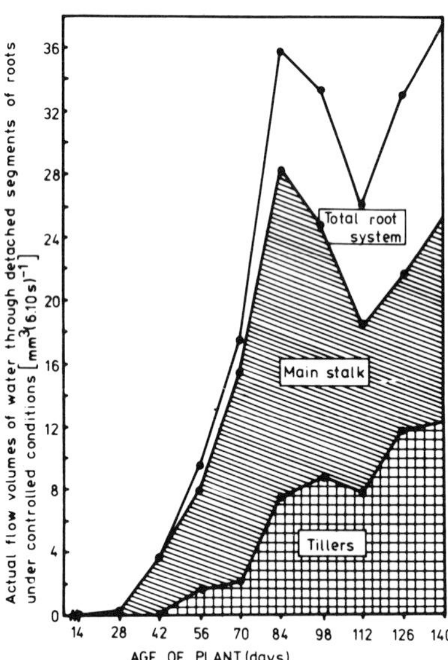

Fig. 4. Total actual flow volumes of water through detached segments of roots of sorghum.

experiments with wheat in which xylem resistance was probably significant and the flow within the xylem vessels appeared to be critical [5].

CONCLUSIONS

The conducting efficiency of roots for the longitudinal flow of water is high on the basis of measurement of the actual flow volumes through the detached segment of roots (Fig. 4). The total cross-section area of conducting tissues is not high but suitable conditions for rapid flow exist.

REFERENCES

1. Kozinka, V. and Luxova, M. 1971 Specific conductivity of conducting and non-conducting tissues of Zea mays root. Biol. Plant (Praha) 13, 257–266.
2. Luxova, M. and Kozinka, V. 1970 Structure and conductivity of the corn root system. Biol. Plant (Praha) 12, 47–57.
3. Luxova, M. and Kozinka, V. 1973 Study of the vascular flow in the root segments of Zea mays. Biologia (Bratislava) 28, 227–234.
4. Mandour, M. S. 1974 Root system conductivity of Sorghum saccharatum L. for longitudinal water flow. PhD thesis Bot. Inst. Slovak Academy Sci., Bratislava.
5. Newman, E. I. 1976 Water movement through root systems. Phil. Trans. R. Soc. Lond. B 273, 463–478.

B. ION UPTAKE AND TRANSPORT

34. Mechanism of ion uptake across barley roots

R. DEJAEGERE, L. NEIRINCKX, J. M. STASSART and V. DELEGHER

Lab. Plant Physiology, Free University, Brussels, Belgium

Since Mitchell showed that ATP-ase acts as a reversible proton carrier in mitochondria, several observations have focused attention on the existence of such proton-pumps in plant membranes [15, 18, 20]. These facts have opened new possibilities regarding the mechanisms of ion uptake and have led us to propose a new model for ion-absorption. We shall first review the main observations upon which the model is based and then check some of its consequences with results of other laboratories.

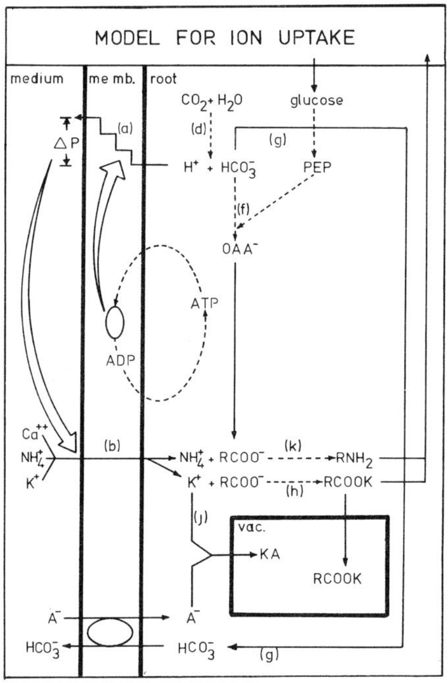

Fig. 1. A model for ion uptake.

R. Brouwer et al. (eds.), Structure and Function of Plant Roots, 173–178.

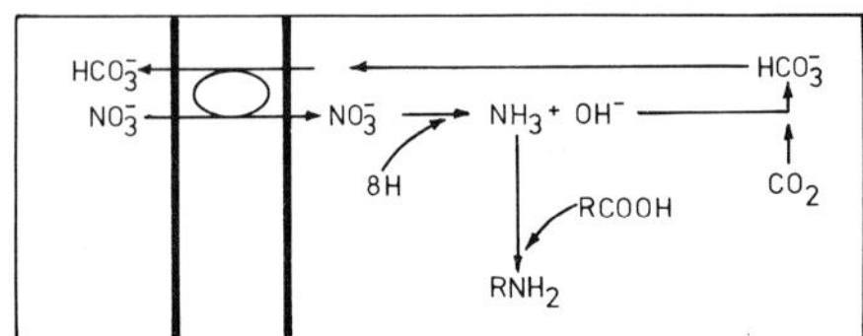

Fig. 2. NO_3^--uptake and assimilation.

THE GENERAL MODEL

The first statement one can make is that H^+-extrusion through the roots is an active mechanism [6, 7].

Second, the H^+-extrusion is linked to the absorption of all the cations tested [5, 6, 7].

Third, when considering not the apparent proton extrusion (which is equal to the amount of base necessary to maintain the pH of the external medium constant) but the net proton efflux (i.e. the apparent proton extrusion corrected for the bicarbonate efflux in the medium), one comes to the conclusion that the proton extrusion is equal to the cation uptake [7].

Fourth, the uptake of the monovalent cations (NH_4^+, K^+, Na^+) is independent of the associated anion (Cl^-, SO_4^{--}, $H_2PO_4^-$), while the reverse is not true [7].

Starting from these results, and from ideas which have been put forward by other people [16, 19], a general model for ion uptake has been built up (cf. Fig. 1). The cation uptake (phase 'b' of the model) is driven by the electrochemical gradient (dP) resulting from the active extrusion of protons ('a'). The absorption of the cations lowers the gradient and allows the further extrusion of protons: the H^+-efflux generates but is at the same time sustained by the cation uptake.

The protons are generated in the equilibrium reaction between CO_2 and H_2CO_3 ('d'). The HCO_3^--ions left over by the H^+-efflux are either used for the

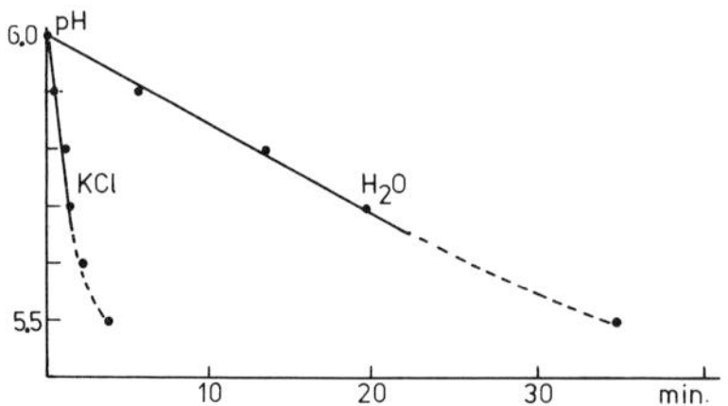

Fig. 3. H^+-extrusion in water as compared with salt-solutions.

Table 1. Ion-uptake by NO_3^-- and NH_4^+-fed plants

Author	Plant	Total uptake (μeg/gFW)	Relative uptake (%)			
			NO_3^-	A^-	NH_4^+	C^+
1. NO_3^--fed plants:						
Neirinckx [17]	Cotton	428	32	19	–	49
Kostic [14]	Wheat	782	53	18	–	29
Kirkby [12]	Tomato	576	47	8	–	45
2. NH_4^+-fed plants:						
Dejaegere [4]	Cotton	464	–	17	29	54
Kostic [14]	Wheat	697	–	15	58	27
Kirkby [12]	Tomato	536	–	14	58	28

A^- = anions other than NO_3^-; C^+ = cations other than NH_4^+.

biosynthesis of organic acids ('f' and 9) or exchanged for other anions in an antiport system ('g').

The cations taken up are neutralized (with the exception of ammonium) partly by the newly synthesized organic acids ('h'), partly by the inorganic anions absorbed ('j').

NITROGEN ASSIMILATION

The assimilation of ammonium ions requires the biosynthesis of keto acids, i.e. the carboxylation of phospho-enol-pyruvate by HCO_3^-. But there are just as

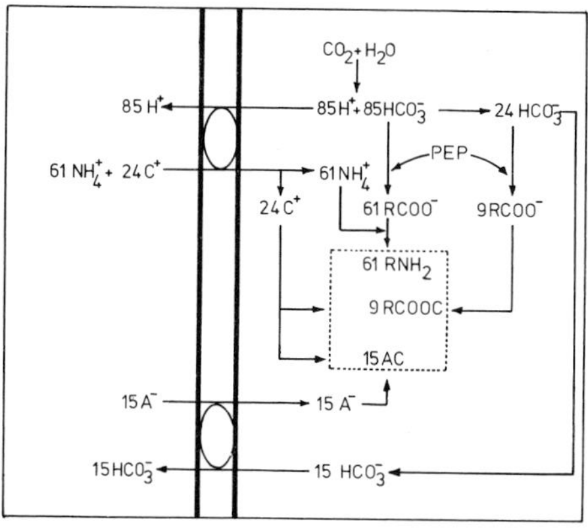

Fig. 4. The model applied to the NH_4^+-nutrition.

many HCO_3^--ions produced as NH_4^+-ions absorbed (because of the 1 : 1 relationship between cation uptake and H^+-efflux and they are all used up in the (usual) case of complete assimilation of NH_4^+ ('k'). A consequence of this fact is that the NH_4^+ uptake, in contrast to all other cations, does not bring about any anion uptake.

The reduction of nitrate to ammonium (Fig. 2) in the plant generates one HCO_3^- which can be used for further NO_3^- uptake. Nitrate sustains its own uptake and is therefore the only anion the absorption of which does not depend on cation uptake.

CONSEQUENCES OF THE MODEL

If, as stated in Section 2, the H^+-efflux and the cation uptake are tightly bound to each other, the proton efflux in distilled water must be negligible. This is indeed what happens (Fig. 3).

If the absorption of all anions but NO_3^- depends on cation uptake, the cationic content of a plant should always be higher than the anionic content (NO_3^- not being taken into account). It is so, as shown in Table 1. The same is true for NH_4^+-fed plants, even if one discards the NH_4^+ uptake because it does not involve any anion uptake, as shown in Section 3 (Table 1).

Competition between cations should only occur when the H^+-efflux becomes limiting. This can happen with a shortage of carbohydrates in the roots (when working with excised roots, for instance) or with a high cation concentration in the medium. Otherwise, it does not occur [1, 8, 21].

In Fig. 4 the model is applied to the ion uptake of NH_4^+-fed plants. The figures are the means of several experiments [3, 10, 11, 12, 14, 17] and may be considered as representative of the ammonium nutrition: for every 100 ions taken up, those plants absorb 61 NH_4^+, 24 other cations and 15 anions. The same rules have been applied to NO_3^--fed plants and some of the data are summarized in Table 2. They show: that the NH_4^+-nutrition leads to an excess of about 70 H (for every 100 ions

Table 2. Theoretical ion-extrusion and biosynthesis by NO_3^-- and NH_4^+-fed plants

N source	Efflux			Biosynthesis		
	H^+	HCO_3^-	Differ.	RNH_2	RCOOC	AC
NH_4^+	85	15	+70	61	9	15
NO_3^-	39	61	−22	47	25	14

Table 3. Comparison of the measured and the theoretical biosynthesis of carboxylates (after results of Kirkby and Knight [13])

Treatm. (NO_3^-, meq/l)	Ion uptake			Bios. carboxylates	
	NO_3^-	A^-	C^+	measured	calculated
0.5	119	121	243	125	122
1.5	180	100	256	159	156
4.0	237	97	282	203	185
6.0	251	98	331	245	233
8.0	258	92	342	246	250

Symbols as in Table 1/ all results: meq100 g D.W.

taken up) in the medium, which accounts for the well known acidification of the NH_4^+-solutions; the NO_3^--nutrition leads on the contrary to a (smaller) deficit of 22 H^+ in the medium, which accounts for the slow alkalinization of most NO_3^--solutions. The biosynthesis of organic acids (RCOOH) in the NH_4^+-fed plants is about three times smaller than in NO_3^--fed plants. This is indeed what is generally observed [2, 3, 11, 12].

Another consequence of the model is that the biosynthesis of carboxylates (RCOOH) is equal to the difference between the cations other than ammonium (C^+) and the anions other than nitrate (A^-) (see for instance Fig. 4). This has been confirmed experimentally (Table 3).

To sum up, we believe that this model, despite some weaknesses has two advantages: it gathers facts which, although well known, were seemingly unrelated and it allows the prediction of the impact of ion uptake on several aspects of the metabolism.

REFERENCES

1. Bange, G. et al. 1965 Interactions in the absorption of potassium, sodium and ammonium ions in excised barley roots. Acta Bot. Neerl. 14, 116–130.
2. Chouteau, J. 1963 Etude de la nutrition nitrique et ammoniacale de la plante de tabac en présence de doses croissantes de bicarbonate dans le milieu nutritif. Ann. Inst. Tabac Bergerac 4, 319–332.
3. Coïc, Y. et al. 1961 Comparaison de l'influence de la nutrition nitrique et ammoniacale sur l'absorption des anions-cations, et plus particulièrement des acides organiques, chez le maïs. Ann. Physiol. Vég. (Paris 3, 141–163.
4. Dejaegere, R. 1967 Effets de la nature de l'alimentation azotée sur la croissance du cotonnier. Ann. Phys. Vég. ULB 12, 93–109.
5. Dejaegere, R. et al. 1978 Comparaison de plantules d'orge riches ou pauvres en sels quant à leur capacité d'excrétion de protons en solution monosaline. Bull. Soc. Bot. Belgique 111, 119–124.

6. Dejaegere, R. and Neirinckx, L. 1978 Proton extrusion and ion uptake: some characteristics of the phenomenon in barley seedlings. Z. Pflanzenphysiol. 89, 129–140.
7. Dejaegere, R. et al. 1979 Relationship between proton extrusion and ion uptake in barley seedlings. In: Kudrev, T. (ed.), Mineral nutrition of plants, vol. 1, 25–38. Sofia, Bulgaria: Publ. House Centr. Cooperative Union.
8. Epstein, E. 1972 Mineral nutrition of plants: principles and perspectives, pp. 1–412, J. Wiley & Sons.
9. Jacoby, B. and Laties, G. 1971 Bicarbonate fixation and malate compartmentation in relation to salt-induced stoichiometric synthesis of organic acids. Plant Physiol. 47, 525–531.
10. Kirkby, E. and De Kock, P. 1965 The influence of age on the cation-anion balance in the leaves of Brussels sprouts. Z. Pflanzenernähr. Düng. und Bodenkunde 111, 197–203.
11. Kirkby, E. 1968 Influence of ammonium and nitrate nutrition on the cation-anion balance and nitrogen and carbohydrate metabolism of white mustard plants grown in dilute nutrient solutions. Soil Sci. 105, 133–141.
12. Kirkby, E. and Mengel, K. 1967 Ionic balance in different tissues of the tomato plant in relation to nitrate, urea or ammonium nutrition. Plant Physiol. 42, 6–14.
13. Kirkby, E. and Knight, A. 1977 Influence of the level of nitrate nutrition on ion uptake and assimilation, organic acid accumulation and cation-anion balance in whole tomato plants. Plant Physiol. 60, 349–353.
14. Kostic, M. et al. 1967 Evaluation of the nutrient status of wheat plants. Neth. J. Agric. Sci. 15, 267–280.
15. Marrè et al. 1974 Evidence for coupling of proton extrusion to K^+ uptake by pea internode segments treated in fusicoccin or auxin. Plant Sci. Letters 3, 365–379.
16. Marrè and Ciferri (eds.) 1977 Regulation of cell membrane activities in plants. Amsterdam: North-Holland Publ. Co.
17. Neirinckx, L. 1964 Influence du milieu alimentaire sur la composition minérale du cotonnier. Ann. Phys. Vég. ULB 9, 57–79.
18. Pitman, M. 1970 Active H^+-efflux from cells of low-salt barley roots during salt accumulation. Plant Physiol. 45, 787–790.
19. Pitman, M. and Schaeffer, N. 1975 Stimulation of H^+-efflux and cation uptake by fusicoccin in barley roots. Plant Sci. Letters 4, 323–329.
20. Raven, J. and Smith, F. 1974 Significance of hydrogen-ion transport in plant cells. Can. J. Bot. 52, 1035–1048.
21. Tromp, J. 1962 Interactions in the absorption of ammonium, potassium and sodium by wheat roots. Acta Bot. Neerl. 11, 147–192.

35. Evidence for an ion uptake controller in *Helianthus annuus*

D. J. F. BOWLING

Department of Botany, University of Aberdeen, Aberdeen AB9 2UD, Scotland, U.K.

In the period 1935–60 many investigations concerning the uptake of ions by roots were carried out on excised roots. It was then believed that excision left the root in its normal condition for many hours after its severance from the rest of the plant. In the 1960s however, a number of experiments were carried out, the results of which challenged this assumption. In 1963 Koster reported that there was a rapid decline in the uptake of nitrate and ammonium ions by soy bean plants a few hours after they were bark ringed [1]. *Ricinus communis* plants were found to show a decline in potassium uptake only 65 min after ringing [2]. Bowling (1968) found that potassium uptake by *Helianthus annuus* fell rapidly if a length of the stem was cooled to just above 0 °C. Potassium uptake was restored when the cooling coil was removed [3].

These results clearly showed that the physiology of the root was much more dependent on the assimilate supply from the shoot than had been previously thought. It was concluded that the main regulatory influence the shoot exerted on the root was through the supply of carbohydrates in the phloem [4]. Plants left in darkness show a decline in salt uptake which can be restored after a period of illumination. There is a lag period between the onset of illumination and the beginning of the rise in salt uptake. Pitman and Cram [5] found the length of this lag in barley to be 1 h. A similar lag was found in transport of radioactive sugars to the root following a pulse of ^{14}C sucrose at the start of the light period.

In our laboratory the problem of the relationship between the root and the shoot has been investigated for a number of years, mainly with the sunflower plant *Helianthus annuus*. As might be expected of other physiological processes in the root besides ion uptake are dependent on the shoot. We found [6] a linear relationship between the translocation of assimilates from the shoot and the respiration rate of the root. Also the membrane electrical potential between the external medium and the vacuole of the root cortical cells has been found to be closely dependent on the shoot [7]. The root cell PD was depolarised by up to 78 mV when the leaves were left in darkness and the normal PD was restored in the light. The normal PD could be restored in the dark by addition of sucrose to the medium surrounding the root.

R. Brouwer et al. (eds.), Structure and Function of Plant Roots, 179–186.

Whilst the root is obviously dependent on the supply of carbohydrate from the shoot in the medium and long term to supply energy to its various physiological processes the rapidity of some of the changes in the root after ringing or excision indicates that perhaps other more subtle controls are exerted on the root by the shoot. With improvements in the technique of measuring ion uptake by large intact plants through the use of flowing solutions and ion selective electrodes we have found that the effect of ringing on potassium uptake by *Helianthus annuus*, for example, is very rapid indeed [8]. In these circumstances it is hard to accept that it is the carbohydrate which is the limiting factor. Also Pitman and Cram [9] have observed that the rate of potassium uptake becomes independent of sugar concentration in plants with high growth rates.

Therefore it is possible that other factors besides carbohydrates, transported to the root in the phloem, control the ion uptake processes in the root. In this paper evidence for the presence of such a factor regulating uptake of the monovalent ions by *Helianthus* roots is presented.

METHODS AND RESULTS

Plants of *Helianthus annuus* (cv Tall single) were grown from seed germinated in moist vermiculite. At one week old the seedlings were transferred to glass jars (2.5 l) containing a balanced nutrient solution. The plants were illuminated continuously by a bank of fluorescent lamps giving 300 $\mu Em^{-2} s^{-1}$ at leaf height in a growth room at 21–23 °C. The nutrient solution which was continuously aerated had the following composition: KNO_3, 0.8 mM; $Ca(NO_3)_2$, 0.6 mM; $MgSO_4$, 0.4 mM; K_2HPO_4, 0.1 mM. The usual trace elements were provided and iron was given as FeEDTA. The pH was adjusted to 6.0. The nutrient solution was changed regularly.

Plants were used for the experiments when they were 3–4 weeks old. They weighed approximately 100 g (F wt) and had a total length of about 50 cm. A plant was left overnight under constant illumination from a mercury vapour lamp (200 $\mu Em^{-2} s^{-1}$) in the laboratory. Nutrient solution was supplied to the root in the growth jar at 400 ml h^{-1} by means of a peristaltic pump (LKB). It was siphoned from the jar and dripped into a beaker containing ion selective electrodes for K^+, Cl^-, NO_3^- or Ca^{++}. The signal from the electrode was fed to a pH meter and chart recorder. Ion uptake was calculated from the difference in concentration of the nutrient solution entering and leaving the growth vessel and the flow rate.

After a steady uptake rate had been achieved overnight the stem of the plant was bark ringed just below the cotyledonary node. A ring of bark approximately

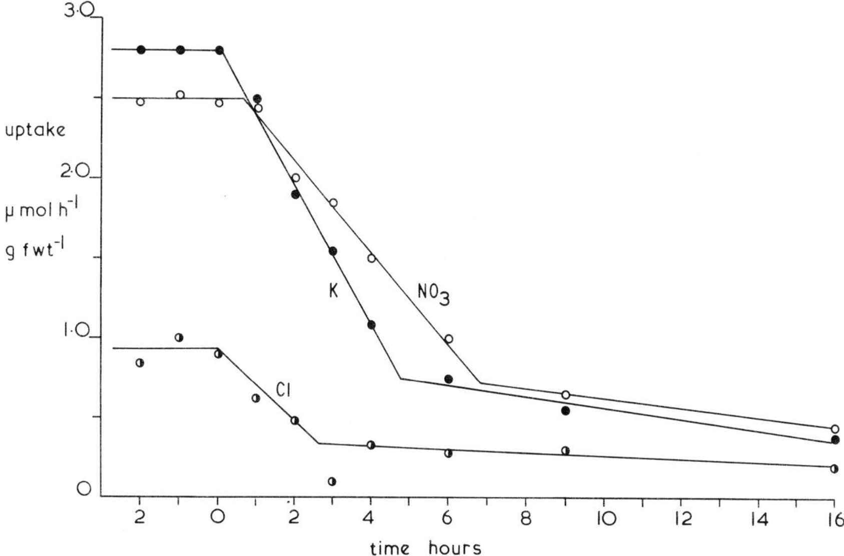

Fig. 1. The effect of ringing a plant of *Helianthus annuus* on the uptake of K^+, NO_3^- and Cl^-. The plant was ringed at time 0.

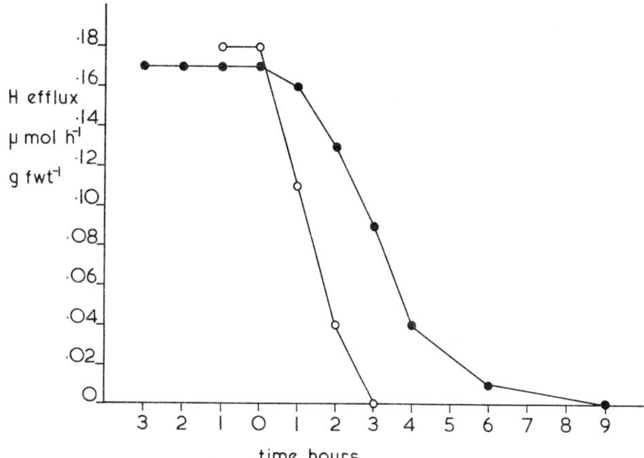

Fig. 2. H^+ efflux by two *H. annuus* plants after ringing. The plants were ringed at time 0.

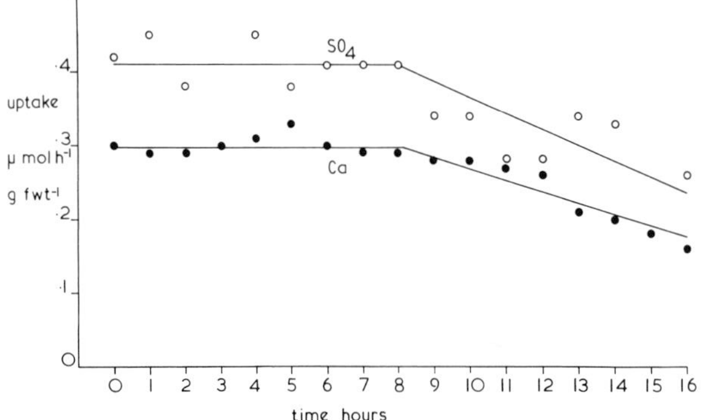

Fig. 3. The effect on the uptake of Ca^{++} and SO_4^{--} of ringing a plant of *H. annuus*. The plant was ringed at time 0.

5 mm wide was removed. Uptake was followed for up to 24 h after ringing. The fresh weight of the root was obtained at the end of the experiment.

Ringing the stem of *H. annuus* had a rapid effect on the uptake of the monovalent ions investigated. The results of one experiment are shown in Fig. 1. Uptake of K^+, NO_3^- and Cl^- all showed a decline within one hour of ringing. Extrapolation of the uptake curves for K^+ and Cl^- to the time of ringing suggested that the decline in uptake set in almost immediately. For NO_3^- the decline appeared to begin 30–60 min after ringing. Uptake of all three ions showed a two phased decline after ringing as can be seen in Fig. 1. There was an initial rapid phase lasting 3–7 h followed by a slower second phase.

H^+ efflux was measured by titrating samples of the solution leaving the root back to the original pH with dilute NaOH (1 mM). Uptake of SO_4^{--} was estimated by precipitation with 5% $BaCl_2$, the turbidity being measured using a

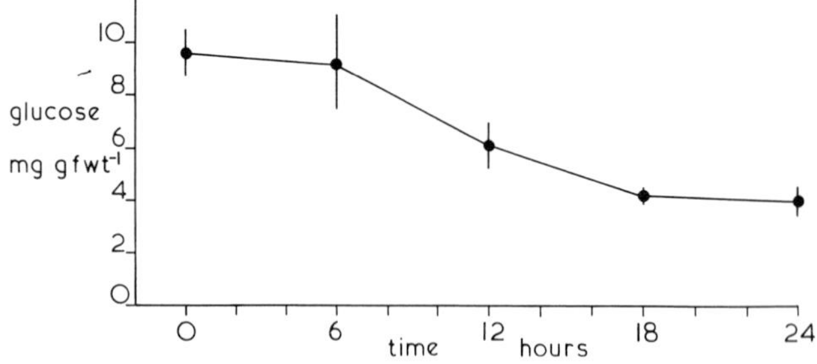

Fig. 4. The trend in glucose level in the roots of *H. annuus* after ringing. Ringing carried out at time 0.

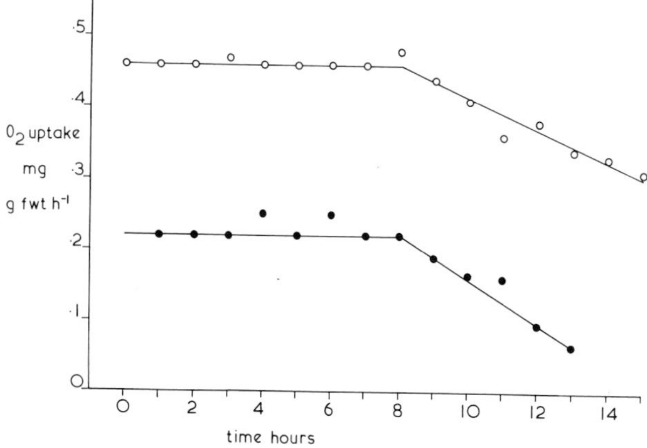

Fig. 5. The trend of respiration of the roots of two *H. annuus* plants after they were ringed at time 0.

nephelometer. Efflux of H^+ was rapidly affected by ringing. It fell to zero within 3–6 h. Fig. 2 shows the results of ringing two plants. In contrast to the behaviour of the monovalent ions the uptake of Ca^{++} and SO_4^{--} was unaffected by ringing for about 8 h. Thereafter there was a decline as can be seen from the results of ringing one plant shown in Fig. 3.

The most likely cause of the rapid decline in the uptake of the monovalent ions and in the efflux of protons appeared to be the cessation of the supply of carbohydrate from the shoot on ringing. Therefore the level of glucose in the root was investigated. It was determined on water extracts using an enzymic analysis

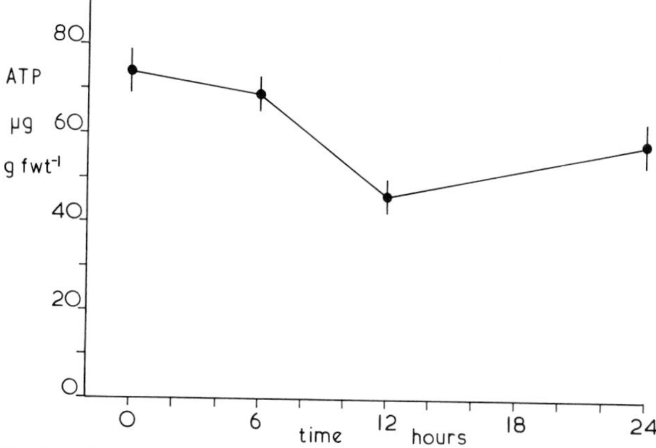

Fig. 6. The effect of ringing on ATP content of the roots of *H. annuus*. Ringing took place at time 0.

(Boehringer test kit 139 041). Fig. 4 shows the trend in glucose content after ringing. The data are means ± standard error of the mean, of values from 8 plants. It can be seen that there was no significant change for 6 h or more after ringing but after 12 h there was a marked decline.

Respiration of the root was measured by sealing the growth vessel and immersing an oxygen electrode (EIL) in the nutrient solution. Respiration appeared to mirror the decline in glucose level. Fig. 5 shows the trend in respiration after ringing for two plants. There was no change for approximately 8 h before decline set in.

ATP content of the roots was determined by following the method of Lin and Hanson [10]. Root extracts in glycine buffer at pH 8.3 were analysed for ATP using firefly lantern extract in an Aminco chem-glo photometer. ATP levels also showed no significant decline until between 6 and 12 h after ringing. Data from 4 plants ± standard error of the mean are plotted in Fig. 6. The curve indicates a slight rise in ATP between 12 and 24 h after ringing.

Ringing had no effect on water uptake by the plants.

DISCUSSION

The rapid decline in uptake of K^+, NO_3^- and Cl^- and in the efflux of H^+ brought about by ringing the stem of *H. annuus* indicated that the root was almost immediately dependent on the supply of some factor from the shoot. The very speed of the reaction suggested that it was not carbohydrate that was the limiting factor. The finding that glucose levels in the root showed no significant decline during the first six hours after ringing strongly supported this notion. As the glucose levels did not decline it was not surprising therefore to observe that respiration of the root also remained steady for up to eight hours after ringing.

Substantial concentrations of ATP have been detected in phloem sap [11] and if it is mobile in the sieve tubes then ringing might be expected to bring about a rapid reduction of ATP in the root. The data in Fig. 6 indicate that this did not occur. Instead the ATP level broadly mirrored the trend in glucose concentration and respiration rate. This indicated that all the ATP in the root was originating from respiration of carbohydrate in the root with none (or at the very most only very small amounts) coming to the root directly from the shoot. Therefore ATP did not appear to be the factor limiting the transport of the monovalent ions.

In contrast to the behaviour of the monovalent ions the uptake of Ca^{++} and SO_4^{--} was not immediately affected by ringing. In fact the eventual decline in uptake of these ions appeared to parallel the decline in respiration rate of the root. The uptake of phosphate by *Helianthus* was found to decline at around

6–8 h after ringing [8]. Uptake of phosphate in this plant is electrogenic with maximum uptake generating approximately 70 mV of the membrane potential in the root cortical cells. Therefore the PD declines in parallel with phosphate uptake. As the PD can be restored by adding sucrose to the medium [7] it would appear that phosphate uptake and also possibly the uptake of Ca^{++} and SO_4^{--} are directly dependent on respiratory energy. Whether this is due to electron flow or ATP supply remains to be determined.

The uptake of the monovalent ions and the efflux of H^+ of course must also be dependent on energy from respiration but the present results suggest the presence of an intermediary controlling factor. This factor must regulate the application of the energy to the active transport systems involved in moving K^+, Cl^- and NO_3^- into the cell and H^+ out. The controller (or controllers) does not appear to be made in the root but has to come from the shoot. This gives the shoot close control over the uptake into the plant of two of the ions taken up from the soil in large quantities. In the situation where carbohydrate supply and the salt supply are adequate the shoot may need to exercise direct control over the transport of these ions into the root. Otherwise on moving into the shoot they may build up large concentrations in the inter-cellular spaces of the leaf which may be deleterious.

Before the evidence for the controller can be proven it must be isolated from the plant and identified. An alternative approach is to try to reverse the decline in monovalent ion transport after ringing by providing compounds in the external solution which might be candidates for the role of controller. So far sucrose, ATP, gibberellic acid (GA_3), vitamins B_1 B_{12} and nicotinic acid have been tried without any positive effect. Of the remaining known candidates IAA, ABA and cytokinins seem unlikely as their effect on ion transport appears to be small and often inhibitory [9]. It is the author's view that the proposed ion uptake controller will turn out to be a highly labile compound of novel structure produced by the leaf in very small quantities.

REFERENCES

1. Koster, A. L. 1963 Changes in metabolism of isolated root systems of soy bean. Nature 198, 709–710.
2. Bowling, D. J. F. 1965 Effect of ringing on potassium uptake by *Ricinus communis* plants. Nature 206, 317–318.
3. Bowling, D. J. F. 1968 Translocation at 0 °C in *Helianthus annuus*. J. exp. Bot. 19, 381–388.
4. Bowling, D. J. F. 1976 Uptake of ions by plant roots. London: Chapman and Hall.
5. Pitman, M. G. and Cram, W. J. 1973 Regulation of inorganic ion transport in plants. In: Anderson, W. P. (ed.), Ion transport in plants, pp. 465–481. London: Academic Press.

6. Hatrick, A. A. and Bowling, D. J. F. 1973 A study of the relationship between root and shoot metabolism. J. exp. Bot. 24, 607–613.
7. Graham, R. D. and Bowling, D. J. F. 1977 Effect of the shoot on the transmembrane potentials of root cortical cells of sunflower. J. exp. Bot. 28, 886–893.
8. Bowling, D. J. F., Graham, R. D. and Dunlop, J. 1978 The relationship between the cell electrical potential difference and salt uptake in the roots of *Helianthus annuus*. J. exp. Bot. 29, 135–140.
9. Pitman, M. G. and Cram, W. J. 1977 Regulation of ion content in whole plants. In: Jennings, D. H. (ed.), Integration of activity in the higher plant, pp. 391–424. Cambridge University Press.
10. Lin, W. and Hanson, J. B. 1974 Increase in electrogenic membrane potential with washing of corn root tissue. Plant Physiol. 54, 799–801.
11. Hall, S. M. and Baker, D. A. 1972 The chemical composition of *Ricinus* phloem sap. Planta 106, 131–140.

36. Efficiency of passive movement of ions in the root

MIROSLAV DVOŘÁK and JANA ČERNOHORSKÁ

Plant Physiology Deptartment, Charles University, 128 44 Praha 2, C.S.S.R.

Electric conductivity of plant tissues results from the presence of freely mobile ions that are discharged at the applied electrodes. The current model of a plant tissue consists of resistance (R) and capacitance (C) branches (with further resistors R_i connected in series with the capacitances). For an analysis of the conductance structure of a plant, a saw-tooth d.c. voltage can be used (U_{max} = +13 V, T = 0.25 ms). The current response to the applied voltage is given by the relationship:

$$i = \frac{U_{max}}{T}\left[C(1 - e^{-t/R_iC}) + \frac{1}{R}t\right]$$

assuming that only a single capacitance branch in parallel is considered. The capacitances are located on various surfaces, above all on cytoplasmic membranes; the conductance is determined by the concentrations of passively mobile ions in the freespace (FS) and the symplast and by the effective cross-section (i.e. the spatial structure) of these systems.

Some current characteristics can be derived from the curves photographed on an oscilloscope screen; these characteristics are independent of the actual current passing through the tissue. In this way, an error dependent on the size of the contact area between the electrode and the tissue is avoided in this method.

The character of the curves is affected by structural elements of plant organs, by their age (the degree of development), nutrition, cultivation conditions and by various operations influencing the actual physiological state of the plant. From this point of view, the conductance of segments of primary roots of *Zea mays* and *Cucurbita pepo* (10 mm long, proceeding from the root tip) was analyzed. A needle-shaped Pt-electrode was inserted into the central cylinder (St) of the basal part of the segment, another electrode was inserted from the side through the rhizodermis into the cortex (Ct), using a micromanipulator. On a further centripetal shift, the curve shape suddenly changed; we assume that the endodermis was penetrated. The electrode entered the stele, but the contact with the cortex was not interrupted. Some data, clearly characterizing the recorded curves, are given in Table 1.

R. Brouwer et al. (eds.), Structure and Function of Plant Roots, 187–189.

Table 1. A comparison of values characterizing the current response to a saw-tooth voltage for root segments of *Cucurbita pepo* and *Zea mays*. The segments are numbered starting from the apex. The 'basic constants' were calculated for T = 1 and a maximum current of each curve equal to unity; the 'reference values' are the data concerning the current and conductance related to the initial quantities for cortex, 1. segment = 1.

		Cortex							Stele						
		Basic const.			Reference values				Basic const.			Reference values			
		a_1	a_2	k	I	C	G	R_i^{-1}	a_1	a_2	k	I	C	G	R_i^{-1}
C. pepo	1	25.5	—	0.85	1	0.15	0.85	1	23.4	—	0.85	1.15	0.17	1.00	1.07
	2	29.6	9.3	0.83	1.76	0.30	1.46	2.42	21.6	6.0	0.72	2.24	0.39	1.62	3.67
Zea	1	35.7	—	0.73	1	0.27	0.73	1	32.3	—	0.68	1.15	0.37	0.78	1.22
mays	2	29.4	—	0.71	1.20	0.35	0.85	1.04	27.8	—	0.66	1.60	0.55	1.05	1.56
	3	29.3	—	0.72	1.23	0.35	0.88	1.04	22.6	—	0.59	2.12	0.88	1.25	2.03

$a = (R_i)^{-1}$; $k = i_{R_{max}} \cdot (i_{max})^{-1}$; $I = i_{R_{max}} + i_{C_{max}}$; C = capacitance; $G = R^{-1}$ = conductance of the R-branch; R_i^{-1} = conductance of the C-branch.

With both species, the maximum current is not substantially changed, on exceeding the boundary Ct → St in the first (apical) segment. The situation is different with the second or third segment: the current and the conductance in the C-branch considerably increase with electrode transfer Ct → St.

The ageing of the root is manifested both for Ct and St by an increase in I, i.e. by an increased ion mobility; this is characteristic of the resistance branch (an increase in G) and sometime of the capacity branch (R_i^{-1}, especially for the stele). The curvature characteristic, a, always slightly decrease on the Ct → St transfer.

It follows from the above data that, on ageing (lengthening) of cells, the tissue capacity increases more than the conductance in the C-branch and the ionic mobility in the R-branch simultaneously increases. The capacitance elements are apparently saturated in another way. On the basis of the present knowledge on the structure of the root tissue and on the small contribution of FS to the total conductance [1] we assume that the symplast plays a role in the R-branch: an increase in $[K^+]$ in vacuolated cells affects its conductance positively. The FS should exert an effect in the C-branch. The comparison of the first segment with the further ones indicates a delayed development of the endodermis as a barrier for passive movement of ions in the FS.

Using a Ca deficient plant we followed the significance of the trophic conditions for the formation of the conductance characteristics of the root. It is

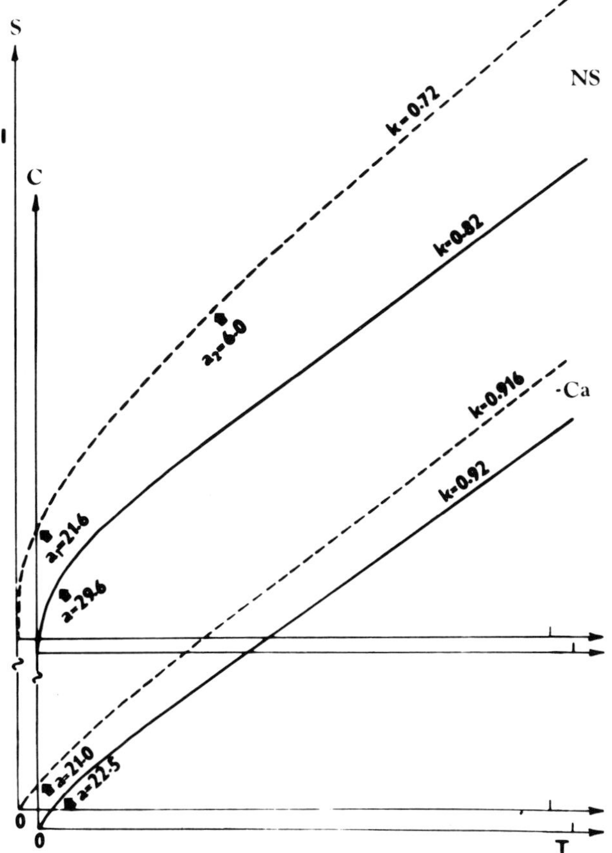

Fig. 1. Changes in the current response measured in the root of *Cucurbita pepo* after breaking the endodermis, in dependence on the Ca nutrition of the plant (7 days cultivation).

Coordinate C – data for the cortex (Ct); Coordinate S – data obtained after electrode passage into the stele (St); NS = the Knop nutritive solution; -Ca = a Ca deficient solution. x – axis = time (t), T = 0.25 ms; y – axis = the relative current.

evident that with '-Ca' the capacitance component is not pronounced and there is no difference between the cortex and stele. Both the effects can be related to plasmalemma faults; if it loses its semipermeability, its capacitance function is also lost.

REFERENCE

1. Spanswick, R. M. 1972 Electrical coupling between cells of higher plants: a direct demonstration of intercellular communication. Planta 102, 215–227.

37. Comparison of some membrane-bound ATPases of glycophytes and halophytes

NATALYA I. TIKHAYA and NATALYA E. MISHUTINA

K. A. Timiriazev Institute of Plant Physiology, Academy of Sciences, Moscow 127276, U.S.S.R.

In our previous studies we have shown that barley roots contain a membrane-bound ATPase that requires Mg and is additionally stimulated with Na + K ions by 34% [6]. If the ATPase is responsible for pumping sodium ions out and potassium ions into the cells, it is logical to suggest that a higher activity of this enzyme will be observed in halophytes grown on saline soils than in glycophytes.

The synergistic effect of sodium and potassium in the presence of Mg and the inhibition of this activity by ouabain were the criteria used for detection of this enzyme.

The membrane ATPase from *Halocnemum strobilaceum* shoots had properties similar to those of the membrane ATPase from *Hordeum vulgare* roots. Both enzymes were synergistically stimulated by Na and K only in the presence of Mg. The additional (Na + K) activation of the enzyme was inhibited almost completely by 10^{-4} M ouabain up to the level of Mg-ATPase activity. Their optimal Na/K ratios were approximately identical (Table 1).

Table 1. Comparison of membrane-bound ATPases from *Hordeum vulgare* L. roots and *Halocnemum strobilaceum* L. shoots. Reaction mixture contained 3 mM ATP, 30 mM tris-HCl (pH 8-glycophyte and pH 6-halophyte) and other additions at concentrations indicated.

Additions	*Hordeum vulgare*	*Halocnemum strobilaceum*
	(ATPase micromoles P_i/mg protein · h)	
none	1.27 ± 0.11	2.03 ± 0.16
NaCl (120 mM)	0.83 ± 0.14	1.92 ± 0.15
KCl (120 mM)	1.39 ± 0.23	2.88 ± 0.18
NaCl (86 mM) + KCl (34 mM)	1.26 ± 0.11	2.16 ± 0.15
$MgCl_2$ (3 mM)	4.58 ± 0.09	3.55 ± 0.31
$MgCl_2$ + NaCl (120 mM)	3.84 ± 0.07	4.58 ± 0.18
$MgCl_2$ + KCl (120 mM)	4.15 ± 0.62	5.57 ± 0.23
$MgCl_2$ + NaCl (86 mM) + KCl (34 mM)	6.22 ± 0.09	11.99 ± 0.30
$MgCl_2$ + NaCl (86 mM) + KCl (34 mM) + ouabain (10^{-4} M)	4.60 ± 0.47	4.02 ± 0.31

R. Brouwer et al. (eds.), Structure and Function of Plant Roots, 191–192.

However these enzymes had two essential differences. The first difference, a qualitative one is the following. The glycophyte enzyme has only one pH optimum at 8 whereas the halophyte enzyme exhibits two optima at 6 and 8. Two pH optima for the enzyme of *H. strobilaceum* and the ability of shoots of this plant to accumulate NaCl in high concentration in their cells, mainly in vacuoles, permit us to suspect that at least two membrane-associated ATPases are being measured. The first enzyme is located at the tonoplast (optimum pH 6) and the other enzyme is at the plasmalemma (optimum pH 8). Thus, the excess of Na can be pumped out from the cells of halophyte both into the intercellular space through the plasmalemma and into vacuoles through the tonoplast.

The second difference, a quantitative one, is more important because the (Na + K) activation of the halophyte enzyme is significantly higher than that of the glycophyte enzyme (137 per cent and 34 per cent over the Mg-ATPase activity level respectively). These results indicate indirectly that the membrane-bound Mg (Na + K)ATPases of plants, like those of animals, are responsible for the functioning of the Na-extruding pump.

Comparing our results with the data of other researchers we have come to the conclusion that the greatest activation by Na and K is observed in halophytes [3, 4, 5] and the lowest – in glycophytes [1, 6]. The intermediate value belongs to a facultative halophyte – a sugar beet [2].

Thus, the higher ATPase activity of the enzyme in halophytes, as compared to that in glycophytes, suggests that Mg(Na + K)ATPase could be involved in the extrusion of Na from the cytoplasm.

REFERENCES

1. Cassagne, C., Lessire, R. and Carde, J. P. 1976 Plasmalemma enriched fraction from leek (*Allium porrum* L.) epidermal cells. Pl. Sciences Letters 7, 127–135.
2. Hansson, G. and Kylin, A. 1969 ATPase activities in homogehates from sugar beet roots, reaction to Mg^{2+} and (Na^+ + K^+)-stimulation. Z. Pflanz. 60, 270–275.
3. Kylin, A. and Gee, R. 1970 ATPase activities in leaves of the mangrove *Avicennia nitida* Jacg. Plant. Physiol 45, 169–172.
4. Mishustina, N. E., Tikhaya, N. I. and Chaplygina, N. S. 1979 (Na^+ + K^+)-ATPase activity of membrane isolated from the halophyte *Halocnemum strobilaceum* shoots. Fiziol. Rast. 26, 541–547.
5. Sullivan, C. W. and Volcani, B. E. 1974 Synergistically stimulated (Na^+ + K^+)-adenosine triphosphatase from plasma membrane of a marine diatom. Proc. Nat. Acad. Sci. 711, 4376–4380.
6. Tikhaya, N. I., Mishustina, N. E., Kurkova, E. B., Vakhmistrov, D. B. and Samoilova, S. A. 1976 Ouabain-sensitive (Na + K)-ATPase activity of cell membranes isolated from barley roots. Fiziol. Rast. 23, 1197–1206.

38. Uptake and translocation of sodium in salt-sensitive and salt-tolerant *Plantago* species

OTTO G. TÁNCZOS, LÁSZLÓ ERDEI* and JAN SNIJDER

Department of Plant Physiology, University of Groningen, Haren (Gr), The Netherlands

The work reported here forms part of a larger research program on the physiological aspects of adaptation of grassland species, in particularly *Plantago* species, to different habitats.

Erdei and Kuiper [3] have already published the results of a comparative study concerning the effects of saline conditions on growth, ion-content and translocation in three ecologically different *Plantago* species. The investigated species were: *Plantago media*, a glycophyte; *Plantago maritima*, a halophyte and *Plantago coronopus*, which occurs in habitats which may vary in salinity.

The most important conclusions of these experiments were that (1) *Plantago media* was sensitive to 25 mM NaCl, while *Plantago coronopus* and *Plantago maritima* could grow in 150 and 300 mM NaCl respectively, and that (2) the three species did accumulate Na^+ in the shoot and maintained a relatively low Na^+ level in the root. K^+, Mg^{2+} and Ca^{2+} levels of shoot and roots decreased with increasing salinity.

Our results correspond with those of Erdei and Kuiper [3] for Na^+ uptake in the roots and further translocation to the shoot in the three species in the presence of 150 mM NaCl (Fig. 1). Na^+ uptake rates in the roots of *Plantago media* and *Plantago maritima* were similar, while *Plantago coronopus* showed a higher rate (Fig. 1A). Translocation of Na^+ to the shoot of *Plantago media* started after a lag period of about two hours (Fig. 1B). Na^+ translocation in *Plantago coronopus* also showed a lag period, although to a lesser extent. Translocation in *Plantago maritima* responded without delay. This difference in translocation capacity is shown in Fig. 1C, which represents the translocated amount of Na^+ as a percentage of the total uptake. This lag period of the translocation system in *Plantago media* resulted in a change in distribution of Na^+ between roots and shoot (Fig. 1D).

The results suggest that the difference between the salt-sensitive and salt-

* Institute of Biophysics, Biological Research Center, Hungarian Academy of Sciences, P.O. Box 521, 6701 Szeged, Hungary.

R. Brouwer et al. (eds.), Structure and Function of Plant Roots, 193–198.

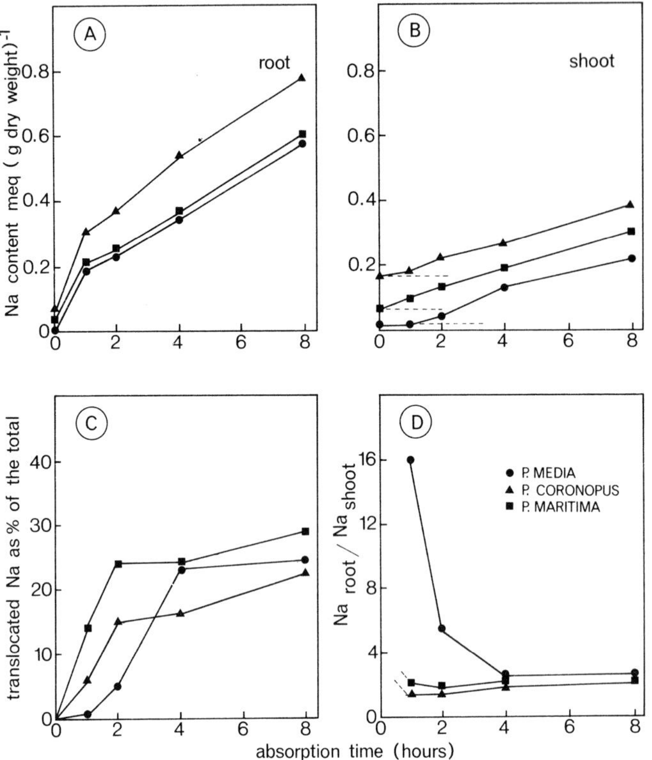

Fig. 1. Uptake and translocation of Na by *Plantago media*, *Plantago coronopus* and *Plantago maritima*. (A) Na content; (B) translocated Na from root to shoot; (C) translocated Na as percentage of total Na uptake of plant; (D) ratio of Na content of roots to Na content of shoots as a function of time. Plants grown in 25 per cent nutrient solution supplemented with 150 mM NaCl. Averages of duplicate samples of six plants. Deviations are less than 10 per cent.

tolerant *Plantago* species may be located in the ion-secretory system of the xylem parenchyma cells next to the xylem vessels.

We continued these experiments in order to obtain more information about the uptake and translocation mechanism of Na^+ in these species and particularly about the lag period found in *Plantago media*.

The plants were grown for six weeks in nutrient solution in a climate chamber at 18 °C, 12 h photoperiod and 60 per cent relative humidity. For the uptake and translocation experiments the plants were transferred to another nutrient solution, complemented with different concentrations of NaCl. For every measurement duplicate samples were taken from six plants of each treatment. The dried

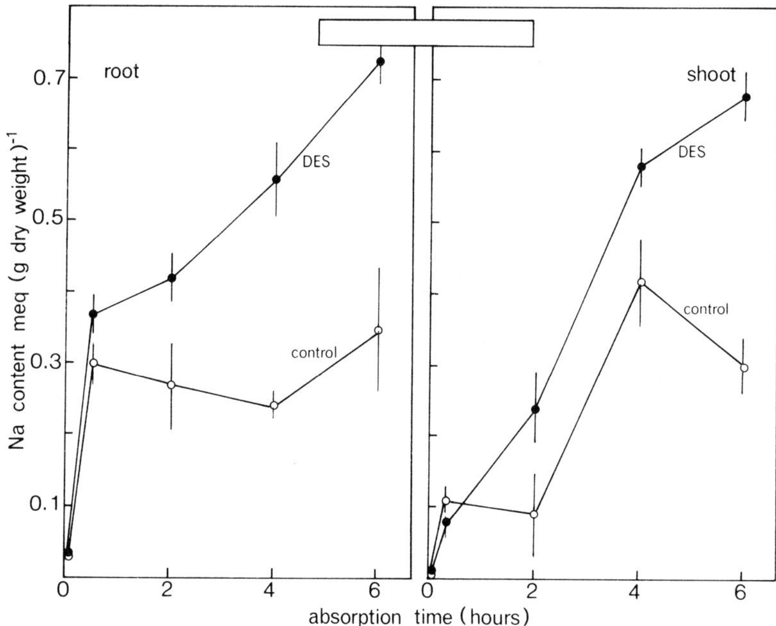

Fig. 2. Effect of diethylstilbetrol (10^{-4} M DES in 0.3 per cent ethanol) on the uptake and translocation of Na^+ by *Plantago media*. Control (○); DES (●); treatment with DES started 8 h before 150 mM NaCl addition. Vertical bars denote SE n = 6.

and measured samples were digested and the Na content was measured by atomic absorption spectrophotometry.

Experiments with an ATPase inhibitor may be helpful to investigate whether ATPase is involved in uptake and translocation of Na^+. Balke and Hodges [1] and more recently Cheeseman et al. [2] reported that diethylstilbestrol (DES) is an effective inhibitor of the plasmalemma ATPase.

10^{-4} M DES in a 0.3% ethanol solution was added to the nutrient solution 8 h before the start of the uptake experiment. The concentration of NaCl was 150 mM. Fig. 2 shows the Na content of the roots and the translocated Na to the shoot of *Plantago media*. The uptake of Na was much higher after DES treatment. The lag period of the control was not so pronounced, perhaps due to the presence of 0.3 per cent ethanol. It completely disappeared after DES treatment and the rate of translocation was higher. Our interpretation is that a membrane-bound ATPase in the roots of the salt-sensitive *Plantago media* pumped out Na^+ and that this system was blocked by DES. Therefore, the 'overloading' of the cortex

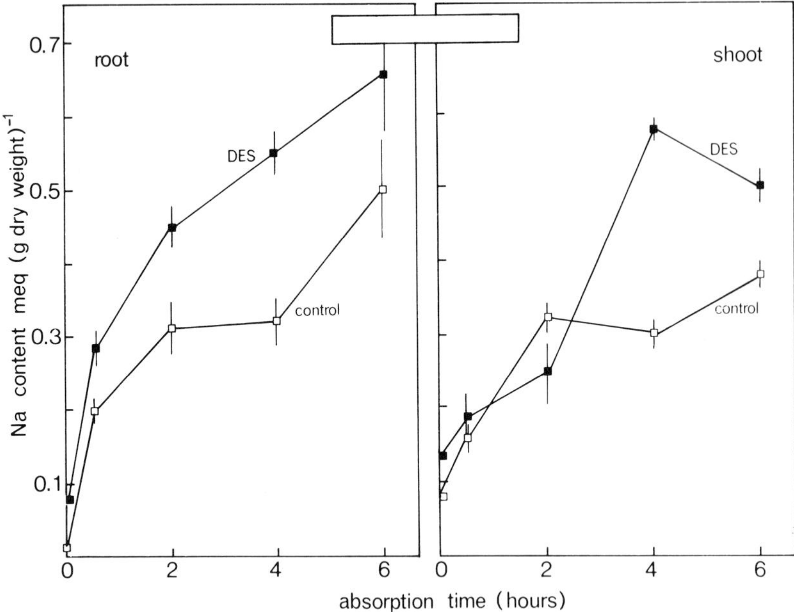

Fig. 3. Effect of diethylstilbetrol (10^{-4} M DES in 0.3 per cent ethanol) on the uptake and translocation of Na^+ by *Plantago maritima*. Control (○); DES (●); treatment with DES started 8 h before 150 mM NaCl addition. Vertical bars denote SE n = 6.

cells by Na was faster and it was followed by translocation to the shoot without delay. Results with *Plantago maritima* after treatment with DES were the same as those in *Plantago media* (Fig. 3). It is possible that at a high external concentration of Na^+ this salt-tolerant plant was also able to pump out Na^+ in order to maintain its ion-balance.

The experiments should be repeated at lower NaCl concentrations to get additional evidence on the role of the membrane-bound ATPase in Na^+ uptake and translocation.

We have completed a number of experiments at lower concentrations of NaCl. Fig. 4 shows uptake and translocation of Na^+ in *Plantago media* at different concentrations of NaCl in the outer solution. Virtually no uptake, and therefore no translocation, was observed in the presence of 1 and 10 mM NaCl, indicating that the root cells of *Plantago media* were able to exclude Na at less till 10 mM NaCl. Na^+ uptake in plants, incubated at 50 mM NaCl was relatively low and translocation to the shoot showed no lag period in comparison with experiments done at 150 mM NaCl.

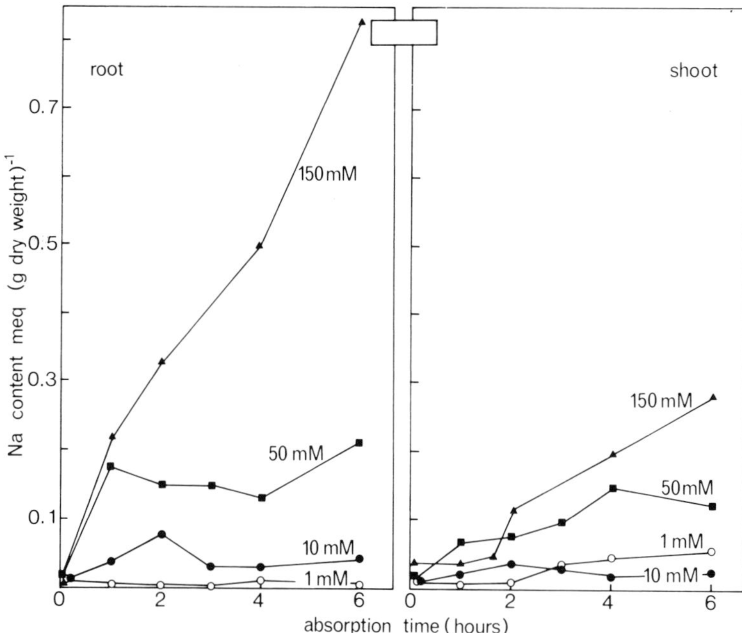

Fig. 4. Uptake and translocation of Na^+ by *Plantago media*. NaCl concentrations: 1mM (○), 10mM (●), 50mM (■), 150mM (▲), Averages of duplicate samples of six plants. Deviations are less than 10 per cent.

Fig. 5 shows the results of such an experiment with *Plantago maritima* at 1 and 150 mM NaCl. The content of the plants exposed to 1 mM NaCl was much higher than in Plantago media and it reached the level of *Plantago media* grown at 50 mM NaCl. No differences in Na^+ content between the two species were observed when grown at 150 mM NaCl.

Conclusions: *Plantago maritima* showed active uptake of Na^+ at low external concentrations of NaCl. At high external concentrations of NaCl the osmotic potential of the shoot could possibly be maintained by accumulation of sorbitol in the cytoplasm [4].

The glycophyte *Plantago media* prevented accumulation Na^+ in the roots until a concentration of 25 mM in the external solution was reached. The lag period of two hours of translocation of Na^+ to the shoot, observed at high external NaCl concentrations, might be caused by diminished transpiration. The increased translocation of Na^+ after the lag period may be caused by damage of the xylem parenchyma cell membranes.

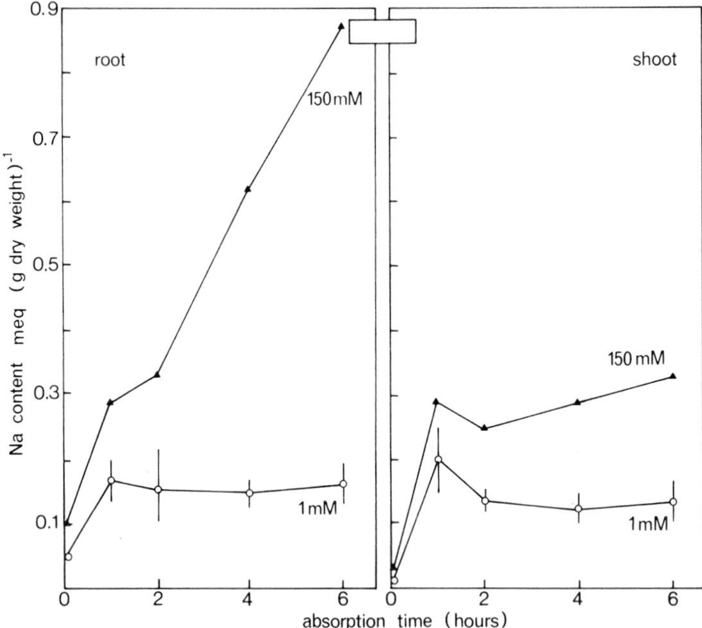

Fig. 5. Uptake and translocation of Na^+ by *Plantago maritima*. NaCl concentrations: 1mM (○), 10mM (●), 50mM (■), 150mM (▲), Averages of duplicate samples of six plants. Deviations are less than 10 per cent.

REFERENCES

1. Balke, N. E. and Hodges, T. K. 1979 Effect of diethylstilbestrol on ion fluxes in oat roots. Plant Physiol. 64, 42–47.
2. Cheeseman, J. M., LaFayette, P. R., Gronewald, J. W. and Hanson, J. B. 1980 Effect of ATPase inhibitors on cell potential and K^+ influx in corn roots. Plant Physiol. 65, 1139–1145.
3. Erdei, L. and Kuiper, P. J. C. 1979 The effect of salinity on growth, cation content, Na^+ uptake and translocation in salt-sensitive and salt-tolerant *Plantago* species. Physiol. Plant. 47, 95–99.
4. Lambers, H., Blacquière, T. and Stuiver, C. E. E. 1980 Photosynthesis and respiration in *Plantago coronopus* as affected by salinity. Interactions between osmoregulation and the alternative pathway. Physiol. Plant. in press.

39. The effect of CCCP on potassium uptake by root cells

MARINA S. KRASAVINA[1], SVETLANA V. SOKOLOVA[1] and OLEG O. LYALIN[2]

[1] *K.A. Timiriasev Institute of Plant Physiology, Moscow*
[2] *Institute of Agrophysics, Leningrad, U.S.S.R.*

Uncouplers of oxidative and photosynthetic phosphorylation are known to inhibit the transport of ions across membranes of plant cells [2]. This effect can be mediated by the influence on mitochondrial and chloroplast membranes or by the dissipation of transmembrane $\Delta\mu H$ at the plasmalemma [1]. The effect of CCCP on K^+ uptake by root cells, membrane potential and electric conductivity of the root cell membranes is studied. The *Cucurbita pepo* L. root segments are used for the K^+ uptake experiments. The membrane potential and membrane resistance are measured in large root hair cells of *Trianea bogotensis* Karst. by means of two intracellular microelectrodes.

^{42}K uptake is inhibited approximately by 40 per cent and the membrane potential is lowered from 150–180 mV to 60–90 mV after 5-min exposure to 10^{-5} M CCCP. These processes are not accompanied by a decrease of the

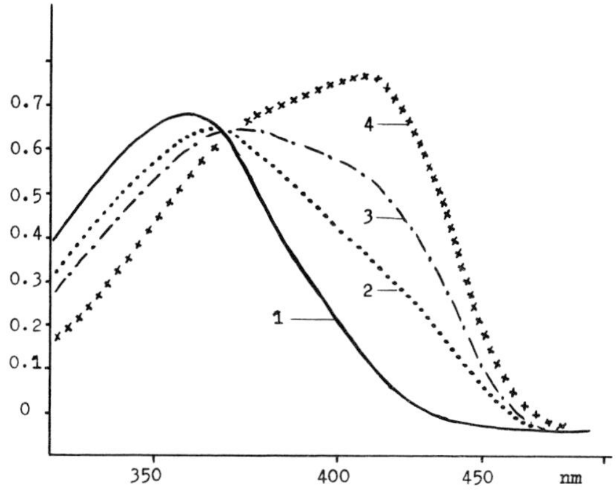

Fig. 1. Effect of dithioerythritol on the absorption spectrum of CCCP. Medium composition: MES–NaOH, pH 5 – 15 mM, CCCP – 30 μM, dithioerythritol – 0, 2, 5, 40 mM (curves 1–4 respectively).

R. Brouwer et al. (eds.), Structure and Function of Plant Roots, 199–201.

Table 1. Effect of thiol reagents on CCCP inhibitory power.

Experimental conditions	Without thiols	Glutathione		Cysteine	DTE
		oxidized	reduced		
		^{42}K uptake, μmol/h.g fr.wt			
– CCCP	2.37	2.36	3.06	2.29	2.93
+ CCCP	0.33	0.30	1.54	1.36	2.63
		Inhibition of ^{42}K uptake by CCCP, per cent			
	86.2	87.4	49.7	40.7	10.1

Medium composition: MEŚNaOH(pH 5)´15 mM, thiol reagents´5 mM, CCCP´0.01 mM. Incubation time´15 min.

membrane resistance which would be expected if the inhibitory effect of CCCP is due to facilitation of proton transfer across cell membranes.

The action of CCCP is little affected by the pH of the medium in the range from 5 to 7 and therefore it seems not to depend on the value of transmembrane ΔpH. The inhibitory power of CCCP decreases in alkaline solutions (pH 9), and the effect correlates with the depression of the inhibitor influx into the cells.

K^+ uptake is depressed by the inhibitor of respiration, KCN, to the same extent as by CCCP. The inhibitory effect of CCCP is not revealed in the presence of KCN, therefore it may be proposed that effect of CCCP is due to uncoupling of mitochondrial ATP production.

The inhibition of transmembrane K^+ transport by CCCP is not reversed after removal of CCCP from the medium. Washing with alkaline solutions is unable to promote the reversibility of the inhibition and CCCP is probably bound irreversibly in the cells.

In the model experiments it is shown that CCCP interacts with different thiol reagents. The interaction is detected as an alteration of the CCCP absorption spectrum, especially pronounced for dithioerythritol (Fig. 1). In the presence of dithioerythritol, CCCP penetrates into the cell without inhibiting K^+ transport. The CCCP inhibitory power is also decreased by cysteine and by reduced (but not oxidized) glutathione (Table 1).

The data presented indicates that the inhibition of transport processes by CCCP is probably mediated by its action on respiratory metabolism and is not due to a protonophore effect on the plasmalemma.

REFERENCES

1. Mitchell, P. 1970 Membranes of cells and organelles: morphology, transport and metabolism. In: Charles, H. P. and Knight, B. C. J. G. (eds.), Organization and control in prokaryotic and eukaryotic cells, pp. 121–139. Cambridge University Press.
2. Pitman, M. G. 1976 Ion uptake by plant roots. In: Luttge, U. and Pitman, M. G. (eds.), Encyclopedia of plant physiology, New ser. 2, part B, pp. 95–128.

40. Specialization of root tissues in ion transport

D. B. VAKHMISTROV

K. A. Timiriazev Institute of Plant Physiology, Academy of Sciences, Moscow 127276, U.S.S.R.

The present communication reviews briefly our experimental results concerned with the contribution of different tissues arranged along the root radius to the lateral transport of ions from the ambient medium to xylem vessels, following some events occurring in a sequence from root epidermis to xylem parenchyma.

The first question concerns the location of the loading site for the ions to be translocated along the symplastic pathway – either in the root epidermis (directly from the external medium) or within the cortex (from the intercellular free space)?

We showed more than ten years ago [7] that potassium ions were absorbed mainly by the epidermal cells of barley seedling roots. As to the cortical free space, it was just an inert volume in terms of ion uptake. This conclusion was confirmed later by a number of workers [1, 4, 11]. Nevertheless, it is still a commonly held view that the bulk of nutrients penetrates a root up to the endodermis *via* apoplasm and is absorbed by the whole surface of cortical cells

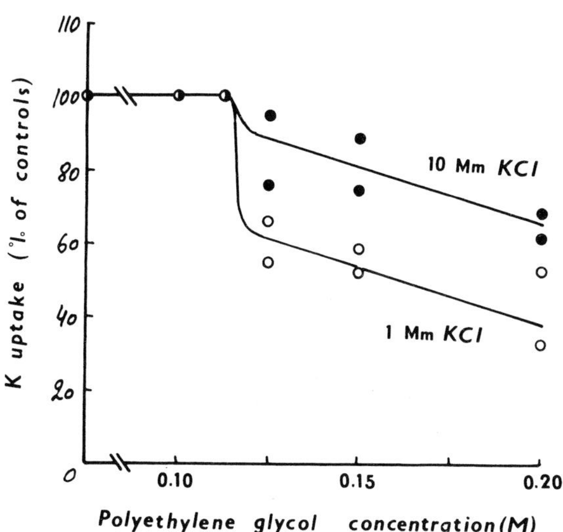

Fig. 1. Rate of K^+ absorption by corn seedling roots after osmotic shock as a function of osmoticum concentration.

R. Brouwer et al. (eds.), Structure and Function of Plant Roots, 203–208.

[3]. For this reason we have recently turned to this subject and have used osmotic shock to investigate it (D. B. Vakhmistrov and Z. V. Titova, unpublished). Intact roots of 7-day old corn seedlings were treated for 2 min with polyethylene glycol solutions of varied concentrations. Just after the treatment the roots were deplasmolysed in 0.25 mM $CaSO_4$ and the kinetics of K^+ absorption was followed during 6 h.

Fig. 1 shows that the rate of K^+ absorption was not affected by osmoticum at the concentration up to 0.11 M. With increasing osmoticum concentration up to 0.12 M the rate of absorption fell abruptly. However, at a higher concentration range the rate of the process decreased more slowly. The first step of the process may be interpreted as resulting from damage of epidermal cells, and the second one – with a spread of shock effect on cortical cells. It should be noted that the first step was more pronounced at low external concentration of K^+ (1 mM). As to the second step, the slope of both concentration curves was the same. Therefore, at a moderate external K^+ concentration, the ions are taken up mainly by epidermal cells. When the external ion concentration is rather high some of them can penetrate the root *via* apoplasm and be absorbed by cortical cells.

If this conclusion is correct, the second question arises: either ions are taken up by all epidermal cells equally or the process is mainly located just within the root hairs?

To answer this question we compared root hairs with adjacent hairless cells in relation to intracellular K^+ activity on the one hand and to the frequency of plasmadesmata in cell walls on the other. K^+ activity was estimated with pin point K^+ sensitive microelectrodes, and plasmodesmata frequency with the electron microscope [8, 9]. The work was performed with two plant species: the higher floating freshwater plant *Trianea bogotensis* [9, 10] and the terrestrial plant radish (*Raphanus sativus*) (D. B. Vakhmistrov, E. B. Kurkova and I. F. Zlotnikova, in press).

Table 1. Intracellular K^+ activity and a number of plasmodesmata in tangential walls of hair and hairless cells of root epidermis

Plant species	Cell types	K^+ activity, mM	Number of plasmodesmata	
			per 1 micron^2	per 1 cell junction
Trianea bogotensis	Hair	133	2.06	10419
	Hairless	74	.11	693
Raphanus sativus	Hair	129	.16	273
	Hairless	124	.07	150

One can see from Table 1 that intracellular K^+ activity in mature root hairs of the water plant is twice as high as that in hairless cells. One can, therefore, reasonably suggest that the concentration gradient, down which ions move inside the symplasm towards xylem vessels, is maintained just by root hairs but not by the whole layer of epidermal cells. If it is really so, the frequency of plasmodesmata in the base of hair cells (on their border with the cortex) should be greater than that in the base of hairless cells. In fact, the number of plasmodesmata in hair cells was 20 times as great as that in hairless ones, when calculated both per 1 square micron of cell wall surface and per 1 cell junction.

However, these patterns were not consistent with those of the terrestrial plant, radish (see Table 1). The levels of intracellular K^+-activities in both cell types of the plant were found to be the same. Frequency of plasmodesmata in the base of hair cells was only twice as high as that in the hairless cells. Thus, the root hairs of the terrestrial plant (radish) apparently do not play as specific role in ion absorption as do the cells of the water plant (*T. bogotensis*).

The role of the cortex in radial ion transport is rather peculiar and it should be noted that the cortex is the only multilayered tissue of the root. The function of the cortex might consist not in promoting ion movement but in hindering it, in order to give the nutrients a chance to interact with organic substrates coming from the phloem.

As a proof of this, I can present the following findings obtained in our laboratory by Dr. D. V. Shtrausberg [5]. Tomato plants were grown in aeroponics and in water culture with continuous aeration or without it. It was observed that the lower the oxygen supply, the higher the ratio of cortex volume to the stele volume; in aeroponics the ratio was equal to 4.7, in water culture with aeration and without it – to 8.1 and 16.5 respectively. We are under the impression that the root compensated the delayed assimilation of ions absorbed for the elongation of their pathway to the vessels, thus increasing the duration of this process.

Whereas the role of the endodermis is well known as far as the function of the pericycle in ion transport is concerned, it has not been studied in detail. To study the direction of symplastic ion movement through the pericycle of 7-day old barley seedlings we counted the plasmodesmata distributed within radial and tangential cell walls of this tissue [8]. It was found that almost a half of the plasmodesmata (43 per cent) were situated at the junction of the endodermis with the pericycle, and about the same portion (45 per cent) – in radial cell walls along the pericycle ring. But only one = tenth of the plasmodesmata (12 per cent) was located at the exit from the pericycle into the stelar parenchyma.

We conclude that the majority of ions moving from the cortex into the stele does not enter the stelar parenchyma but is directed immediately to xylem vessels *via* the shortest way along the pericycle ring. Moreover, such a distribution of plasmodesmata allows us to propose that the pericycle is also responsible for the uniform distribution of photosynthates moving from the phloem to the cortex [8].

Any discussion of root tissue functions in ion transport would be incomplete without mentioning xylem parenchyma. So, the last question is whether ions are passively liberated from xylem parenchyma cells into the vessels, or they are pumped out actively?

According to the widely accepted Bowling's model [2] the ions are liberated passively. This view is based partially on measurements of electropotential difference between root exudate and surrounding solution. The potential is believed to reflect the electric gradient along root symplasm. As to xylem vessels, they play a role only of salt bridges with low electric resistance. So, if we cut the root off under solution level, the potential would fall down to zero – because of direct electric shunt through cut xylem vessels.

This interpretation was checked experimentally by Dr V. A. Soloviev in our laboratory [6]. He measured the electropotential difference between root exudate of decapitated sunflower plants and surrounding nutrient solution at steady state. Then, he cut the root under solution level and continued the measurement. It may be seen from Fig. 2 that the potential did not disappear after

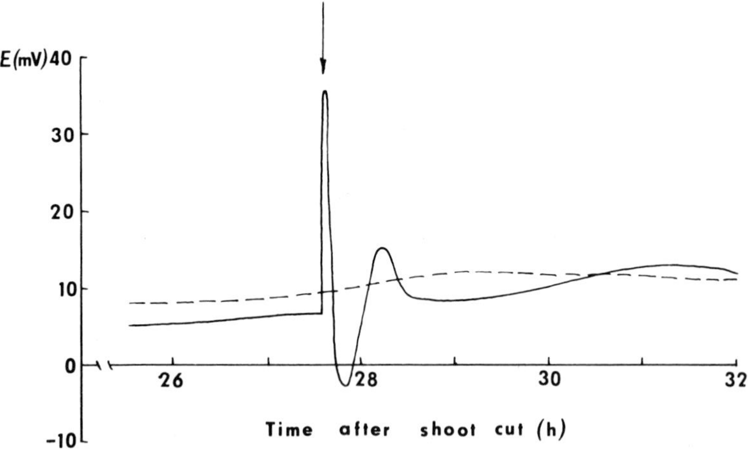

Fig. 2. Kinetics of electropotential difference between an exudate of excised sunflower root after root cutting under solution level (solid line, arrow) and without the cutting (dashed line).

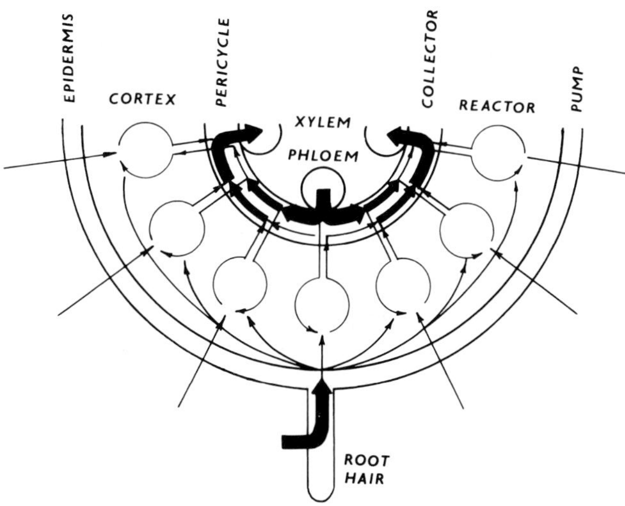

Fig. 3. General scheme of root tissue functions in the transport of ions (centripetal arrows) and photosynthates (centrifugal arrows).

root cutting, but returned back to the initial level after some fluctuations. Therefore, the electropotential difference, measured between root exudate and external solution, apparently does not reflect just the symplastic potential of the root. This result, of course, does not answer the question discussed. But it indicates, that the question is still open. All the data presented are summarized schematically in Fig. 3.

At moderate external concentrations of ions the root epidermis serves as a pump, that is as a site of initial entrance of ions into the symplasm. The roots of the water plant absorb ions mainly through root hairs whereas the roots of the terrestrial plant – through all epidermal cells uniformly. The cortex performs a function of a metabolic reactor. Its purpose is to elongate the pathway of ions absorbed to xylem vessels and to give them an opportunity to interact, with organic compounds coming from the phloem. The endodermis is omitted in this picture: its function as an osmotic barrier is widely known. And at last, the pericycle serves as a distributive collector. A centrifugal flow of carbohydrates from the phloem goes first along the pericycle ring, gradually being released into the cortex. For this reason the carbohydrates themselves can not reach xylem vessels, a carbohydrate shunt between phloem and xylem of the root is impossible to occur. On the other hand, a centripetal flow of ions and products of their primary assimilation moving from the cortex into the stele is concentrated in the pericycle and progressively expanding is directed *via* the shortest arc way immediately into xylem vessels. As to the function of xylem parenchyma in ion liberation into the vessels, the question is still open.

REFERENCES

1. Bange, G. G. J. 1973 Diffusion and absorption of ions in plant tissue. III. The role of the root cortex cells in ion absorption.
2. Bowling, D. J. F. 1976 Uptake of ions by plant roots. New York: Chapman and Hall.
3. Clarkson, D. T. 1974 Ion transport and cell structure in plants. London: McGraw-Hill.
4. Grunwaldt, G., Ehwaldt, R. and Göring, H. 1978 Suitability of osmotic shock procedure for the analysis of membrane transport in root tips of *Zea mays* L. J. Exp. Bot. 29, 97–106.
5. Shtrausberg, D. V. 1973 Changes of tomato root structure in connection with aeration conditions. Sovjet Plant Physiol. 20, 477–483.
6. Solov'ev, V. A. 1978 Nature of escape of ions into xylem vessels of the root. Sovjet Plant Physiol. 25, 843–848.
7. Vakhmistrov, D. B. 1967 On the function of the apparent free space in plant roots. A study of the absorbing power of epidermal and cortical cells in barley roots. Sovjet Plant Physiol. 14, 103–107.
8. Vakhmistrov, D. B., Kurkova, E. B. and Solov'ev, V. A. 1972 Characteristics of plasmodesmata and lomasome-like formations in barley roots in connection with transport of substances. Sovjet Plant Physiol. 19, 808–817.
9. Vakhmistrov, D. B., Mel'nikov, P. V. and Vorob'ev, L. N. 1974 Differences in absorption of potassium by root hairs and hairless cells of the root epidermis in *Trianea bogotensis*. Sovjet Plant Physiol. 23, 448–454.
10. Vakhmistrov, D. B. and Kurkova, E. B. 1979 Symplasmic connections in the rhizodermis of *Trianea bogotensis* Karst. Sovjet Plant Physiol. 26, 763–771.
11. Van Gren, F. and Boers-van der Sluijs, P. 1980 Symplasmic and apoplasmic radial ion transport in plant roots. Planta 148, 130–137.

41. Sugar beet root as an organ for sucrose accumulation

VALENTINA P. KHOLODOVA, YULIYA P. BOLYAKINA, ANATOLI B. MESHCHERIAKOV and MARINA S. ORLOVA

K. A. Timiriasev Institute of Plant Physiology, Academy of Sciences, Moscow, U.S.S.R.

In parenchyma cells of the sugar beet root, sucrose accumulates in large central vacuoles formed at a very early stage of root development. The growth of parenchyma cells is mainly the result of vacuole volume increase, small individual vacuoles being replaced by a single central vacuole. For apoplastic solute movement, cell walls and intercellular spaces may be used, the latter being well developed in young roots [2]. Later, an intensive thickening of cell walls takes place followed by an alteration of their structure [1]. In growing roots, plasmodesmata may also serve for sucrose movement. Plasmodesmata between parenchyma cells are located in large groups in relatively thin regions of the cell walls. They are branched in two directions with large median cavities which seem to unite all the plasmodesmata of a zone (Fig. 1). Although sucrose is really a principal reserve product in the sugar beet root, starch accumulates in plastids in large amounts (Fig. 2). Therefore, the sugar beet root can transform sucrose arrived from leaves into starch, however starch content is very low since plastids are small and sparse.

Sucrose and monosaccharides are actively taken up by different transport systems in storage cells, as is proved by the absence of their interaction in the course of transport from mixed solutions [3]. The rate of membrane transport of sugars significantly decreases during plant growth but may be highly activated by washing excised tissue in aerated solutions ('ageing') (Fig. 3). Acceleration of sugar transport is also provided by using citrate buffer in comparison with phosphate or acetate buffers (Fig. 4).

Sucrose seems to be retained in storage cells for a long time due to a low rate of its efflux out of vacuoles, especially in mature roots or during the period of winter storage. Preliminary experiments show that the pH of bathing solutions may influence the rate of sucrose leakage. Small pH decrease (on 0.2–0.3 pH unit only) results in a two-fold increase of sucrose efflux. However the process appears to be strongly dependent on the chemical nature of substances used for pH establishing (Fig. 5). As for absorbing roots [4], it is assumed that solute efflux out of sugar beet root tissue is controlled by changes in cytoplasmic pH which are effectively caused by weak lipophilic organic acids.

R. Brouwer et al. (eds.), Structure and Function of Plant Roots, 209–213.

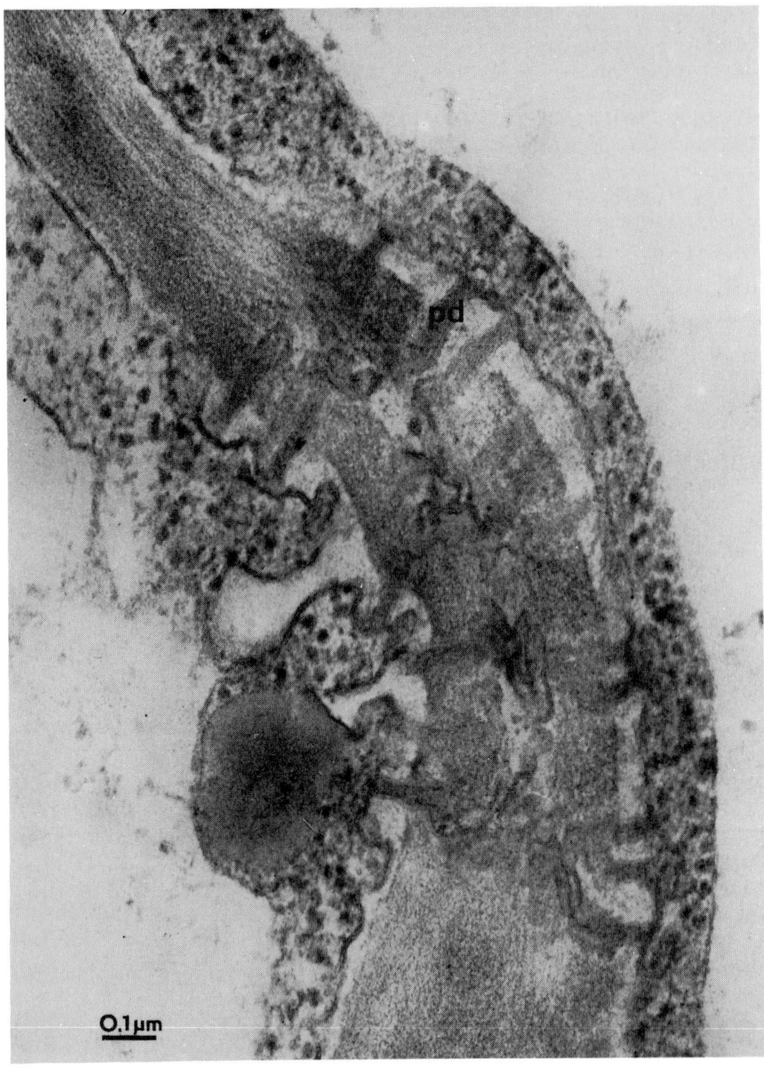

Fig. 1. A group of numerous plasmodesmata (pd) with extend median cavities between two root parenchyma cells.

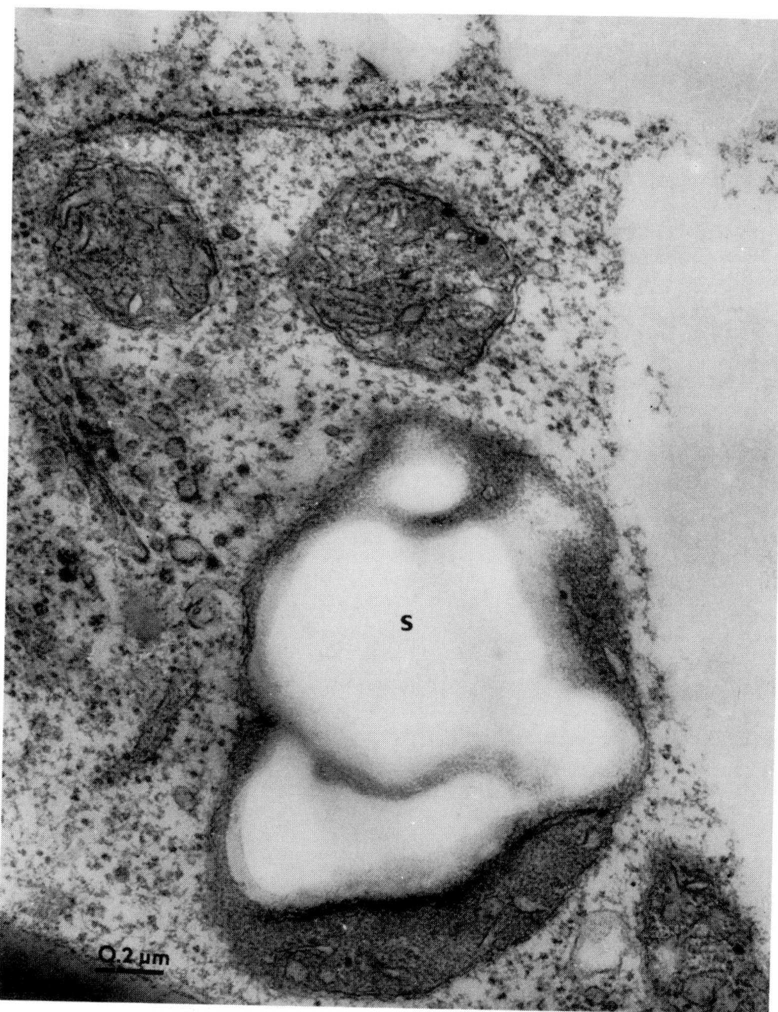

Fig. 2. Young root parenchyma cell with a plastid rich in starch(s).

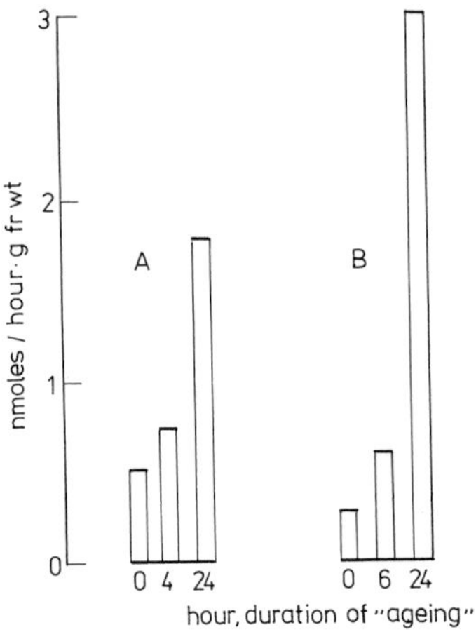

Fig. 3. Fructose transport in root parenchyma of young (A-62 days) and mature (B-128 days) sugar beet plants and an effect of 'ageing'.

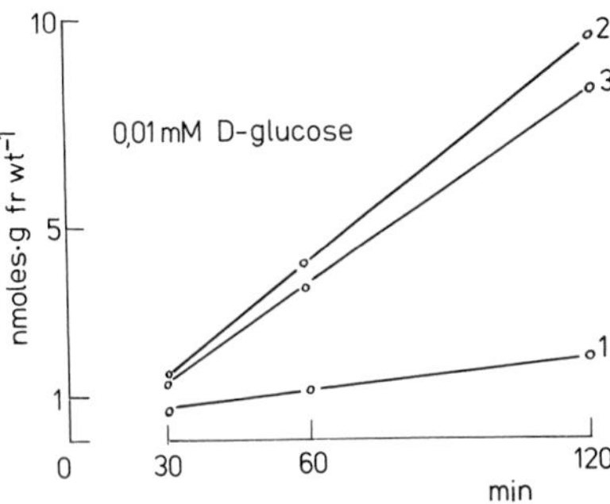

Fig. 4. Effect of acetate (1), phosphate (2) or citrate (3) buffers, on D-glucose uptake by parenchyma cells of root.

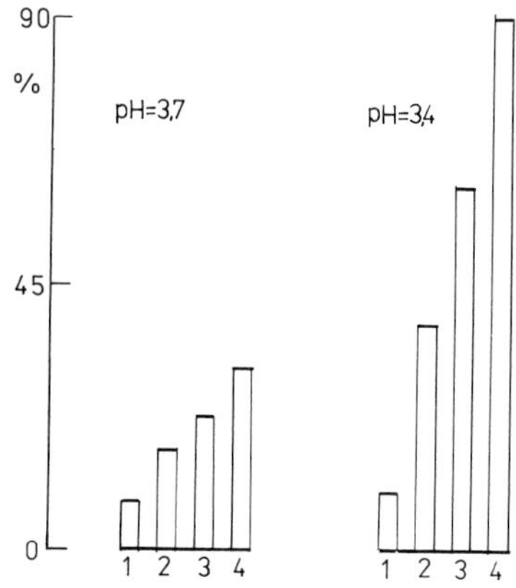

Fig. 5. Comparison of the effect of hydrochloric (1), formic (2), acetic (3) and butyric (4) acids on sucrose leakage (per cent of total sucrose content) out of sugar beet root tissue.

REFERENCES

1. Bolyakina, Yu. P., Kholodova, V. P. 1974 Some peculiarities of the ultrastructure of storage parenchyma in sugar beet roots. Phiziologiya rastenii 21, 573–577.
2. Kholodova, V. P., Buzulukova, N. P. and Bolyakina, Yu. P. 1980 Storage tissues. In: Danilova, M. F. and Kozubov, G. M. (eds.), Ultrastructure of plant tissues, pp. 213–226. Petrozavodsk, Kareliya.
3. Kholodova, V. P., Sokolova, S. V., Turkina, M. V. and Meshcheryakov, A. B. 1976 Transport and accumulation of di- and monosaccharides in sugar beet tissues. Wiss. Zeitschr. der H-U zu Berlin, Math-Nat. Reie XXV, 1.
4. Lee, R. B. 1977 Effect of organic acids on the loss of ions from barley roots. J. Exp. Bot. 28, 578–587.

42. Influence of IAA and ABA on compartmentation in red beet cells (*Beta vulgaris* L. ssp. *vulgaris* L. *var. rubra*)

G. FRANÇOIS, G. J. BOGEMANS and L. NEIRINCKX

Laboratorium voor Plantenfysiologie, Vrije Universiteit Brussel, St. Genesius-Rode, Belgium

The influence of the growth substances IAA and ABA on the sodium compartment capacities (free space, cytoplasm and vacuole) of red beet cells (*Beta vulgaris* L.ssp *vulgaris* L. *var.rubra*) was investigated. Evidence of a hormonal regulation of ion transport was given by other authors [3]. A comparison was made between the rate constants of the individual sodium movements across plasmalemma and tonoplast.

Cylindrical disks (diameter = 6 mm, thickness = 1 mm) of aged (7 days) red beet storage tissue [2] were loaded in a labeled (^{22}Na)NaCl solution (0.5 mM) during 24 h. Two series of experiments were carried out: or the efflux of radioactivity was measured in a saccharose solution (2.5 mM) or in an unlabeled NaCl solution (0.5 mM). Three different cases were considered: loading and elution without phytohormones, in the presence of IAA (10^{-5} M) or in the presence of ABA (10^{-5} M). Each time compartment sizes and unidirectional sodium fluxes were calculated by means of compartmental analysis [1]. The compartments were supposed to be arranged in series. All sodium movements were considered as being first order phenomena.

RESULTS

1. Efflux of ^{23}Na in saccharose (2.5 mM). The rate constants k_{co}, k_{cv} and k_{vc} of the sodium movements are given in Table 1, while the sodium capacities Q_o, Q_c and Q_v are shown in Fig. 1.

2. Efflux of ^{22}Na in ^{23}NaCl (0.5 mM). The rate constants k_{oc}, k_{co}, k_{cv} and k_{vc} of the ^{22}Na movements are presented in Table 2. Fig. 2 shows the capacities Q_o, Q_c and Q_v and Fig. 3 gives the individual sodium fluxes.

DISCUSSION

These results clearly show an influence of both hormones on sodium fluxes and capacities.

R. Brouwer et al. (eds.), Structure and Function of Plant Roots, 215–218.

Table 1. Influence of IAA(10^{-5} M) and ABA(10^{-5} M) on k_{co}, k_{cv} and k_{vc} (min^{-1}) during efflux in saccharose (2.5 mM).

Rate constant	Treatment		
	No hormones	IAA	ABA
k_{co}	0.02438	0.02862	0.04802
k_{cv}	0.00153	0.00029	0.00361
k_{vc}	0.00204	0.00275	0.00246

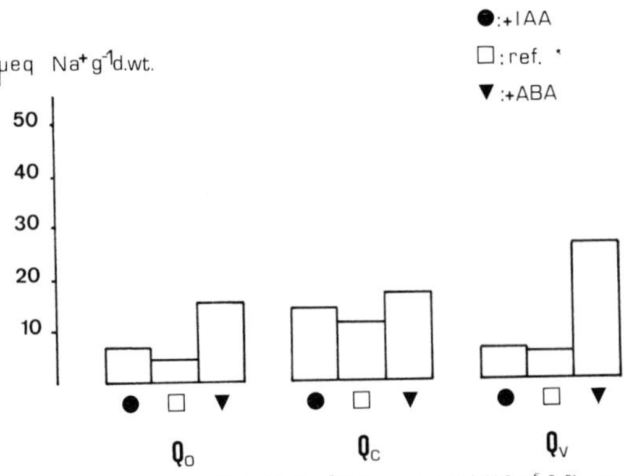

Fig. 1. Influence of IAA(10^{-5} M) and ABA(10^{-5} M) on Q_o, Q_c and Q_v (µeq Na.g^{-1}d.wt.) during efflux in saccharose (2.5 mM).

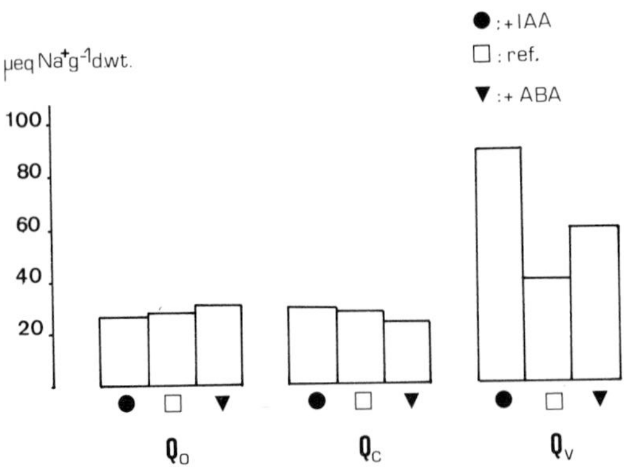

Fig. 2. Influence of IAA(10^{-5} M) and ABA(10^{-5} M) on Q_o, Q_c and Q_v (µeq Na.g^{-1}d.wt.) during efflux in NaCl (0.5 mM).

Table 2. Influence of IAA(10^{-5} M) and ABA(10^{-5} M) on k_{oc}, k_{co}, k_{cv} and k_{vc} (min^{-1}) during efflux in NaCl (0.5 mM).

Rate constant	Treatment		
	No hormones	IAA	ABA
k_{oc}	0.03609	0.02384	0.00472
k_{co}	0.03891	0.03663	0.01683
k_{cv}	0.00050	0.00029	0.00357
k_{vc}	0.00213	0.00458	0.00550

They indicate that:

1.IAA:
- enlarges the exchangeable sodium amount in the vacuole;
- retards the influx across the plasmalemma and accelerates the outflux across tonoplast in quasi steady state;
- stimulates a net sodium efflux in quasi steady state.

2.ABA:
- enlarges the washable sodium amount in free space, cytoplasm and vacuole;
- enlarges the exchangeable sodium in the vacuole;
- retards all fluxes across plasmalemma in quasi steady state, but mainly the influx;

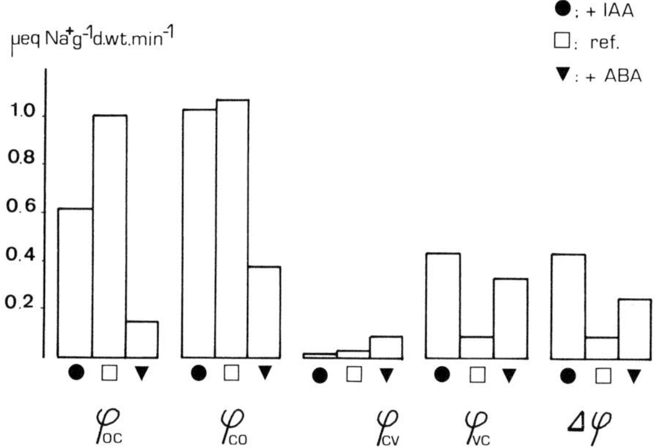

Fig. 3. Influence of IAA(10^{-5} M) and ABA(10^{-5} M) on φ_{oc}, φ_{co}, φ_{cv}, φ_{vc} and $\Delta\varphi$ (µeq Na.g^{-1}d.wt.min^{-1}) during efflux in NaCl (0.5 mM).

- accelerates all fluxes across tonoplast in quasi steady state, but mainly the outflux;
- stimulates a net sodium efflux in quasi steady state.

REFERENCES

1. Grignon, C. 1974 Etude des flux et de la distribution endocellulaire du potassium chez les cellules libres d'Acer pseudoplatanus L. Thèse de doctorat es-sciences naturelles Université de Paris VII, pp. 62 and 136–158.
2. Poole, R. J. 1971 Development and characteristics of sodium-selective transport in red beet. Plant Physiol. 47, 735–739.
3. Van Steveninck, R. F. M. 1974 Hormonal regulation of ion transport in parenchyma tissue. In: Zimmermann, U. and Dainty, J. (eds.), Membrane transport in plants, pp. 450–456. Berlin-Heidelberg-New York: Springer-Verlag.

43. The effect of calcium on the accumulation and transport of sodium into different parts of barley roots

J. M. STASSART and L. NEIRINCKX

Laboratorium voor plantenfysiologie, Instituut voor molekulaire biologie, Vrije Universiteit Brussel, Sint Genesius-Rode, Belgium

The accumulation of sodium (1 mM NaCl) was measured in different root tissues, the cell wall, the cortex and the stele. The effect that calcium could have on this process in the separated parts of the root was also investigated, since it is known [1] calcium effects the sodium uptake at different levels. It is only in recent years that some workers have been trying to find out what relative importance each different part of the root could have in the whole process of accumulation and transport of ions through the root system [2, 3]. The difference between cortex and stele was already stated [4, 5]. Cell walls were prepared from whole roots by a Triton X-100 detergent treatment [6]. The separation of the other parts was obtained by stripping the roots.

The role and importance of each part of the root in the process of transport was

Table 1. Sodium accumulation from a 1 mM NaCl solution in cell wall, cortical cells and stele compared with intact roots. Results expressed in μeq/g fr wt.

Time (h)	1	2	3	4	5	6
cell wall	1.0	1.0	1.0	1.0	1.0	1.0
cortex	1.2	2.0	2.5	3.0	3.5	4.0
stele	2.8	3.2	3.7	4.0	4.0	4.0
sum	5.0	6.2	7.2	8.0	8.5	9.0
intact root	5.5	7.2	8.3	9.8	10.5	11.2

Table 2. Sodium accumulation from a 1 mM NaCl + 0.5 mM $CaCl_2$ solution in the cell wall, the cortex and the stele compared with intact roots. Results expressed in μeq/g fr wt.

Time (h)	1	2	3	4	5	6
cell wall	0.5	0.5	0.5	0.5	0.5	0.5
cortex	0.7	0.8	1.0	1.2	1.4	1.6
stele	0.8	1.2	1.5	1.6	1.7	1.9
sum	2.0	2.5	3.0	3.3	3.6	4.0
intact root	3.0	4.8	6.2	7.9	8.2	8.3

R. Brouwer et al. (eds.), Structure and Function of Plant Roots, 219–220.

Table 3. Sodium transport, measured as the xylem exudation on intact roots and stripped roots. The effect of the presence of calcium in the uptake solution. Results are expressed in μeq/g fr wt.

roots	Ca^{2+}	time	1 h	2 h	3 h	4 h	5 h	6 h
intact	−		0.2	0.6	1.0	1.2	1.4	1.6
intact	+		0.1	0.2	0.5	0.7	0.9	1.0
stripped	−		0.5	0.9	1.3	1.5	1.9	2.3
stripped	+		0.4	0.8	1.2	1.4	2.0	2.6

tested by measurements of xylem exudation, using intact roots or partly stripped roots. We can summarize the present findings as follows:

- A rapid equilibrium is reached between Na^+ and Ca^{2+} in the cell wall, this Na/Ca ratio seems to influence the subsequent uptake into the cells.
- The stele is very selective in sodium accumulation and sensitive to the presence of calcium.
- It is rather the cortex which regulates the amount of sodium transferred to the stele for further transport in intact roots.

REFERENCES

1. Neirinckx, L. and Stassart, J. M. 1979 The effect of calcium on the uptake and distribution of sodium in excised barley roots. Physiol. Plant 47, 235–238.
2. Richter, Ch. and Marschner, H. 1974 Verteilung von K^+, Na^+ und Ca^{2+} zwischen Wurzelnrinde und Zentralzylinder. Z. Pflanz. Physiol. 71, 5, 95–105.
3. Stassart, J. M. and Neirinckx, L. 1980 The accumulation and transport of sodium in barley roots: a comparative study between: cortex, stele and intact roots. An. Bot., in press.
4. Pitman, M. G. 1972 Uptake and transport of ions in barley seedlings. Evidence for two active stages in transport to the shoot. Aust. J. Biol. Sci. 25, 243–257.
5. Baker, D. A. 1973 The radial transport of ions on maize roots. Ion transport in plants, pp. 511–517. London: Academic Press.
6. Cathala, N. 1975 Absorption du cuivre par des racines de maïs (Zea mays L.) et de tournesol (Helienthus Annuus L.) Doctoral thesis, Academie de Montpellier, France, p. 20.

44. Permeability and conductance properties of the maize root epidermis

JOZEF MICHALOV

Institute of Experimental Biology and Ecology, Slovak Academy of Sciences, 885 34 Bratislava, Czechoslovakia

Permeability properties of the epidermis have been studied separately for seven elements [1], [2], [3], [4], [5]. Potential difference E_T (Fig. 1A, B) across an epidermis, experimentally measured for single ions according to the percentage of the relevant ion present in litre of solution, shows the sequence:

$$E_T(K^+) > E_T(Na^+) > E_T(Mn^{2+}) > E_T(Ca^{2+}) > \\ > E_T(Mg^{2+}) > E_T(Fe^{3+}) > E_T(Zn^{2+}). \quad (1)$$

Potential difference E_K (Fig. 1C, D), measured for single elements by the same method as E_T when, two solutions were in direct contact, conforms to the sequence:

$$E_K(K^+) > E_K(Na^+) \gg E_K(Ca^{2+}) > E_K(Mg^{2+}) > \\ > E_K(Mn^{2+}) > E_K(Fe^{3+}) > E_K(Zn^{2+}). \quad (2)$$

The calculated Nernst potential E_N for these ions is governed by the sequence:

$$E_N(Na^+) \geqq E_N(K^+) \gg E_N(Mg^{2+}) \geqq E_N(Ca^{2+}) > \\ > E_N(Zn^{2+}) \geqq E_N(Mn^{2+}) > E_N(Fe^{3+}). \quad (3)$$

A similar inequality order applies to membrane potential E_m calculated from a modified Goldman Equation:

$$E_m(K^+) \geqq E_m(Na^+) \gg E_m(Ca^{2+}) \geqq E_m(Mg^{2+}) > \\ > E_m(Mn^{2+}) \geqq E_m(Zn^{2+}) > E_m(Fe^{3+}). \quad (4)$$

The situation with the calculated diffusion potential E_D for the elements has been investigated by sequence:

$$E_D(K^+) \geqq E_D(Na^+) \gg E_D(Mg^{2+}) \geqq E_D(Zn^{2+}) > \\ > E_D(Ca^{2+}) \gg E_D(Mn^{2+}) \ggg E_D(Fe^{3+}). \quad (5)$$

R. Brouwer et al. (eds.), Structure and Function of Plant Roots, 221–224.

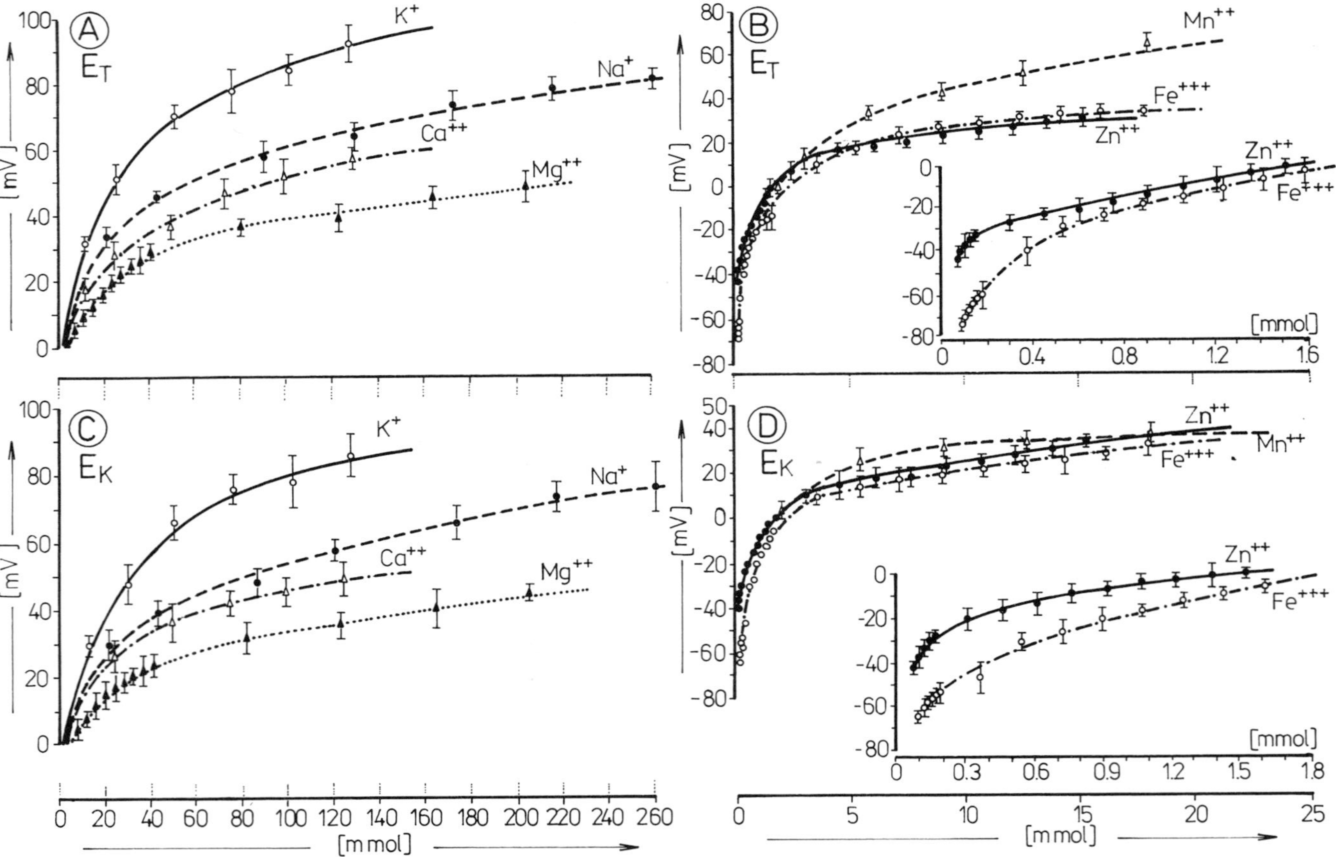
A
E_T
K+
Na+
Ca++
Mg++
[mV]
B
E_T
Mn++
Fe+++
Zn++
[mmol]
C
E_K
D
E_K
[mmol]

Comparing the 1st sequence with the sequence characterizing ionic mobility values

$$u_{Na} < u_K < u_{Ca} < u_{Mg} < u_{Fe} < u_{Zn}, \qquad (6)$$

we see that ions with small mobility induce a large potential difference on the epidermis and vice versa. This is also supported by the sequence characterizing the hydration entropy

$$S_H(K^+) < S_H(Na^+) \lll S_H(Ca^{2+}) < S_H(Fe^{3+}) \doteq \doteq S_H(Zn^{2+}) < S_H(Mg^{2+}). \qquad (7)$$

Here the ions inducing high transepidermal potentials have a small hydration entropy and a great effective ionic radius given by the sequence:

$$r_{ef}(Mn^{2+}) < r_{ef}(Mg^{2+}) < r_{ef}(Fe^{3+}) \doteq \doteq r_{ef}(Zn^{2+}) < r_{ef}(Na^+) < r_{ef}(Ca^{2+}) < r_{ef}(K^+). \qquad (8)$$

Quite significant is the fact that the calculated Nernst potential (E_N) and membrane potential (E_m) have almost the same values for calcium, iron, manganese, magnesium and zinc. From the E_N and E_m of calcium, iron and manganese we can verify the empirical relation $E_T > E_N \doteq E_m$ and the potassium, sodium, magnesium and zinc hold reversed inequality $E_T < E_N \doteq E_m$.

On the strength of inequalities and the relationship between E_T, E_K, E_N and E_m as well as on other data, it may rightly be supposed that fixed bounded electric charges exist in the rhizodermis structure. Furthermore, based on the permeability coefficients changes, we may conclude that, from the biophysical viewpoint of transport, the epidermis is a porous barrier regulating ion transport from the outside medium into the inside of the root and come back.

According to our findings, the fixed electric charges of the epidermis seem to be mostly positive, inhibiting the movement of cations, or their possible carriers. However, this does not mean that all cations, have to penetrate the epidermis by active transport.

Fig. 1. Resultant transepidermal potentials (E_T) in variation with external concentrations of macro-elements (Part A) and of micro-elements (Part B). 2. Liquid-contact potentials (E_K) in variation with the external concentrations of macro-elements (Part C) and micro-elements (Part D). Internal concentrations were constant. E_K^+ – potassium potential, E_{Na}^+ – sodium potential, E_{Ca}^{++} – calcium potential, E_{Mg}^{++} – magnesium potential, E_{Fe}^{+++} – iron potential, E_{Mn}^{++} – manganese potential and E_{Zn}^{++} – zinc potential, respectively.

It should be added that, in general, hydration and the quality of bound ions plays a very important role in this case.

REFERENCES

1. Michalov, J. 1975 Biophysical aspect of the transport phenomenon. Research report UEBE SAV Bratislava.
2. Michalov, J. 1977a Sodium permeability and conductance of maize primary root epidermis. Z. Pflanzenphysiol. 84, 1–11.
3. Michalov, J. 1977b Magnesium permeability and conductance of maize primary root epidermis. Z. Pflanzenphysiol. 84, 13–23.
4. Michalov, J. 1977c Potassium ion permeability and conductance properties of maize primary root epidermis. Z. Pflanzenphysiol. 84, 377–390.
5. Michalov, J. 1977d Mn^{2+} ionic permeability and conductance properties of maize primary root epidermis. Z. Pflanzenphys. 85, 189–199.

45. Absorption and transport of nutrient cations and anions in maize roots

WILLEM G. KELTJENS

Department of Soil Science and Plant Nutrition, Agricultural University, Wageningen, The Netherlands

In most plant species rates of absorption of nutrient cations and nutrient anions are unequal on an equivalent basis; consequently, processes within the plant are active to keep both the plant interior as well as the external root medium electrically neutral. Although excess cation over anion absorption initiates synthesis and accumulation of organic anions in the plant [4], reversal conditions are much more complicated. For these conditions Dijkshoorn et al. [3] and Ben Zioni et al. [1] proposed a K^+ recirculation hypothesis in the plant. Part of the K^+, transported by xylem upward to the upper plant parts, will be redirected via the phloem to the roots in association with malate ions synthetised by nitrate reduction in the upper plant parts. In the roots malate will be decarboxylated and excreted from the root as OH^-, while K^+ will be transferred into the xylem again. According this K^+ cycle excess absorbed negative charge (mainly NO_3^-) can be eliminated.

However, with nitrate reduction in the plant root hydroxyl ions, synthetized by nitrate reduction, can directly be excreted from the root and a NO_3^-/OH^- exchange mechanism can exist without K^+ recirculation in the plant. The final goal of this work was to see how the maize plant regulates electroneutrality under different ion uptake regimes.

MATERIALS

For exudation experiments maize plants were grown for 5 weeks on a complete nutrient solution. One week before the absorption-exudation experiment plants were transferred to a 0.5 mmol l^{-1} $Ca(H_2PO_4)_2$ solution. Subsequently, plants were decapitated and transferred to different single salt solutions of KCl, K_2SO_4, $CaCl_2$ and $Ca(NO_3)_2$, all 20 meq l^{-1}. Absorption, accumulation and transport was measured during a 24 h period.

For split-root experiments maize plants were grown for 4 weeks continuously on a complete nutrient solution (with NO_3^-). ^{86}Rb was used as a physiological substitute for ^{42}K.

R. Brouwer et al. (eds.), Structure and Function of Plant Roots, 225–232.

RESULTS

During the exudation experiment absorption of nutrient cations (C_a) exceeded absorption of nutrient anions (A_a) with both KCl and K_2SO_4 nutrition (Fig. 1). Externally, electroneutrality was maintained by a H^+ efflux of the root equal to the difference in absorbed cat- and anions. This proton flux was measured by the titration method described by Breteler [2]. Internally, accumulation and transport of absorbed cat- and anions differed on charge basis. Particularly in the case of K_2SO_4 nutrition xylem transport of potassium exceeded $SO_4^=$ transport by far. Chemical composition of plant exudates (Table 1) indicated that differences in contents of total inorganic cations (C) and total inorganic anions (A) were compensated by organic anions or carboxylates. Qualitatively, malate seemed to be the predominant organic anion in the K_2SO_4 exudate (Table 2). Thus by production of organic anions, mainly malate, electroneutrality of xylem sap

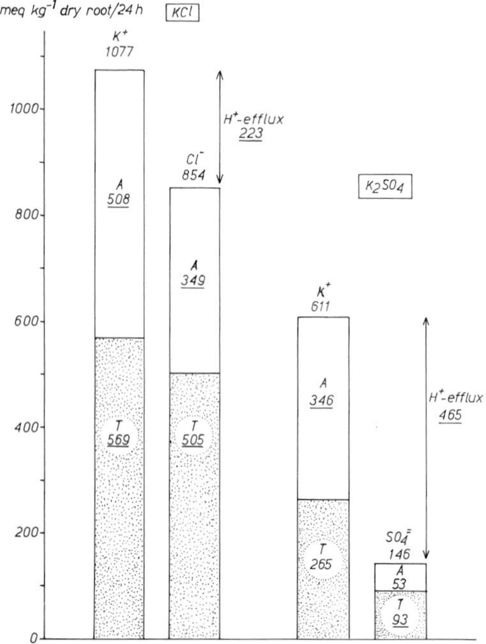

Fig. 1. Absorption, accumulation and xylem transport of K^+, Cl^- and $SO_4^=$ in decapitated maize plants transferred to single salt solutions of 20 meq l^{-1} KCl and K_2SO_4. Experimental period 24 h. Total length of the columns represents absorption, while T and A stand for the fractions of xylem transport and root accumulation, respectively.

Table 1. Contents of C, A, C–A and sum carboxylates in exudates collected during two time intervals after transfer of decapitated maize plants to single salt solutions of KCl and K_2SO_4 both 20 meq l^{-1}. Contents in meq l^{-1}.

Solution	Time interval (h)	C	A	C–A	Sum carboxylates
KCl	0–16	46.1	42.2	3.9	4.1
	16–24	52.7	47.4	5.0	4.4
K_2SO_4	0–16	29.9	14.9	15.0	19.0
	16–24	36.7	14.7	22.0	26.0

(exudate) and root tissue (results not presented) was guaranteed. Because of the difference in ion absorption pattern, both production and final content of carboxylates in xylem sap and root tissue were rather different in KCl and K_2SO_4 fed plants.

Reversal nutritional conditions were created by transferring decapitated plants to single salt solutions of $CaCl_2$ and $Ca(NO_3)_2$. As shown in Fig. 2, under these conditions, absorption of nutrient anions exceeded absorption of nutrient cations by far. Processes involved in regulating and maintaining electroneutrality of root interior and exterior were studied.

As far as the root exterior is concerned electroneutrality was achieved by an excretion of OH^- ions from the root equivalent to $(A_a - C_a)$. With $CaCl_2$, xylem transport of Ca^{++} and Cl^- were equivalent. Excess absorbed Cl^- was stored in the root and electrically compensated, partly by Ca^{++} ions, partly by a release of negative charge from the root by way of decarboxylation of organic anions present in the root. Excretion of hydroxyl ions from the root as indicated before is the result of this decarboxylation process. Processes involved in the $Ca(NO_3)_2$

Table 2. Contents of the different organic anions in the K_2SO_4 exudate collected during the 16–24 h time period. Absolute contents in meq l^{-1}.

Organic anion	Absolute content	Relative content
succinate	0.33	1
malonate	1.50	6
malate	24.50	91
citrate	0.20	1
total	26.03	100

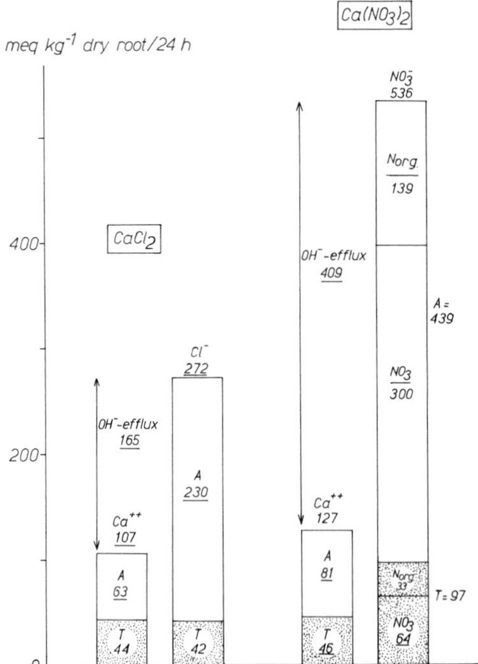

Fig. 2. Absorption, accumulation and xylem transport of Ca^{++}, Cl^- and N in decapitated maize plants transferred to single salt solutions of 20 meq l^{-1} $CaCl_2$ and $Ca(NO_3)_2$. Experimental period 24 h. Total length of the columns represents absorption, while T and A stand for the fractions of xylem transport and root accumulation, respectively.

system were much more complicated. As shown in Fig. 2 part of absorbed NO_3^- has been accumulated or transported in the root as such. However, about one third of the nitrogen in both fractions, accumulated or transported, has been reduced already within the root and was present as organic nitrogen. This confirms the findings of Ivanko and Maxianová [5] that the process of NO_3^- reduction is active also in the roots of maize plants. Nitrate reduction also contributes to the internal/external charge balance of the plant root by a direct NO_3^-/OH^- exchange mechanism (Kirkby [6]). However, in this experiment NO_3^- reduction in the root did not completely neutralize excess absorbed nutrient anions; part of excess absorbed negative charge was also neutralized by the process of decarboxylation of organic anions as in the $CaCl_2$ system.

Analysis of maize roots before and after the exudation period showed indeed a loss of carboxylates of 183 and 182 meq kg^{-1} dry root with $CaCl_2$ and $Ca(NO_3)_2$

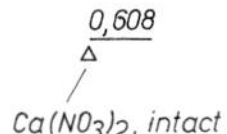

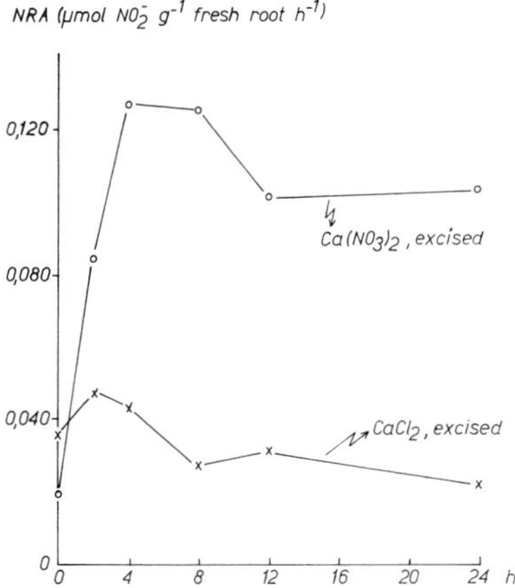

Fig. 3. Nitrate reductase activity (NRA) in maize roots. On time zero roots are excised and transferred to single salt solutions of $CaCl_2$ (×) and $Ca(NO_3)_2$ (○), both 20 meq l^{-1}. NRA at t = 24 h in roots of intact plants fed with $Ca(NO_3)_2$ is also indicated (△).

nutrition, respectively. Although with $Ca(NO_3)_2$ xylem transport of NO_3^- exceeded transport of Ca^{++} (Fig. 2), (C–A) of the xylem sap had values of almost zero to slightly negative. This implies an additional xylem transport of inorganic cations already present in the root before the experiment. Negative values of (C–A) of the exudates can be explained or by the presence of positive charged organic compounds or analytical errors.

Fig. 3 shows the activity of the enzyme nitrate reductase (NRA) in roots during the exudation experiment for both $CaCl_2$ and $Ca(NO_3)_2$ nutrition. It indicates very clearly the adaptive character of the enzyme by a strong increase of its activity during the first 4 h. Subsequently during the next 20 h activity stabilized or slightly decreased. NRA in roots of intact plants measured at the end of the exudation period showed a 5 fold higher value than similar roots of decapitated plants at the same time (Fig. 3). This supports the assumptions that:

– removal of the plant shoot blocks further energy supply from shoot to root; thus assimilates are needed for maintaining enzyme activity [7].

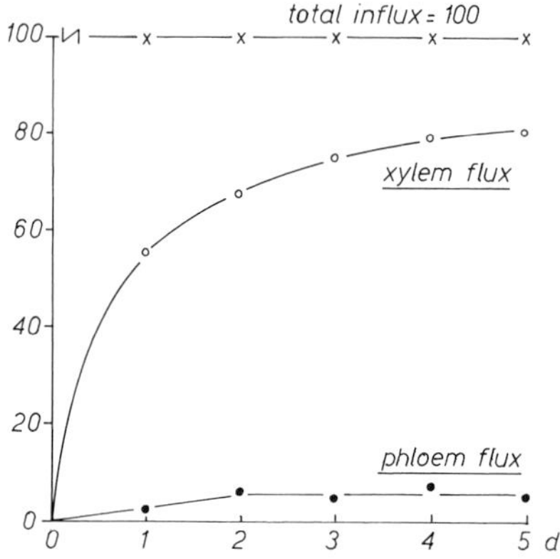

Fig. 4. Total influx (×), xylem flux (○) and phloem flux (●) of potassium in intact maize plants. All fluxes were measured in a split-root experiment with ^{86}Rb as a tracer for potassium and expressed relative to the total influx (= 100).

- NO_3^- reduction as measured in the exudation experiment gives an underestimation of the real NO_3^- reduction in roots of intact maize plants.

If so, actual nitrate reduction in roots of intact maize plants will be higher than 30% of the total absorbed NO_3^- as measured in the exudation experiment. Moreover, the necessity of a downward phloem transport of K^+-malate as proposed by Ben Zioni et al. [1] does not exist for the maize plant if excess absorbed negative charge (NO_3^-) will be reduced in the root.

In a split-root experiment with intact maize plants phloem flux of K^+ was measured. Labelled K^+ (^{86}Rb) was added to the one root half while all other nutrients were equal for both root compartments. During a 5 days period distribution of the radio-isotope inside the plant was monitored and phloem flux of potassium was estimated by measuring ^{86}Rb influx into the other root half. Results of this experiment (Fig. 4) illustrate that at equilibrium phloem flux of K^+ was only about 5–7% of total K^+ influx. This corresponds with a K^+ phloem transport in the plant of 60.9 µmol K^+ $plant^{-1}$ day^{-1} at a K^+ influx rate of 870 µmol K^+ $plant^{-1}$ day^{-1} as was measured during steady state (day 3–5). Simultaneous measurement of OH^--efflux by maize roots in a similar split-root experiment showed values of 1281 µmol OH^- $plant^{-1}$ day^{-1}. This indicates that under these conditions (NO_3^- nutrition) supply of roots with K^+-malate via phloem is far from sufficient to cover total hydroxyl excretion by the root. According to the Ben Zioni model hydroxyl efflux and downward phloem flux of (K^+ + Mg^{++}) carboxylates should be stoichometric in the absence of NO_3^-

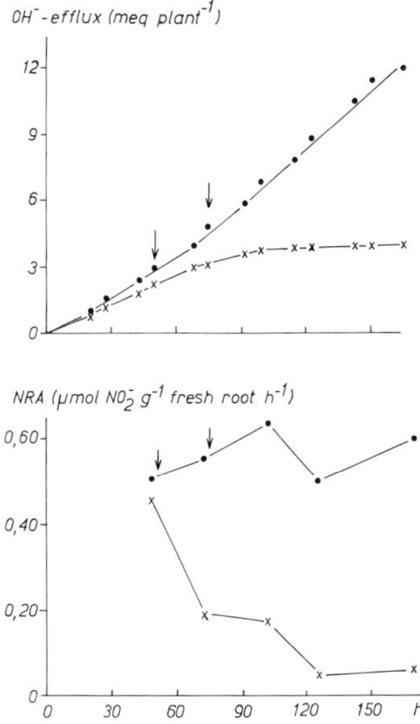

Fig. 5. Hydroxyl efflux and NRA in maize roots with (×) and without (•) addition of tungstate to the nutrient solution. Addition of 100 and 200 μmol tungstate at times indicated by arrows. Experiment carried out by split-root technique.

reduction in the root. Considering that K^+ exceeds Mg^{++} in phloem, the conclusion will be justified that almost total OH^--efflux by the root is a direct consequence of NO_3^- reduction in the root. Hydroxyl ions produced within the root by the reduction process can be exchanged directly with external NO_3^-.

That NO_3^- reduction within the root is the driving force for the OH^--efflux was proved in another split-root experiment by adding tungstate, an inhibitor of the enzyme nitrate reductase, to one root half. Fig. 5 shows that within a short time after adding this inhibitor to the one side both NRA and OH^--efflux in this root half almost stopped, while activity of the other root part was not affected at all.

Finally, the suggestion that nitrate reduction in the roots of maize plants is very active, regulates electroneutrality of the root system and eliminates the necessity of K^+ phloem transport is further supported.

REFERENCES

1. Ben Zioni, A., Vaadia, Y. and Lips, S. H. 1971 Nitrate uptake by roots as regulated by nitrate reduction products in the shoot. Physiol. Plant 24, 288–290.
2. Breteler, H. 1973 A comparison between ammonium and nitrate nutrition of young sugar-beet plants grown in nutrient solutions at constant acidity. I. Production of dry matter, ionic balance and chemical composition. Neth. J. Agric. Sci. 21, 227–244.
3. Dijkshoorn, W., Lathwell, D. and de Wit, C. T. 1968 Temporal changes in carboxylate content of ryegrass with stepwise change in nutrition. Plant and Soil 29, 369–390.
4. Hiatt, A. J. 1967 Relationship of cell sap pH to organic acid change during ion uptake. Plant Physiol. 42, 294–298.
5. Ivanko, S. and Maxianová, A. 1974 Nitrogen uptake and assimilation by maize roots. In: Kolek, J. (ed.), Structure and function of primary root tissues, pp. 461–469. Veda, Publishing House of the Slovak Academy of Sciences, Bratislava.
6. Kirkby, E. A. 1974 Recycling of potassium in plants considered in relation to ion uptake and organic acid accumulation. Proc. 7th Int. Coll. Plant. Anal. Fertilizer Problems, pp. 557–568.
7. Radin, J. W., Parker, L. L. and Sell, C. R. 1978 Partitioning of sugar between growth and nitrate reduction in cotton roots. Plant Physiol. 62, 550–553.

46. Absorption and transport of anions by different roots of *Zea mays* L.

M. HOLOBRADÁ, I. MISTRÍK and J. KOLEK

Institute of Experimental Biology and Ecology, Slovak Academy of Sciences, 885 34 Bratislava, Czechoslovakia

Despite earlier investigations of ion absorption by different parts of the intact root system [1, 3], only a very incomplete picture is available of the relative extent to which the different roots participate both in the uptake and the transport of ions [2]. In our earlier work, studying the uptake and metabolism of anions by intact maize root system, the question of the participation of morphologically different types of roots has arisen.

In the present paper some of the above mentioned problems are discussed. The participation of primary seminal (PS), adventitious seminal (AS) and nodal (N) roots in the uptake and upward transport of ions was followed with two anions, sulphate and phosphate. The roots were incubated in a nutrient medium labelled with ^{35}S-sulphate or with ^{32}P-phosphate. The capacity for absorption of all types of roots after a short, 15 min uptake time is not very different neither for ^{32}P-

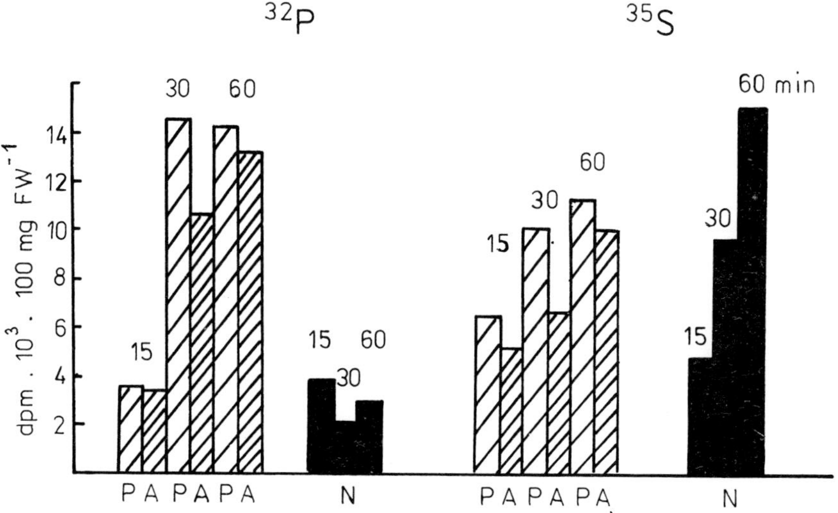

Fig. 1. The uptake of ^{32}P-phosphate and ^{35}S-sulphate by different types of maize roots during 15, 30 and 60 min incubation. (P – seminal primary root, A – adventitious seminal roots, N – nodal roots).

R. Brouwer et al. (eds.), Structure and Function of Plant Roots, 233–236.

phosphate nor for ^{35}S-sulphate (Fig. 1). Primary roots have a higher capacity for sulphate uptake than the other roots. By prolonging the uptake time to 30 and 60 min the pattern for both anions is altered (Fig. 1). Seminal roots did not exceed nodals in phosphate uptake in 15 min, but later the phosphate uptake capacity of seminal roots increased strongly with some reduction at the end of the experiment. Nodal roots exhibit a low capacity for phosphate uptake through the whole experiment.

Sulphate uptake, shown in the second part of Fig. 1, increases with time in all roots. In contrast to phosphate uptake, nodals also exhibit increase of absorbed sulphate, to an even greater extent than that of seminals. The differences suggest that nodal roots develop different capacities for different anions. All types of roots are more equal in the extent of sulphate uptake than was observed with phosphate.

By incubation of different roots with the labelled anion, e.g. those of seminals only, there was a very good possibility of determining the rate of ion transport between the roots and also their participation in the upward transport. When incubating the seminal and nodal roots separately in labelled nutrient solution, higher phosphate absorption occurred in seminals than nodals. For ^{35}S-sulphate uptake the opposite pattern was found.

Transport of ^{32}P-phosphate from seminal roots after the first 15 min is highest into nodal roots. Upward transport after 15 min is predominantly into the first and into the youngest leaf – the fifth one (Fig. 2, top). Later, after 60 min uptake,

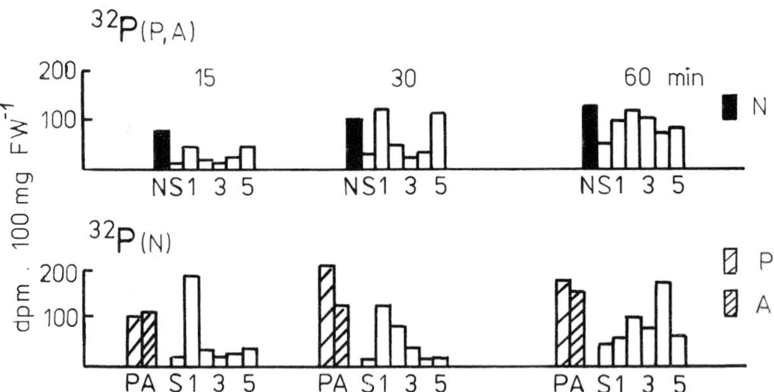

Fig. 2. The transport of ^{32}P-phosphate from roots incubated in labelled nutrient solution into unlabelled roots and leaves. top: seminal roots incubated with ^{32}P-phosphate, bottom: nodal roots incubated with ^{32}P-phosphate. (P – primary seminal root, A – adventitious seminal roots, N – nodal roots, 1, 2, 3, 4 and 5 – leaves).

the distribution of ^{32}P is more balanced in all 'unlabelled' parts of plants, with higher accumulation in the older, more developed leaves.

When nodal roots were incubated in labelled ^{32}P medium (Fig. 2, bottom), most of the phosphate is transported into the first leaf, but later, after 60 min more is found in younger leaves. The rate of ^{32}P transport into seminal roots also is high and transport into primary roots is greater than into adventitious roots.

Comparing ^{35}S-sulphate (Fig. 3) and ^{32}P-phosphate (Fig. 2) from the aspect of transport from the 'labelled' roots, some differences are obvious. More sulphate than phosphate is released for transport into leaves. Transport of ^{35}S into the first and the second leaf is large when the nodal roots are labelled (Fig. 3, bottom). Sulphate transport between roots is smaller than was noted for phosphate. In contrast to phosphate, where most of this anion is transported from nodal roots into seminal ones (Fig. 2, bottom), the rate of sulphate transport between roots is smaller when compared with that into leaves (Fig. 3).

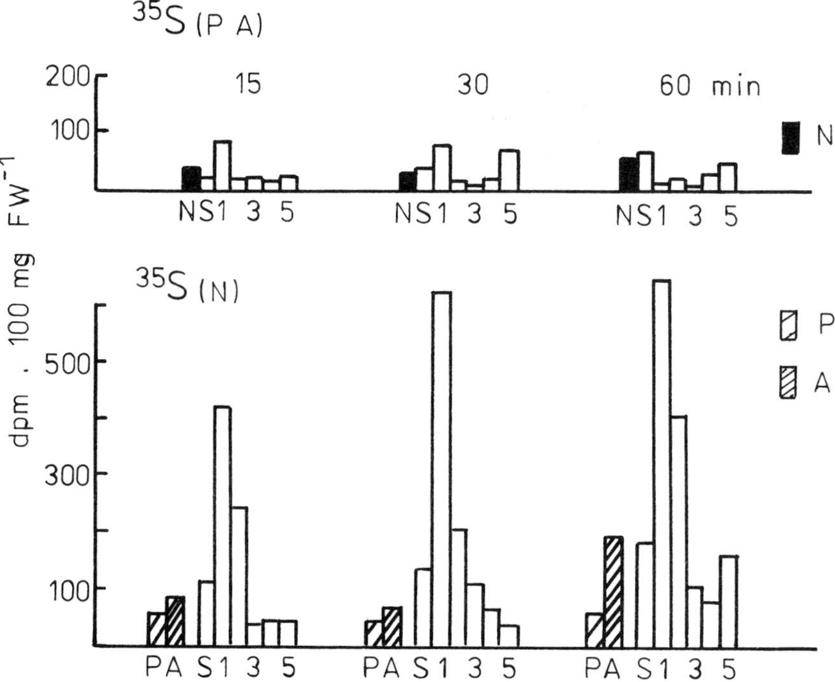

Fig. 3. The transport of ^{35}S-sulphate from roots incubated in labelled nutrient solution into unlabelled roots and leaves. top: seminal roots incubated with ^{35}S-sulphate, bottom: nodal roots incubated with ^{35}S-sulphate. (P – primary seminal root, A – adventitious seminal roots, N – nodal roots, 1, 2, 3, 4 and 5 – leaves).

It is evident that morphologically different maize roots (PS, AS, N), are characterized by different absorption capacities for sulphate and phosphate anions (Fig. 1). There are also considerable differences between seminal and nodal roots with respect to their participation in the root-shoot anion transport. The participation of the various types of roots in the processes here followed depends also on the duration of uptake.

It seems, that for sulphate uptake and upward transport the nodal roots take over the absorption function of seminal roots more rapidly than was observed for phosphate. Seminal roots are concerned both in ion uptake and transport, but the increasing participation of nodals in sulphate absorption and transport processes is important and develops earlier for sulphate than for phosphate. This difference may be closely connected with the metabolic function of sulphur and occurs in response to increased requirements of developing, intensively metabolising young parts of plants.

REFERENCES

1. Clarkson, D. T. and Sanderson, J. 1970 Relationship between the anatomy of cereal root and the absorption of nutrients and water. Agr. Res. Counc. Letc. Lab. Annu. Report, pp. 16–25.
2. Holobradá, M., Mistrík, I. and Kolek, J. 1980 The effect of root temperature on the uptake and metabolism of anions by the root system of Zea mays L. I. Uptake of sulphate by resistant and non-resistant plants. Biológia (Bratislava), 35, 259–265.
3. Russell, R. S. and Sanderson, J. 1967 Nutrient uptake by different parts of the intact roots of plants. J. exp. Bot. 18, 491–508.

47. Metabolic aspects of the transport of ions by cells and tissues of roots

BRIAN C. LOUGHMAN

Department of Agricultural Science, University of Oxford, Parks Road, Oxford OX 1 3PP, U.K.

There is no clear cut viewpoint concerning the mechanism of transport of solutes into and within roots that is applicable to all plants and all solutes. Most attention has been given to ionised species, particularly those that are not incorporated into organic molecules, e.g. K^+, Na^+, Ca^{++} and Cl^-. Ions that are rapidly metabolised such as NO_3^-, SO_4^{--} and $H_2PO_4^-$ give rise to special problems as do uncharged solutes such as sugars that become charged during entry by phosphorylation.

The first two are rapidly involved in reductive processes after entering the plant and both organic and inorganic forms are transported to the xylem. Phosphate is incorporated via oxidative phosphorylation into a wide range of organic forms prior to being released into the xylem for upward transport as inorganic phosphate. Although nitrate and sulphate appear to be transported as such across the plasmalemma of root cells prior to reduction, the incorporation of phosphate into ATP is an integral part of the entry process. The factors that affect the utilization of phosphate are discussed in this paper as an example of the problems inherent in studies of ions involved in rapid metabolic processes.

Orthophosphate ions have a much lower mobility in soil than in free solution and their concentration normally lies in the range 0.1–10 μM. The entry of the ion into the plant root is largely limited by physical rather than metabolic factors, and it is clear that a very efficient biochemical mechanism operates to enable significant quantities of phosphate to be absorbed from these very low concentrations in the soil solution. The $H_2PO_4^-$ ion is the preferred form and the overall process is controlled by the pH of the soil solution, a factor that is itself modified by the biological activity of the root and associated microorganisms.

AVAILABILITY TO PLANTS

Apart from weathering, the mineralisation of insoluble inorganic forms can be brought about by biological mechanisms. Microorganisms are concerned in solubilising apatite and related compounds in addition to accelerating phosphate release *via* the decay of organic matter. A more direct effect of the metabolic

R. Brouwer et al. (eds.), Structure and Function of Plant Roots, 237–245.

activity of the plant root on the availability of phosphate is provided by the development of phosphatases which hydrolyse a broad spectrum of organic phosphates. The synthesis of these enzymes is often activated under conditions of low availability of inorganic phosphate. The organic forms are usually present as insoluble salts and include small amounts of all the common phosphorylated components of living plants such as nucleic acids, sugar phosphates, and phospholipids. In many soils over half is present as inositol phosphates. The major component, myo-inositol hexaphosphate (phytic acid) is an important storage form in many seeds and recent suggestions by Biswaṣ et al. [2] that it holds a key metabolic role in germination may lead to revision of ideas on its utilization in soils.

MECHANISM OF ENTRY

The concentration of available inorganic orthophosphate ions in normal soils is around 1 μM, and a highly efficient mechanism of absorption is therefore required by the cells of the roots. When the root system of an actively growing barley seedling is placed in a radio-active phosphate solution (10 μM $KH_2{}^{32}PO_4$), within three seconds a third of the phosphate associated with the root can be identified chromatographically as being in the form of adenosine triphosphate and other nucleoside phosphates [10]. Subsequent incorporation of absorbed phosphate into other organic forms appears to be a necessary step before transfer to the shoot can occur. The form in which phosphate is transported to the shoot in a number of species appears to be inorganic orthophosphate, implying that a dephosphorylative step occurs prior to transfer to the xylem (Fig. 1).

The process of entry into the root is affected by temperature, pH, oxygen tension and the presence of other essential nutrient elements such as calcium or boron. In the presence of metabolic inhibitors such as 2.4. dinitrophenol where no metabolic conversion occurs some inorganic phosphate enters the root and is transported to the shoot leading to the conclusion that a small proportion of the total phosphate reaching the leaves may do so *via* a physical process [9].

The role of oxygen is not as clear as in the accumulation of other ions which are normally present at much higher concentrations than phosphate. Even under complete anaerobiosis young barley plants can absorb phosphate from 1 μM solutions and transport it to the shoots [7]. These results suggest that glycolytic metabolism produces sufficient energy to maintain membrane function and to absorb phosphate from low concentrations in the complete absence of oxygen. Absorption of phosphate shows normal temperature response and significant

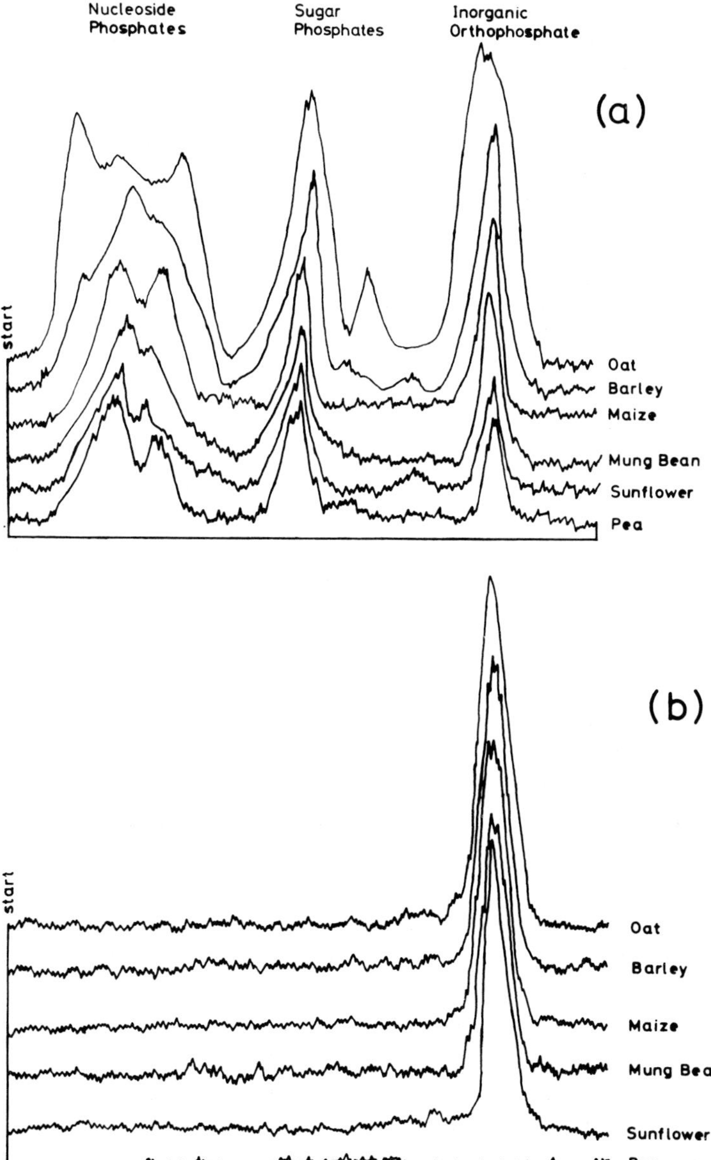

Fig. 1. The forms of phosphate in roots and transport fluid of six species after 1 h in 10 μM $KH_2{}^{32}PO_4$ at 20 °C, pH 5.5 and 12 700 lx.

Chromatogram scans of (a) extracts of roots and (b) exudation fluid from base of shoots of 12-day old seedlings. Root extraction in 0.1 M $HClO_4$ at 0 °C, neutralized to pH 5.0 after centrifugation. Shoot exudate removed as 5 μl samples immediately after removing shoot 5 mm above seed. Chromatogram developed in *tert* butanol/water/picric acid, 80 ml/20 ml 2.0 g.

amounts can be absorbed by barley roots at 0 °C. The phosphorylative processes are still active at this temperature. Most evidence suggests that some stage of the process of oxidative phosphorylation is coupled to the inward transport of orthophosphate ions from the soil solution.

The evidence for the actual mechanism of phosphate entry across the plasmalemma is inconclusive although some experiments with yeast are interpreted by Cockburn et al. [4] as showing that entry is associated with a H^+ influx at least in the region of pH 5, whereas the active transport of phosphate in *Lemna gibba* appears to be driven by a H^+ extrusion pump. This pump maintains a H^+ gradient which provides the energy for a proton-phosphate cotranport system [13]. Evidence is also presented in support of the view that the active transport is regulated by the endogenous level of phosphate.

It has been possible to detect the absorption of non-exchangeable inorganic phosphate into potato tuber tissue in the absence of phosphorylation at 0 °C. Subsequent transfer of the tissue to 20 °C causes immediate incorporation into the normal pattern of organic forms [8]. Such evidence is consistent with the view that a cotransport system of the type described in *Lemna* is reponsible for the movement across the plasmalemma and that this is immediately followed by incorporation of the phosphate into nucleoside phosphate *via* mitochondrial or glycolytic processes. Subsequent transfer of phosphate to sugar phosphate may well be concerned in bringing absorbed phosphate to the stelar region where dephosphorylation occurs to release inorganic phosphate into the xylem. Confirmation that the xylem transports only inorganic orthophosphate has been achieved by nuclear magnetic resonance studies with intact seedlings. A valid criticism of the experiments involving extraction of the xylem contents immediately after cutting at the base of the shoot described in Fig. 1 is that any organic forms present in the xylem might be instantaneously hydrolysed by phosphatases. However, sugar phosphates added to the xylem exudate are not significantly hydrolysed during the period of a few minutes between severing the stem and termination of enzyme action by contact with the acid chromatographic solvent. The support given by NMR studies strengthens the view that all the phosphate entering the root is incorporated into organic forms and that one of these forms, probably a hexose phosphate, is hydrolysed during the transfer of inorganic phosphate to the transport stream.

MECHANISM OF UPWARD TRANSPORT

After the initial process of phosphorylative absorption by the cytoplasm of the cells of the root epidermis and cortex, inorganic phosphate is then released into the xylem and is transported to the leaves. The successive processes of absorption and transport are readily separable by suitable experimental methods. At a concentration of 10 μM $H_2PO_4^-$ the transport process is more sensitive to anaerobiosis than is the initial absorption.

Interference with the normal glycolytic pathway of some species by the addition of mannose to sequester incoming phosphate as mannose-6-phosphate causes complete cessation of transport to the shoot but no reduction of entry into the root [9]. This mechanism of selection depends on the relatively low activity of phosphomannoisomerase in sensitive species such as barley and other cereals when compared with mung bean where the activity of the enzyme is much higher. It is possible to distinguish between consecutive components in the overall process of phosphate utilization with this experimental approach. If roots are treated simultaneously with mannose and phosphate, transport to the shoot begins but stops completely after 30 min and recovers after transfer to water. If mannose is added to the root environment transport is again blocked and can also be reversed. It appears that phosphate available in the cytoplasm for export to the shoot can always be sequestered by mannose entering the root (Fig. 2).

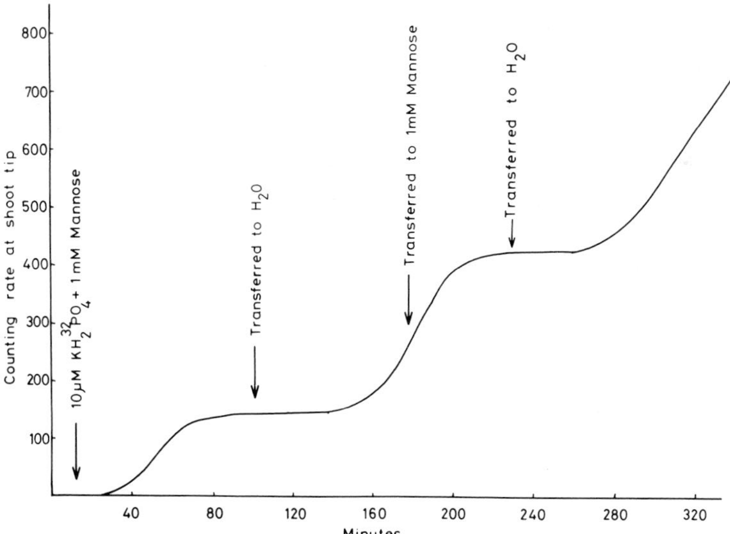

Fig. 2. The effect of mannose on phosphate movement in maize seedlings.

Maize seedlings (12 day) were raised and treated at 20 °C and 12 700 lx. 0.50 cm² of the blade 1 cm from the tip of the 1st leaf was continuously monitored with a GM tube attached to a recording rate meter.

Long distance transport in marine algae such as *Laminaria digitata* also appears to be *via* a metabolic system in which $KH_2{}^{32}PO_4$ applied to the lamina is incorporated into organic forms and it is claimed that the transported form in the conducting tissue is inorganic orthophosphate [10].

SPECIES DIFFERENCES

Reference has already been made to the way in which species differ in their response to environmental changes designed to modify phosphate utilization. Although capacities for absorption by roots are similar for a wide range of species of monocots and dicots the capacity for upward transport in young seedlings can vary by 50-fold (Table 1). Clarke and Brown [11] showed that capacities for phosphate absorption differ markedly between inbred strains of maize and the more efficient strains grow better on low phosphate soils. The low transport capacity in peas and field beans can be rapidly increased by deseeding and it appears that the mechanism is functional in these species but is maintained at a low rate by the presence of the seed.

INTERACTION WITH OTHER NUTRIENTS

An adequate supply of the other essential plant nutrients is required for efficient absorption of phosphate. The effects of divalent ions such as calcium and of minor nutrients such as boron illustrate this point. With regard to the total amount of nutrients it has been shown by Petterson [12] that the ionic strength of the medium in which the root grows can have significant effects on the amount of phosphate absorbed. The specific requirement for calcium in maintaining the integrity of cell membranes is shown by experiments of Hyde [13] where the rate

Table 1. The capacity for absorption and transport of phosphate in 12-day old plants of seven species treated with 10 μM $KH_2{}^{32}PO_4$ at pH 5.5 at 20 °C and 12700 lx.

	Absorption rate $\mu gP/g^{-1}/h^{-1}$	Per cent transported to shoot in 4 h
Oat (var. Blenda)	8.32	30
Barley (var. Proctor)	7.37	30
Maize (var. INRA 200)	3.37	18
Sunflower (var. Polestar)	6.78	24
Mung bean	6.17	20
Pea (var. Alaska)	2.48	1.2
Field bean (var. Equina)	5.30	0.45

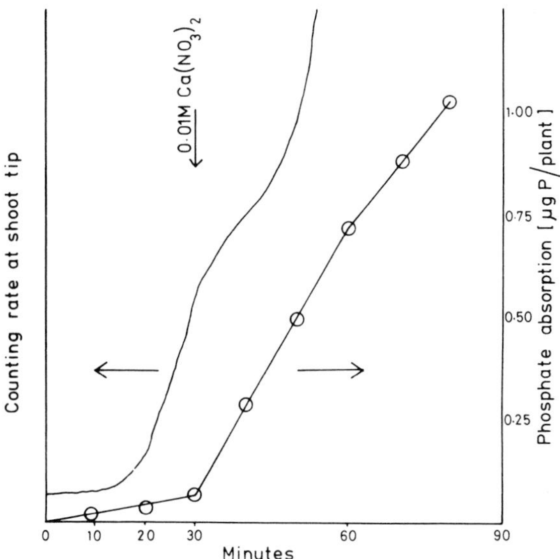

Fig. 3. The effect of calcium ions on the absorption and transport of phosphate by 12 day-old maize seedlings.

Plants raised in full culture solution for 10 days followed by 2 days in Ca-free solution. After 30 minutes in 1 µM $KH_2{}^{32}PO_4$ at pH 5.5, 20 °C and 12700 lx, 0.01 M $Ca(NO_3)$ was added to the solution. Simultaneous measurements of absorption and transport were carried out, the former by removal of samples of culture solution and the latter by continuous recording of the arrival of ^{32}P at the shoot tip.

of absorption by barley plants of phosphate from solutions of low concentrations (0.1–0.01 µM) in the absence of calcium can be rapidly increased on addition of 1 mM $Ca(NO_3)_2$. This stimulation results from a number of effects, both physical and metabolic, operating at the membrane and cell wall but it represents a classic interaction between the two ions. The four fold activation of absorption by the roots of young maize plants is not accompanied by similar effects on the transport to the xylem (Fig. 3).

The progressive loss of capacity for phosphate absorption is one of the first detectable results of witholding boron from the environment of the roots of a number of species. Almost immediate restoration of this capacity can be achieved by the addition of 10 µM H_3BO_3. Boron appears to be involved in the mechanism by which phosphate is absorbed and the rapidity of the response suggests that the membrane transport system is directly affected. Pollard et al. [14] showed that maize plants that have been maintained on adequate supplies of boron still respond to the addition of 10 µM H_3BO_3 to the medium from which they take up phosphate. The results indicate that normal plants operate at only about two thirds of their potential capacity for absorption and that this capacity is decreased by boron deficiency and increased by addition of H_3BO_3.

The environment within the cell is also important and the endogenous levels of both organic and inorganic phosphates are important factors in the control of metabolism. It is probable that in normal cells the greater part of the inorganic phosphate is confined to the vacuole. The control of cytoplasmic reactions may therefore be brought about by an amount of inorganic phosphate representing less than 1 per cent of the total in the cell [8, 16]. Large changes in the overall level of phosphate brought about by starvation or luxury feeding have relatively little effect on the cytoplasmic level thus confirming the importance of the role of this cytoplasmic component. The control of transport of orthophosphate into the chloroplast after reaching the leaf cells is an important example of the cytoplasmic level exerting influence over the functioning of a specific organelle and a number of processes including starch synthesis are governed in this way.

CONCLUSIONS

Fertilizers are becoming increasingly expensive and although agricultural soils in general are well supplied with reserves of phosphate it is necessary to examine the mechanism of utilization by crops if the most efficient use is to be made of applied fertilizers. The fact that capacities for absorption and transport of phosphate appear to be genetically controlled opens up possibilities for producing varieties that absorb more phosphate from low concentrations, under alkaline or acid conditions, at low temperature or in the presence of subnormal or toxic amounts of other elements. The important role of mycorrhizal fungi and other microorganisms is becoming increasingly clear both in terms of making phosphate available in the soil and in the process of active absorption. The fact that the function of metabolic control within the cell by inorganic orthophosphate is provided for by a very small proportion of the total ion intensifies the importance of its contribution to plant growth and development. Information concerning the control of transfer in both directions across the plasmalemma, tonoplast, chloroplast and mitochondrial membranes is slowly emerging and it is to be hoped that agronomic use may eventually be made of at least some of the findings.

REFERENCES

1. Bieleski, R. L. and Laties, G. G. 1963 Turnover rates of phosphate esters in fresh and aged slices of potato tuber tissue. Plant Physiol. 38, 568–594.

2. Biswas, B. B., Biswas, S., Chakravarky, S. and De, B. P. 1978 A novel metabolic cycle involving *myo* inositol phosphates during formation and germination of seeds. In: Eisenberg, F. and Wells, W. W. (eds.), Cyclitol and Phosphoinositides, New York, Academic Press.
3. Clark, R. B. and Brown, J. C. 1974 Differential phosphorus uptake by phosphorus-stressed corn inbreds. Crop Sci. *14,* 505–508.
4. Cockburn, M., Earnshaw, P. and Eddy, A. A. 1975 The stoichiometry of the absorption of protons with phosphate and L-glutamate by yeasts of the genus Saccharomyces. Biochem. J. 146, 705–712.
5. Floc'h, J,-Y. and Penot, M. 1976 Etude comparative du transport a longue distance de differents radioelements dans le thalle de *Laminaria digitata.* C.R. seances Acad. sci., serie D, Paris, 282, 989–992.
6. Hyde, A. H. 1966 Nature of the calcium effect in phosphate uptake by barley roots. Plant & Soil 24, 328–332.
7. Larkum, A. W. D. and Loughman, B. C. 1969 Anaerobic phosphate uptake by barley plants. J. Experim. Botany 20, 12–24.
8. Loughman, B. C. 1960 Uptake and utilization of phosphate association with respiratory changes in potato tuber slices. Plant Physiol. 35, 418–424.
9. Loughman, B. C., 1978 Metabolic factors and the utilization of phosphorus by plants. In: Porter, R. and Fitzimons, D. W. (eds.), Phosphorus in the environment: its chemistry and biochemistry, pp. 155–174. Amsterdam and New York: Elsevier/Excerpta Medica/North Holland.
10. Loughman, B.C. and Russell, R.S. 1957 The absorption and utilization of phosphate by young barley plants. J. Experim. Botany 8, 280–293.
11. Pettersson, S. 1975 Effects of ionic strength of nutrient solutions on phosphate uptake by sunflower. Physiologia Plantarum 33, 224–228.
12. Pollard, A. S., Parr, A. J. and Loughman, B. C. 1977 Boron in relation to membrane function in higher plants. J. Experim. Botany 28, 831–839.
13. Ullrich, C., Novacky, A., Fischer, E. and Luttge, U. 1978 Driving forces of phosphate transport in *Lemna gibba,* presented at SEB/FESPP Inaugural Meeting, Edinburgh, July 9–14.

48. Accumulation of cations and the functional activity of mitochondria from roots

M. G. ZAITSEVA

K.A. Timiriazev Institute of Plant Physiology, Academy of Sciences, Moscow 127 276, U.S.S.R.

The close relationship existing between ion transport and metabolic states of mitochondria has been demonstrated recently with intact renal tubules specialized for this process in mammalian tissues [1]. Ion transport in plants is carried out by roots and mitochondria are known as organelles potent for ion transport and ion accumulation. Therefore the accumulation and the loss of cations related to changes in the metabolic activity of mitochondria isolated from the roots of wheat (*Triticum aestivum* L.) have been studied [4].

The data obtained indicate that the metabolic state transitions of mitochondria are accompanied by cation movements. In the absence of substrate there was a loss of calcium and magnesium from mitochondria, whereas the

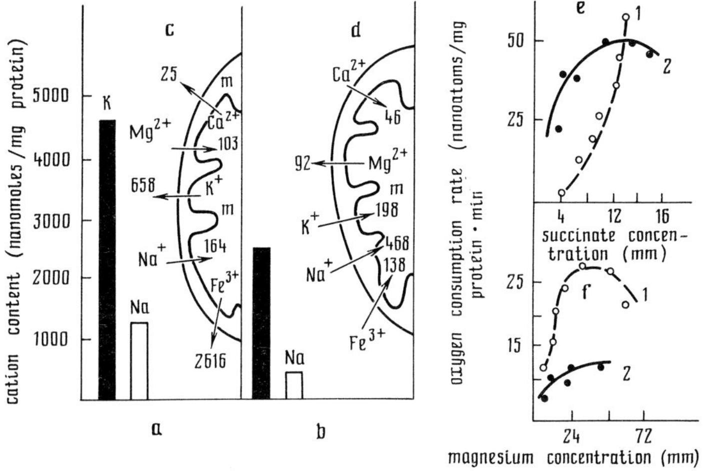

Fig. 1. Cation movements and respiration in mitochondria with different content of the monovalent cations. a, b – potassium and sodium content; c, d – the cation movements related to the state 3; m – mitochondria.

Figures represent the amounts of cations in nanomoles/mg of protein transported in and out of mitochondria. e,f – the effect of succinate and magnesium concentration on the respiratory rate; – mitochondria with high (1) and low (2) monovalent cation content.

R. Brouwer et al. (eds.), Structure and Function of Plant Roots, 247–248.

addition of succinate (substrate state 4) and substrate with ADP (State 3) [2] induced the movements of cations (K^+, Na^+, Ca^{2+}, Mg^{2+} and Fe^{3+}) either from the medium to mitochondria or in the opposite direction. The direction of the cation flow as well as the effect of external magnesium and the substrate concentration on mitochondrial activity seemed to be dependent on the intramitochondrial pool of monovalent cations (Fig. 1).

These data suggest the possible role of cations previously accumulated by mitochondria in the regulation of cation flow which may exist between the organelles and cytosol.

The ion movements may be influenced as well by chelating substances. Added EDTA affected the direction as well as the average rates of cation movements (Ca^{2+}, K^+ and Fe^{3+}). These observations raise the possibility that the accumulation of chelating substances such as organic acids and nucleotides in a cell may induce an increased exchange of cations between the respiring mitochondria and cytoplasm.

It should be mentioned that the movements of cations were rapid enough for as much as $\frac{3}{4}$ of the mitochondrial pool to be transported in 5–10 min. Since mitochondrial calcium and magnesium make up the bulk of the intracellular pool [3] it seems likely that the movements of cations related to metabolic state transitions of mitochondria could play a significant role in the mechanisms mediating the interaction of various calcium- and magnesium-dependent processes. It is also possible that ion transport into the cell and respiration itself are affected by these movements of cations.

REFERENCES

1. Balaban, R. S., Mandel, L. J., Soltoff, S. P. and Storey, J. M. 1980 Coupling of active ion transport and aerobic respiratory rate in isolated renal tubules. Proc. Natl. Acad. Sci. U.S.A. 77, 447–451.
2. Chance, B. and Williams, G. R. 1956 The respiratory chain and oxidative phosphorylation. In: Advances in Enzymology 17, pp. 65–134. New York, Interscience Publishers.
3. Thiers, R. E. and Vallee, B. L. 1957 Distribution of metals in subcellular fractions of rat liver. J. Biol. Chem. 226, 911–920.
4. Zaitseva, M. G. and Zubkova, N. K. 1979 Accumulation and losses of cations by mitochondria in changing metabolic states. Fiziol. Rasten. 26, 1085–1092.

49. Absorption and transport of sulphate in relation to its metabolism

M. HOLOBRADÁ

Institute of Experimental Biology and Ecology, Slovak Academy of Sciences, 885 34, Bratislava, Czechoslovakia

Ion uptake and accumulation by the root system as well as upward transport depend to a large degree on the metabolic state of cells and on their capacity to take up various ions. Sulphate uptake by the root is evidently controlled by the level of sulphur and its intermediates in plant cells. The close relation of sulphate transport and the rate not only of sulphur metabolism, but of the cell metabolism as a whole becomes a factor determining to a considerable extent the uptake of sulphur, its distribution in the roots and transport to the photosynthesizing organs. The changes in sulphur nutrition of plants manifest itself not only in changes in root morphology, but also in cellular metabolism.

A lower content of sulphur and of its metabolites in S-deficient cells of cotton and maize plants induces an intensive uptake of ^{35}S-sulphate. S-deficient cotton plants, 7 weeks old, increased their ^{35}S uptake after 30 min of ^{35}S-sulphur treatment (Fig. 1). Most absorbed sulphur was retained in roots. S-deficient maize plants, 3 weeks old, also exhibit a marked stimulation of uptake of ^{35}S (Fig. 2), and retain it in the root.

In contrast to S-starvation, a higher concentration of sulphate in the medium (0.3 per cent Na_2SO_4 for maize and 0.6 per cent Na_2SO_4 for cotton) resulted in a reduction of uptake and of the distribution of sulphur. In cotton only leaves, organs with the most intensive sulphur metabolism, do not differ in sulphur accumulation between control and −S-plants (Fig. 1A). In roots, the retention of absorbed ^{35}S is much lower than that of control and S-deficient plants as well (Fig. 1). High concentrations of sulphur in plant cells resulted also in maize plants in an inhibition of ^{35}S uptake, but the transport into leaves (Fig. 2) was severely reduced.

The present results show evident differences between the sensitivity of maize and cotton to nutrient stress by adding sulphate to the medium. It has been noted that in maize plants a 2-week treatment with sulphate induces a decline in the uptake and transport of sulphur (Fig. 2), while in cotton a 6-weeks treatment with a double sulphate concentration is needed to decrease uptake and transport of added ^{35}S-sulphate (Fig. 1A).

Reduction of ^{35}S-sulphate to soluble S-amino acids in cotton is stimulated by

R. Brouwer et al. (eds.), Structure and Function of Plant Roots, 249–252.

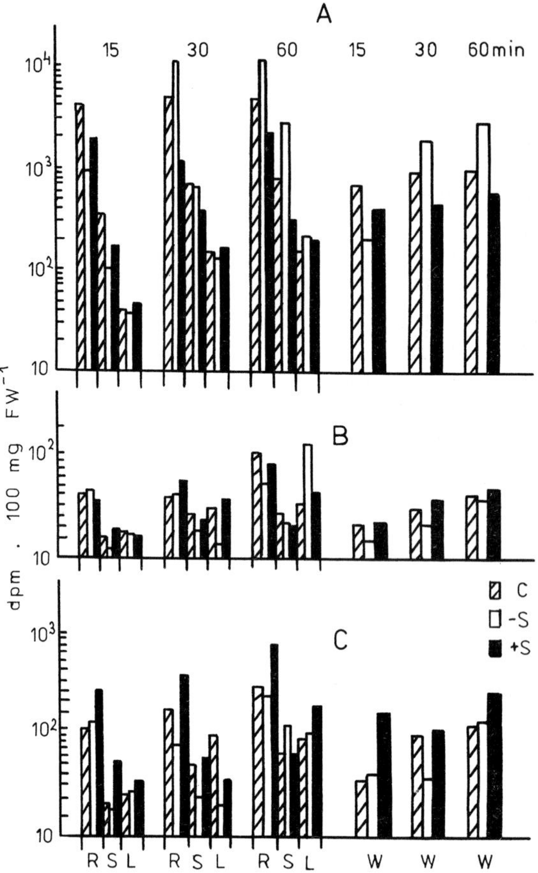

Fig. 1. Uptake and assimilation of ^{35}S-sulphate in cotton plants after 15, 30 and 60 min incubation on ^{35}S labelled nutrient medium. A – total uptake of ^{35}S, B – sulphate reduced into soluble S-amino acids, C – sulphate reduced into S-proteins.

Symbols: C: control plants, –S: S – deficient plants, +S: high sulphur plants, R – roots, S – shoots, L – leaves, W – whole plants.

the excess of sulphate (Fig. 1B), but reduction in maize is inhibited (Fig. 3B). The pattern of synthesis of S-amino acids after S-deficiency varies; stimulation occurs in maize (Fig. 3B) and a depression in cotton (Fig. 1B). Whereas in maize plants

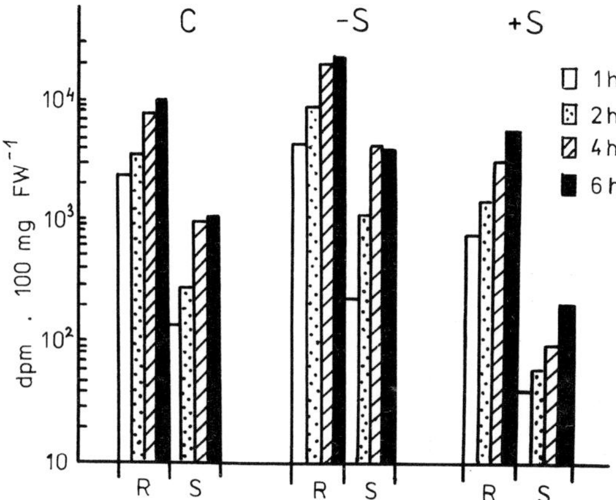

Fig. 2. Uptake and distribution of ^{35}S-sulphate in maize plants after 1, 2, 4 and 6 h incubation on ^{35}S-sulphate.

Symbols: C: control plants, − S: S – deficient plants, + S: high sulphur plants, R – roots, S – shoots.

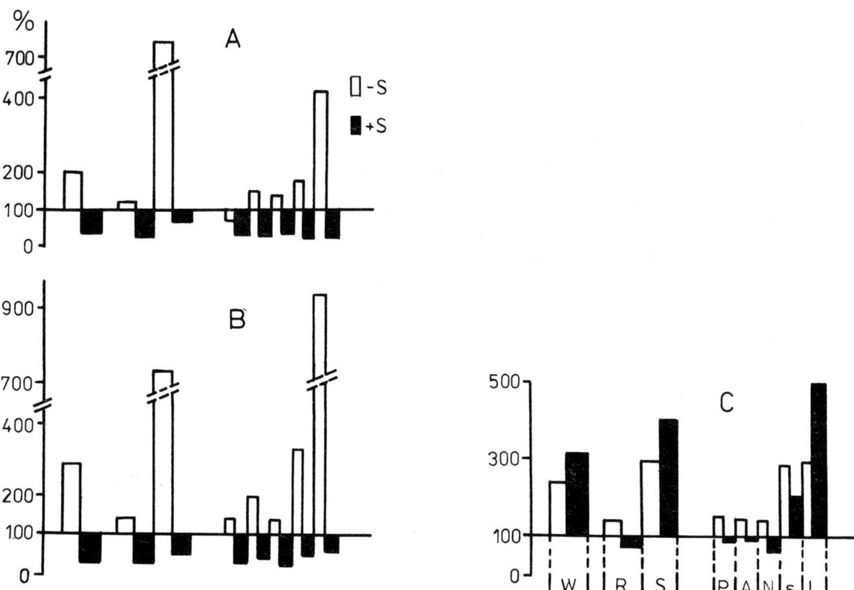

Fig. 3. Uptake and assimilation of ^{35}S-sulphate in maize plants stressed by S-deficiency (− S) and high concentration of sulphur (+ S) in relation to control plants. A – total uptake of ^{35}S, B – sulphate reduced into soluble S-amino acids, C – sulphate reduced into proteins.

Symbols: − S: S-deficient plants, + S: high sulphur plants, R – roots, S – shoots, P – primary seminal root, A – adventitious seminal roots, N – nodal roots, s – shoots without leaves, L – leaves.

both deficit and excess of sulphur induce ^{35}S-protein synthesis, apart from a depression in +S roots (Fig. 3C), cotton plants increase their protein synthesis only in cells with high levels of sulphur (Fig. 1C).

Differential uptake of sulphur by maize and cotton and the range of its metabolic assimilation as a response to deficit or excess of sulphate demonstrate, the close relation between metabolism, the processes of absorption and transport of ions.

50. The effect of phenolic acids on metabolism and nutrient uptake of roots

F. POSPÍŠIL and M. ŠINDELÁŘOVÁ

Institute of Experimental Botany, Czechoslovak Academy of Sciences, 160 00 Praha 6, Czechoslovakia

During the microbial degradation of lignin and humus phenolic acids are formed. Vanillic, p-hydroxybenzoic and p-coumaric acids occur most frequently in soil while ferulic and protocatechuic acids occur to a lesser extent. Phenolic acids are assimilated by plants and influence plant metabolism. Vanillic and protocatechuic acids inhibit the activity of glucose-6-phosphatedehydrogenase in the excised roots of pea whereas gallic acid stimulates the activity of this enzyme in the roots of whole plants.

We investigated the effect of phenolic acids on the growth, nitrate uptake, induction of nitratereductase (NR), and glutaminsynthetase (GS) activity in the roots of pea seedlings. Polarograph methods were used to follow oxidation of phenolic acids.

Three-day old germinating pea seeds (*Pisum sativum*, cv. Raman) were transferred to Richter's solution either with nitrate when growth was investigated or without nitrate when induction of NR and GS activity and nitrate uptake was estimated. After 4 days the nutrient solution was replaced by a new one containing both nitrate and phenolic acids (10^{-3} M). The seedlings were grown under constant conditions (light period 16 h, illumination 7000 lx, tem. $21 \pm 25\,^{\circ}$C).

As is shown in Fig. 1a, the inhibitory effect of phenolic acids on root growth is evident with gallic, p-hydroxybenzoic acid and vanillic acid being most effective. Nitrate uptake was estimated as increasing content of the nitrate in the roots after the nitrate had been added to the solution. All acids except protocatechuic acid decrease nitrate uptake (Fig. 1b), induction of NR (Fig. 1c) and GS activity (Fig. 1d). The results show that phenolic acids inhibit the nitrate uptake and the overall metabolic activity.

Figs. 2 and 3 show the dependence of the half-wave potentials of gallic and protocatechuic acids on pH. The anodic oxidation of phenolic acids on a rotating platinum disc electrode (RPDE) yielded two waves at pH 2.9–6.5 (Figs. 4, 5). This is caused by strong adsorption of the substances on the electrode. The uptake of gallic and protocatechuic acids from the nutrient solution and the formation of

R. Brouwer et al. (eds.), Structure and Function of Plant Roots, 253–257.

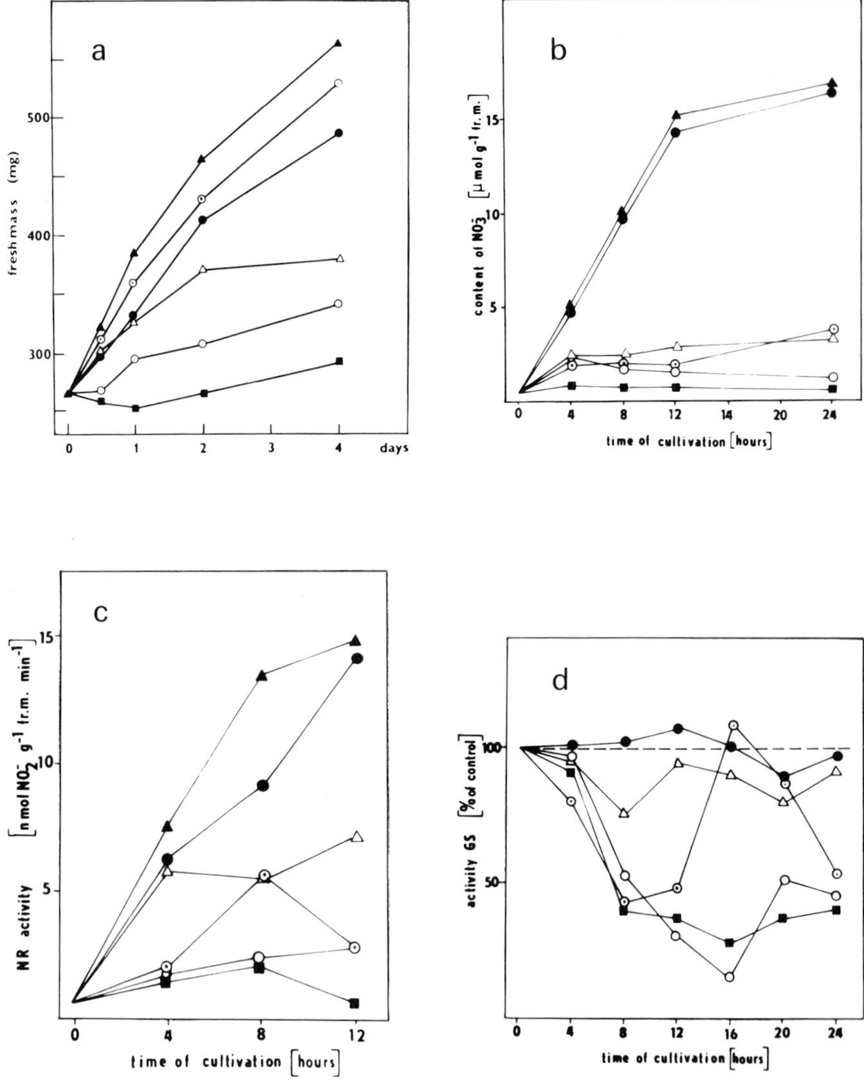

Figs. 1a–d. The effect of phenolic acids (10^{-3} M) on the growth of the roots (a) on the nitrate uptake, (b) on the induction of NR, (c) and on the GS activity, (d) in the roots of pea seedlings. ▲ control, ● protocatechuic acid, △ gallic acid, ⊙ 2,4-dihydroxybenzoic acid, ○ p-hydroxybenzoic acid, ■ vanillic acid.

products (quinones, dimers) on the surface of the roots was investigated. The polarogram at a RPDE (Figs. 6, 7) shows the reduction of the product formed.

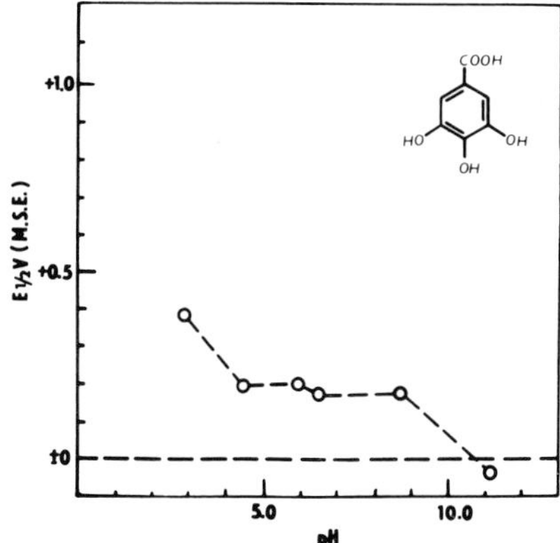

Fig. 2. Dependence of the half-wave potential of gallic [2] and protocatechuic acid [3] on pH. RPDE (vs. M.S.E.), Britton-Robinson buffers.

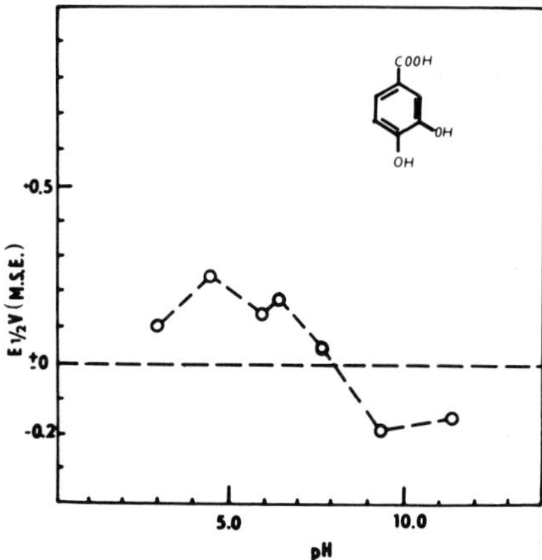

Fig. 3. For caption see Fig. 2.

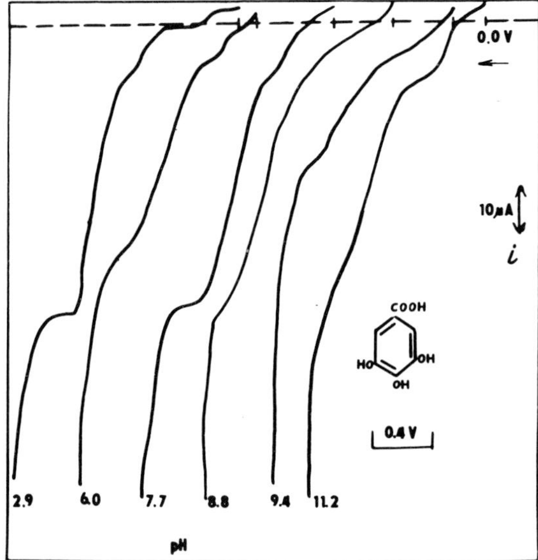

Fig. 4. Polarogram of gallic [4] and protocatechuic acid [5] at pH 2.9–6.5. Britton-Robinson buffers. RPDE (vs. M.S.E.).

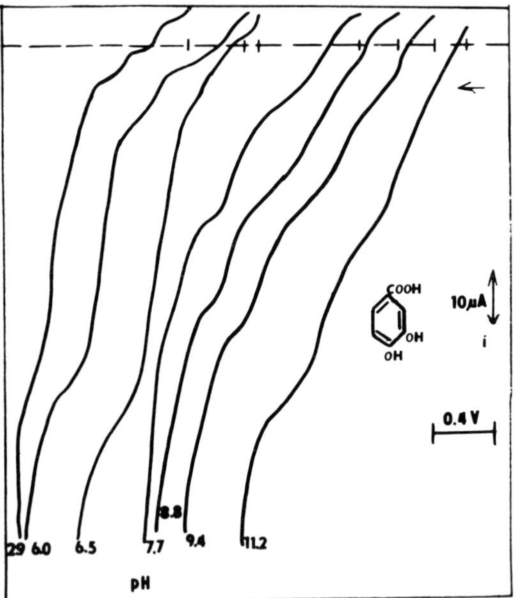

Fig. 5. For caption see Fig. 4.

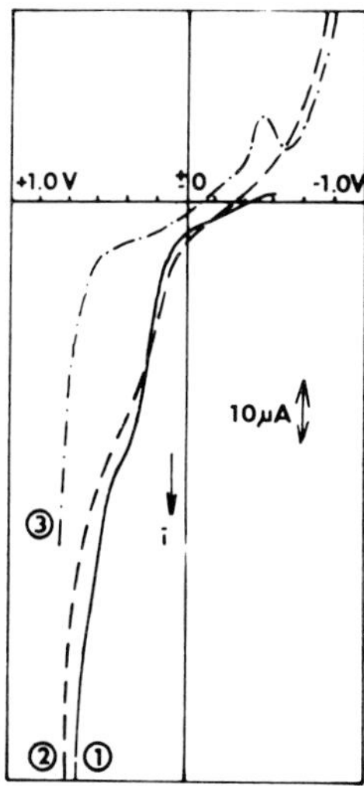

Fig. 6. Polarogram of gallic acid in the B.R. buffer at pH 6.5 (6.1) in nutrient solution, in the presence of pea roots (6.2) and polarogram of 0.1 N KOH extract of roots (6.3). Polarogram of protocatechuic acid at pH 11.2 (7.1) and polarogram of 0.1 N KOH extract of roots (7.2).

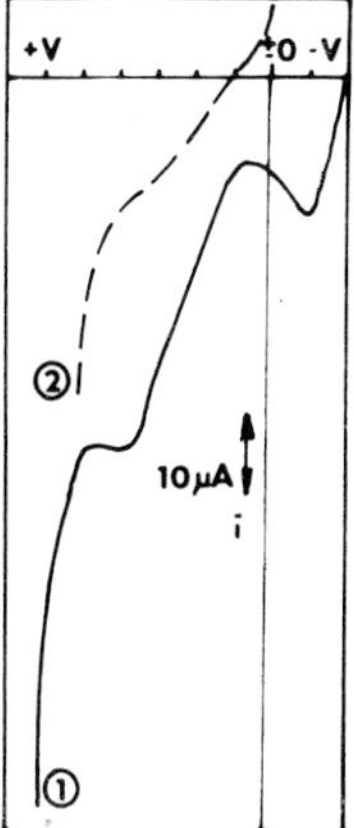

Fig. 7. For caption see Fig. 6.

51. Effects of endomycorrhizal infection on phosphate and cation uptake by *Trifolium subterraneum*

SARAH E. SMITH, F. ANDREW SMITH and D. J. DONALD NICHOLAS

Departments of Agricultural Biochemistry and Botany, University of Adelaide, Adelaide 5001, Australia

Vesicular-arbuscular mycorrhizal (V.A.M.) fungi improve the growth of many plants when the supply of phosphate (P) from the soil is low. At high soil P levels, when the P supply no longer limits plant growth, mycorrhizal growth increases do not usually occur [1, 6]. However, alterations in root: shoot dry weight ratio and, occasionally, reductions in plant dry weight are observed [7]. In the first part of this chapter we describe measurements of P inflow (moles P absorbed per unit root length per unit time [5]) into mycorrhizal (+ M) and non-mycorrhizal (− M) *Trifolium subterraneum* cv. Mt. Barker, over a range of P concentrations. In contrast to the large amount of information about effects of V.A.M. fungi on P nutrition, little is known about the effects on levels of inorganic cations [3, 7]. In the second part of the chapter we describe measurements of K and Na concentrations in + M and − M plants grown under a variety of conditions.

PHOSPHORUS INFLOW INTO MYCORRHIZAL AND NON-MYCORRHIZAL *T. SUBTERRANEUM* AT DIFFERENT LEVELS OF SOLUBLE PHOSPHATE

Root and shoot dry weight, root length, phosphate uptake, nodulation, nitrogenase activity (C_2H_2 reduction) and mycorrhizal infection of pot-grown *T. subterraneum* plants were measured at 3 harvests up to 5 weeks from planting. Phosphate was supplied as Na_2HPO_4 at 0, 0.1, 0.27, 0.6 and 1.0 mequiv. P. per kg soil.

Mycorrhizal plants were raised in a mixture (1 : 10) of non-sterile soil and steamed sand. Indigenous V.A.M. fungi (chiefly *Glomus mosseae*) colonized the roots rapidly, and by 5 weeks there were 1–2 m mycorrhizal roots per plant, irrespective of P fertilization. The % of the root system infected varied considerably between P treatments, as a function of increased root growth at high soil P. Non-mycorrhizal plants were grown in a mixture of γ-irradiated soil and sand

R. Brouwer et al. (eds.), Structure and Function of Plant Roots, 259–265.

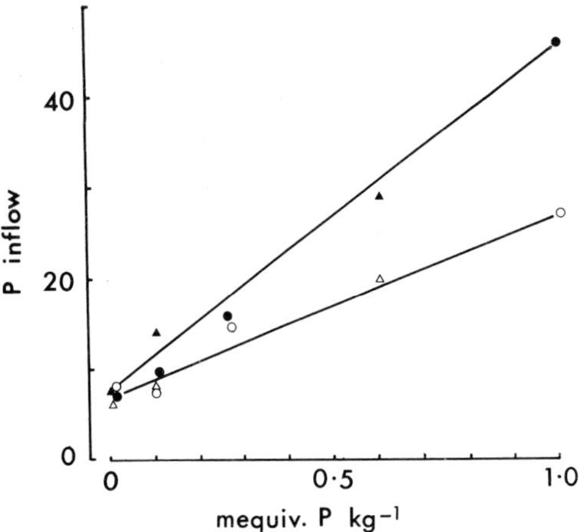

Fig. 1. Phosphate inflow (mol P $cm^{-1}s^{-1} \times 10^{-15}$) into +M (closed symbols) and −M (open symbols) roots of *T. subterraneum* as a function of added Na_2HPO_4. Phosphate content was measured at 22 days in two experiments (circles and triangles).

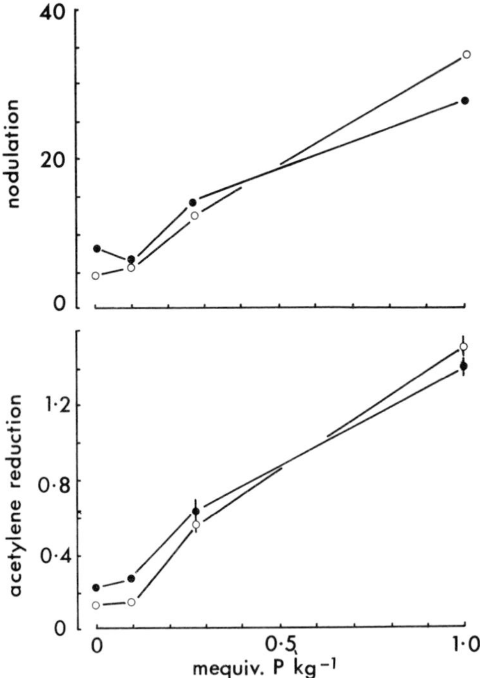

Fig. 2. Upper: nodulation (mm^3 nodule per plant); lower: nitrogenase activity (μmol C_2H_2 reduced per plant per hour); both as a function of added Na_2HPO_4 after 5 weeks growth. Values are means ± standard errors of the means for +M (●) and −M (○) plants.

(1 : 10). All plants were inoculated with *Rhizobium trifolii* and regularly received nutrient solutions (−N and P). Plant growth was increased by P fertilization and +M plants grew better than −M plants at the lower levels of soil P (0, 0.1 and 0.27 mequiv. kg^{-1}) but had fewer leaves and lower dry weight at 0.6 and 1.0 mequiv. kg^{-1} (S.E. Smith, unpublished).

Fig. 1 shows that the inflow of P into clover roots had a linear relationship with soil P supply between 0 and 22 days. There was no evidence that uptake saturated over the concentration range used and +M roots had higher inflows over the whole of this range. P inflow was lower at later harvests, probably because a large proportion of soil P had been absorbed by the plants. Increased P inflow did not necessarily lead to increased shoot P concentrations, as reductions in root : shoot ratio in +M plants and at high soil P led to lower total plant uptake. Retention of P in +M root systems also occurred.

Nodulation and nitrogenase activity are shown in Fig. 2. Fertilization with P increased both these processes, but mycorrhizal effects were relatively small; significant increases in nitrogenase activity were only apparent at 0 and 0.1 mequiv. P kg^{-1}, i.e. where shoot growth was also increased.

EFFECTS OF ENDOMYCORRHIZAL INFECTION ON K AND Na LEVELS IN *T. SUBTERRANEUM*

T. subterraneum was grown as above, except that in some experiments soil was sterilized by autoclaving to give −M plants [6]. Measurements of root and shoot dry weight, per cent mycorrhizal infection, and K and Na (by flame photometry after digestion in HNO_3) were made at each harvest. Large positive growth responses in +M plants normally occurred after 3–4 weeks growth. Sometimes there was a temporary growth depression during the early stages of mycorrhizal infection, preceding the positive response. Addition of small amounts of soluble phosphate (NaH_2PO_4 or superphosphate at only 0.6 mequiv. P kg^{-1}) greatly increased growth of both −M and +M plants without decreasing mycorrhizal infection. In one experiment combined nitrogen as $(NH_4)_2SO_4$ or $NaNO_3$ was added to pots at 5 mequiv. N kg^{-1}; this is sufficient to decrease N_2 fixation [2]. There were 1 or 2 harvests: after $3\frac{1}{2}$–$4\frac{1}{2}$ weeks and after $5\frac{1}{2}$–9 weeks, and the results given below are grouped accordingly.

There were no consistent differences between roots of +M and −M plants with respect to either K or Na (Figs. 3A and 3B). At the first harvest, roots of +M plants grown in soil containing $(NH_4)_2SO_4$ had higher concentrations of K than the corresponding −M roots, but it is possible that these treatments are not comparable, due to slower nitrification in the sterilized soil (despite the addition

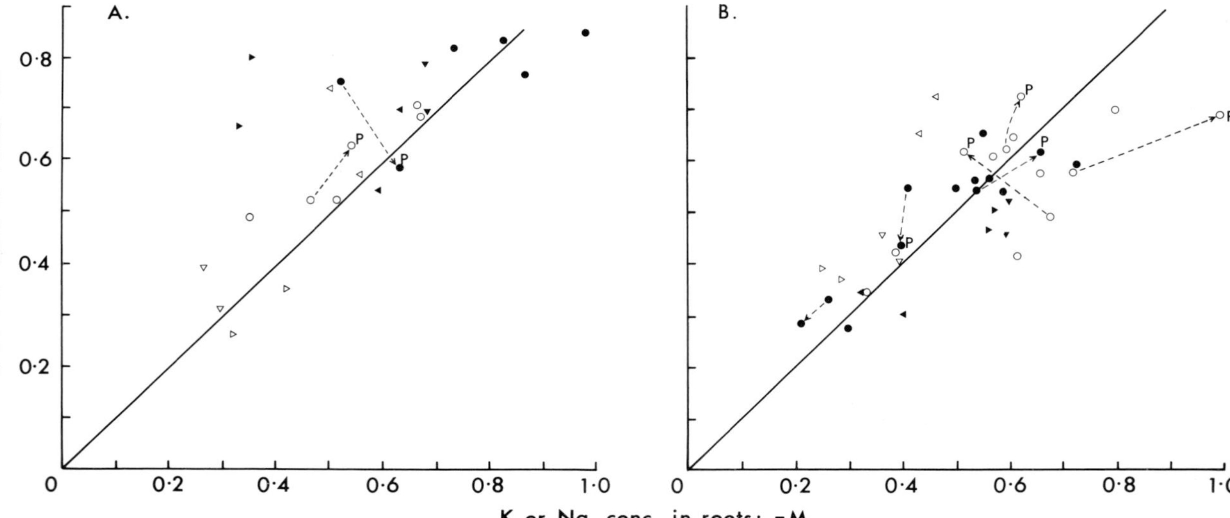

Fig. 3. Concentrations of K (closed symbols) and Na (open symbols), both as mmol g^{-1}, in roots of +M and −M plants at the first harvest (A) and second harvest (B). Circles represent data from different experiments. Triangles are results from an experiment where $NaNO_3$ (◀ ◁) or $(NH_4)_2SO_4$ (▶ ▷) was added to soil, with controls with no added combined N(▲ △). Points marked "P" are for plants grown in soil with added phosphate; dashed arrows start from the comparable treatments without added P.

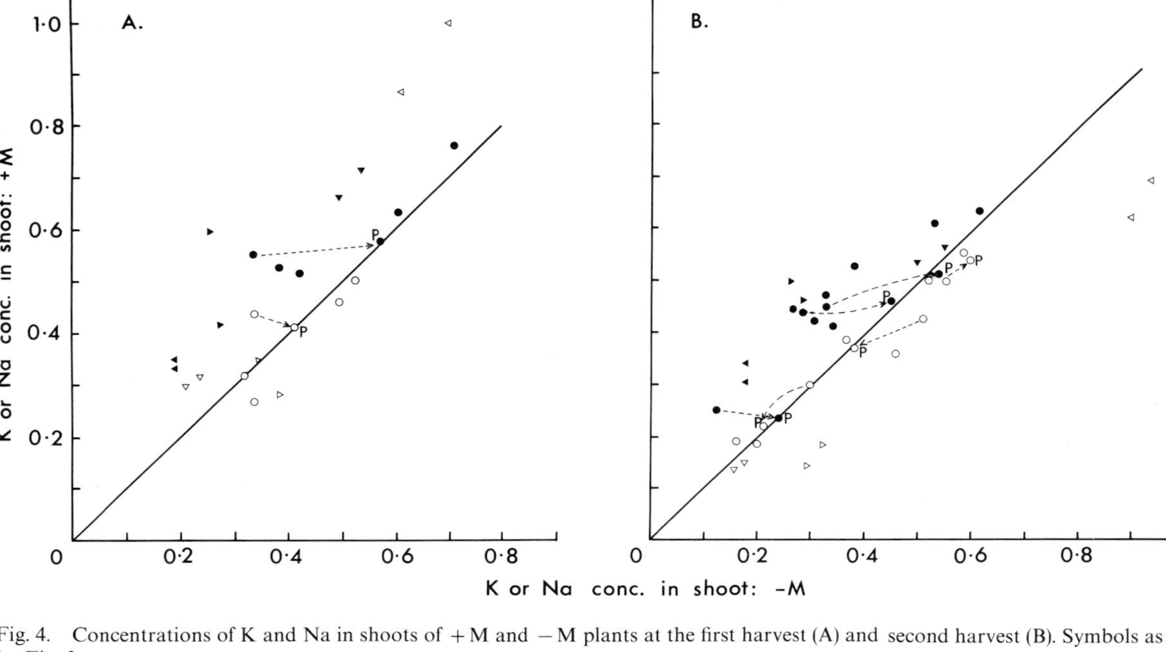

Fig. 4. Concentrations of K and Na in shoots of +M and −M plants at the first harvest (A) and second harvest (B). Symbols as for Fig. 3.

of soil filtrate containing bacteria). In other words, NH_4 might have remained relatively high in −M soil, resulting in an inhibition of K uptake [4]. This difference between +M and −M plants had disappeared by the second harvest at about 9 weeks (Fig. 3B). Addition of available P had no consistent effects on the K or Na concentrations in roots. In contrast, shoots of +M plants often had higher concentrations than the corresponding −M plants (Figs. 4A and 4B), particularly at the lower end of the range. When shoot K was high the difference usually disappeared. Addition of P increased the K concentration in shoots of −M plants to the level found in the corresponding +M plants, which were affected very little by addition of P. The Na concentrations in shoots of +M and −M plants were usually very similar (Figs. 4A and 4B). The main exception was the experiment in which $NaNO_3$ was added to the soil. This greatly increased the shoot Na concentration and lowered shoot K, especially at the first harvest, and at this harvest the shoots of +M plants contained more Na than those of the −M plants. At the second harvest, this latter situation was reversed. Average concentrations of K or Na were also calculated (mmol K or Na in shoots plus roots, divided by weight of shoots plus roots). On this basis, K concentrations in +M plants were still higher than those in −M plants, except where internal concentrations were high, or where P was added, in which cases there was no difference between +M and −M plants. Average Na concentrations were, however, little affected by mycorrhizal infection, as expected from Figs. 3 and 4.

These results show that mycorrhizal infection can improve the K nutrition of *T. subterraneum*, when internal K concentrations are generally low. The improved K nutrition appears mainly in the shoot tissues (Fig. 4A and 4B). In all the experiments the plants were grown in sand/soil mixtures with low available K (about 2 mmol kg^{-1}). The Na varied from 3 to 8 mmol kg^{-1}, depending upon the supply of sand used. This caused the wide range of Na concentrations and rather low K/Na ratios in the tissues. K/Na ratios were also affected by the addition of combined N, particularly $NaNO_3$ [2, 4].

The improved K nutrition associated with mycorrhizal infection may appear analogous to the effect of mycorrhizas on P nutrition: in both cases internal concentrations in +M plants are higher than in −M plants when external levels are low. In the case of P, uptake and translocation by the fungus followed by transfer to the root is thought to be important in maintaining rapid uptake rates. With K, however, there is as yet no need to postulate direct involvement of the fungus in absorption and translocation. Increased K uptake may be an indirect result of improved P nutrition. In our experiments plants with higher P concentrations (either +M or +P) were 'healthier' and showed less senescence of cotyledons and spade leaves. They had higher growth rates and could also have had higher rates of K uptake into the roots, leading to the observed increase in tissue concentrations. This was true even at the early ($3\frac{1}{2}$–4 week) harvest, before +M plants showed a positive increase in dry weight. In other words, comparison of K uptake into 'matched' +M and −M plants could not be made in the absence of added P. A positive growth response to added P in both +M and −M plants always occurred by the first harvest. These plants were better matched and differences in K and Na nutrition were usually negligible.

CONCLUSIONS

Mycorrhizal fungi increase the rate of phosphate uptake by roots (P inflow) over a range of soil P levels even when mycorrhizal growth increases no longer occur. It is likely that the fungi play a direct part in uptake and translocation of P to the roots. V.A.M. effects on nodulation and N_2 fixation are largely indirect, probably resulting from improved P nutrition and growth at low soil P levels. Work on inorganic cation nutrition is much less advanced, but it is already clear that there are interactions between P nutrition and cation uptake which may also be indirect. The pattern of N assimilation (N_2 fixation vis-à-vis $NaNO_3$ or $(NH_4)_2SO_4$ uptake) may modify cation/P interactions. Further work is required to distinguish cause and effect and to clarify the role played by V.A.M. fungi.

ACKNOWLEDGEMENTS

Financial support for this work is provided by grants from the Australian Research Grants Committee, the Rural Credits Development Fund (Reserve Bank of Australia) and the Wool Research Trust Fund.

REFERENCES

1. Abbott, L. K. and Robson, A. D. 1978 Growth of subterranean clover in relation to the formation of endomycorrhizas by introduced and indigenous fungi in a field soil. New Phytol. 81, 575–585.
2. Chambers, C. A., Smith, S. E. and Smith, F. A. 1980 Effects of ammonium and nitrate ions on mycorrhizal infection, nodulation and growth of *Trifolium subterraneum*. New Phytol. 85, 47–62.
3. Powell, C. L. 1975 Potassium uptake by endotrophic mycorrhizas. In: Mosse, B. and Sanders, F. E. (eds.), Endomycorrhizas, pp. 460–468. London and New York: Academic Press.
4. Raven, J. A. and Smith, F. A. 1976 Nitrogen assimilation and transport in vascular land plants in relation to intracellular pH regulation. New Phytol. 76, 415–431.
5. Sanders, F. E., Tinker, P. B., Black, R. L. B. and Palmerley, S. M. 1977 The development of endomycorrhizal root systems: 1. Spread of infection and growth-promoting effects with four species of vesicular-arbuscular endophytes. New Phytol. 78, 257–268.
6. Smith, F. A. and Smith, S. E. 1981 Mycorrhizal infection and growth of *Trifolium subterraneum*: use of sterilized soil as a control treatment. New Phytol., in press.
7. Smith, S. E. 1980 Mycorrhizas of autotrophic higher plants. Biol. Revs., in press.

IV. FUNCTIONAL INTEGRITY OF THE ROOT SYSTEM

52. Co-ordination of growth phenomena within a root system of intact maize plants

RIENK BROUWER

Botanical Laboratory, State University of Utrecht, Utrecht, The Netherlands

An intact higher plant represents an integrated system in which growth and function of various parts appear to be well-co-ordinated. Despite large efforts up to now our knowledge of the nature of the controlling systems is far from complete. Both nutritional and hormonal control mechanisms are assumed to play a part in the co-ordination. It has been shown repeatedly that in the intact plant any individual part grows and functions at rates which are less than their possible rates (capacities). Partial defoliation results in an increased photosynthesis of the remaining leaves as well as in an increased relative growth rate of leaf tissue often accompanied by an increase in absolute growth rate of growing leaves. Partial derooting leads to an enhanced absorption rate of the remaining roots and an increased relative root growth rate accompanied by an increase in absolute growth rate of individual root members. These results suggest that in the intact plant a constraint of growth and functioning induced by comparable parts or organs is a common phenomenon. We are studying these mutual effects of various parts of the root system of intact maize plants on each other by growing seminal roots and crown roots in separate compartments. Basically all compartments contain normal strength Hoagland-solution. Special treatments, such as different temperatures, plus or minus aeration and partial removal of part of the system are described in the next section.

RESULTS

Growth of the seminal roots

In undisturbed growth the seminal roots make a quick start but after emergence of the subsequent whorls of crown roots their growth rate slows down. If,

R. Brouwer et al. (eds.), Structure and Function of Plant Roots, 269–276.

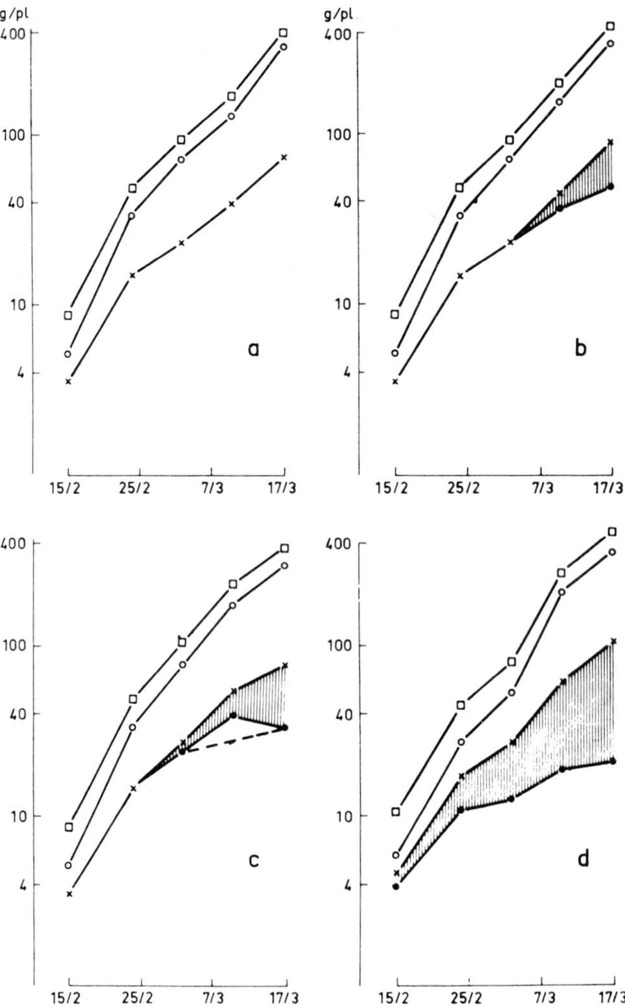

Fig. 1. Time course of weight increase of primary seminal roots (black dots), all roots (crosses), shoots (open dots) and total plants (squares) of maize plants from which the adventitious roots were cut at emergence during different periods: a throughout; b until March 3; c until February 24; d not at all.

however, the tips of the crown roots are cut immediately after emergence, the growth of the seminal roots continues resulting in an almost equal weight as the total weight of seminal and crown roots in the control plants (Figs. 1a and d). The inhibitory effect of crown root growth on the growth of the seminal roots can be evoked at any time during the plant's development (Figs. 1b and c) and the degree of inhibition depends on the number of crown roots which are allowed to grow. In all these treatments shoot growth is not affected significantly neither is the number of crown roots initiated [2]. These growth patterns fit with the assumption that a fixed amount of dry matter is available for root growth provided shoot growth is not affected. The privilege of the crown roots to be a more successful sink may just be a matter of position.

Rivalry between individual crown roots

When after germination the seminal roots are gradually removed in 3–4 days the growth of the seedlings is slightly reduced and the appearance of crown roots is

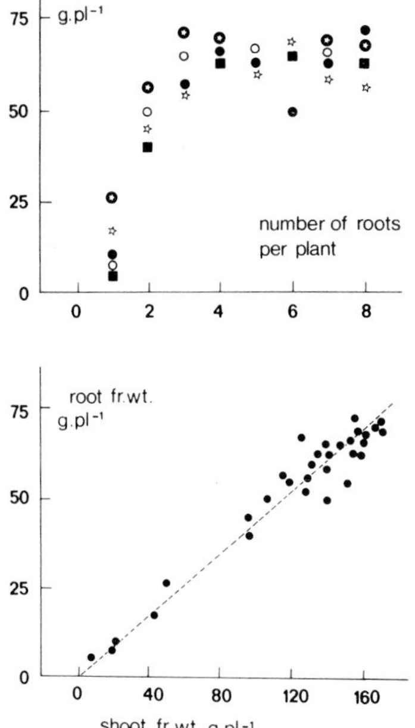

Fig. 2. Top: Weight of the root system of maize plants plotted against the number of crown roots which were left intact. The seminal roots and the remainder of the crown roots were cut at emergence. Bottom: The root weights plotted against the weights of the corresponding shoots.

accelerated [9]. Thereafter growth of the plants is resumed readily reaching quite normal relative growth rates (0.25–0.28 g g^{-1} d^{-1} at 22 °C). By cutting the tips of some of the emerging crown roots in the same way as described in the section on 'Growth of seminal roots' we obtained plants growing with 1 to 8 crown roots (Fig. 2). In all cases a linear relation between shoot weight and root weight exists. Both weights stay behind when only 1 or 2 crown roots are left to the plant but from 3 to 8 crown roots both total root weight and shoot weight are equal. Again a constant proportion of total weight is available for root weight and the data suggest that the distribution is merely a question of competition between the individual crown roots.

Effects of aeration

Root growth in maize is affected by aeration of the nutrient solution. In our experiments we varied the degree of aeration for all of the crown roots or for 50% of the crown roots (Fig. 3 bottom). The most serious reduction (no aeration at all) (Fig. 3 top) reduced root weight to about $\frac{1}{3}$ of the control and shoot weight to about $\frac{1}{2}$ of the control. When only half of the crown roots is treated in that way the reduction of root weight in the treated half is slightly more serious and the

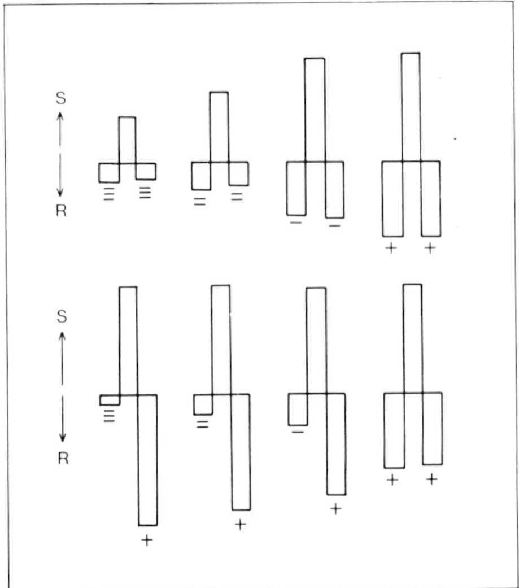

Fig. 3. Top: Shoot and root weight of maize plants grown with from right to left increasingly reduced aeration of both halves of a split root system. Bottom: The same but only one half of a split root system with reduced aeration, the other half normally aerated.

aerated half compensates completely for this reduction so that total root weight is comparable with that of the control plants (both compartments aerated). It is self-evident that shoot growth is not affected when compensatory root growth is possible. These results indicate that in the control plants roots are growing at a rate that is much slower than their possible growth rate. A lack of supply from the shoot may be considered as the reason for the reduction.

Effects of local temperatures

Maize plants grown on a range of temperatures of the root medium tend to show optimal shoot growth at a slightly higher root temperature than optimal root growth [8]. The low concentration of soluble sugars in the root tissue suggests that carbohydrates are limiting root growth [8], certainly at the relatively high root temperatures of 30 °C and 35 °C. Using the split root technique we compared the growth rate of root halves which were exposed to a range of root temperatures from 5 °C to 40 °C. Either both halves were exposed to the same

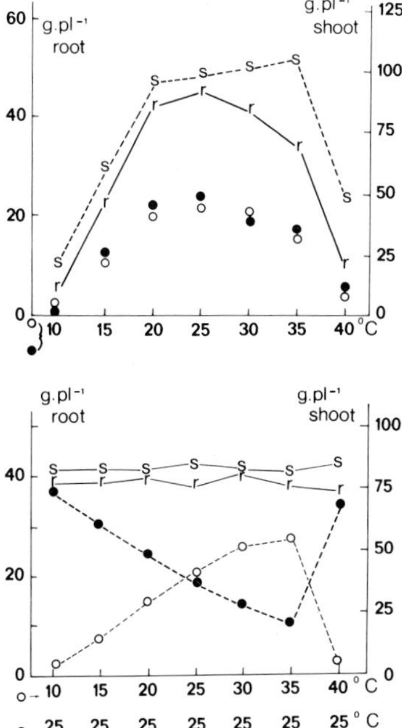

Fig. 4. Shoot and root weight of maize plants grown on a range of root temperatures. Top: Both halves of the split root system kept at the root temperatures indicated. Bottom: One half of the split root system kept at 25 °C, the other half treated with a range of root temperatures as indicated.

temperature or one half to 25 °C (supposed to be optimal) and the other half to the whole range. As might be expected from the results presented, when one half at 25 °C succeeds in draining the more dry matter in its direction the worse the condition in the other compartment is. It appears that 30 °C is slightly more favourable for root growth than 25 °C (Fig. 4).

Effects of the nitrogen supply on crown root growth

For these experiments we used an assembly in which the seminal roots and 4 individual crown roots could be supplied with solutions with or without nitrogen [3]. Before the crown roots started to grow, we had plants with N-deficiency (– N) and plants well-supplied with nitrogen (+N). In both kinds of plants one of the crown roots got the reverse treatment. Hence we have minus nitrogen plants with plus or minus nitrogen roots (Fig. 5 bottom) and plus nitrogen plants with plus or minus nitrogen roots (Fig. 5 top). The results indicate that the rate of root elongation depends on the nitrogen status of the plants and not on the supply of the individual root. The weight distribution along the axis, corresponding with the branching pattern, depends only on the N-supply in the medium of the individual root in case of a nitrogen deficient plant. If the plant is well-supplied with nitrogen both elongation rate and branching pattern of a root which lacks nitrogen nutrition in its medium are identical to those of a root with its own supply. Obviously these roots receive sufficient nitrogen from the internal supply [9].

CONCLUSIONS

The experimental results presented show that within the intact plant growing undisturbed a serious competition exists between various root members. When as a consequence of partial derooting or local inhibitory treatments shoot growth is not affected the remaining root members compensate completely for the reduction elsewhere. Competition for a limited supply of carbohydrates covers most of the observations [1]. However, since it has been shown that with the phloem stream various hormones are transported to the roots, some of which may affect root growth [4, 5], the ultimate solution needs additional experimentation [10]. In case of the differences in the rate of root elongation in nitrogen deficient plants and in plants well-supplied with nitrogen, the higher auxin content in the latter may be responsible for the lower elongation rate [5].

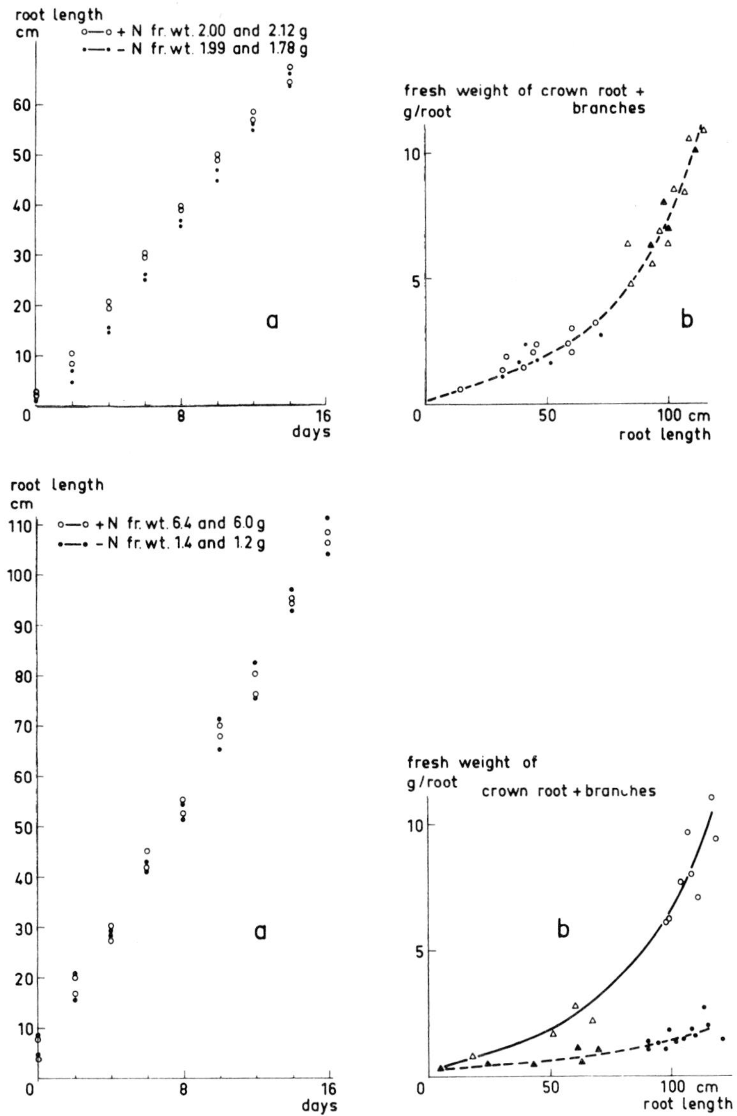

Fig. 5. Effect of nitrogen nutrition on growth of crown roots. Top: a. Length of crown roots plotted against time. Open dots refer to roots in Hoagland solution, filled dots to roots grown in a Hoagland solution in which nitrate has been replaced by chloride. Seminal roots in normal Hoagland solution. b. The relationship between fresh weight and length of individual crown roots plus their branches as dependent on the nitrogen supply in their environs. The plant as a whole is well-supplied with nitrogen (seminal roots). Some of the crown roots under investigation (closed symbols do not find nitrogen in their environs, others do. Bottom: a. As top a, but crown roots of a nitrogen deficient plant. b. As top b, but crown roots of a nitrogen deficient plant.

REFERENCES

1. Bar-Yosef and Lombert, J. R. 1979 Corn and cotton root growth in response to osmotic potential and oxygen and nitrate concentrations in nutrient solutions. In: Harley and Scott Russell (eds.), The soil root interface, pp. 287–299. London: Academic Press.
2. Brouwer, R. and Locher, J. Th. 1965 The significance of seminal roots in growth of maize. Jaarb. I.B.S. Wageningen 1965, 21–28.
3. Brouwer, R. and Loen, E. A. 1962 Growth and uptake of individual crown roots of *Zea mays* L. Jaarb. I.B.S. Wageningen 1962, 19–25.
4. Bruinsma, J. 1977 Root hormones and overground development. In: Plant regulators and world agriculture, pp. 35–47. Plenum Publ. Corp.
5. Burström, H. G. and Svensson, S. B. 1972 Hormonal regulation of root growth and development. In: Hormonal regulation and development. Proc. Adv. Study Inst. Izmir, Verlag Chemie, Weinheim.
6. Drew, M. C. and Saker, L. R. 1975 Nutrient supply and the growth of the seminal root system in barley. II. Localized compensatory increases in lateral root growth and rates of nitrate uptake when nitrate supply is restricted to only part of the root system. J. Exp. Bot. 26, 79–90.
7. Gile, P. L. and Carrero, J. O. 1917 Absorption of nutrients as effected by the number of roots supplied with the nutrient. J. Agric. Res. 9, 73–95.
8. Grobbelaar, W. 1963 Responses of young maize plants to root temperatures. Meded. L.H. Wageningen 63(5), 1–71.
9. Jager, A. de 1979 Localized stimulation of root growth and phosphate uptake in *Zea mays* L. resulting from restricted phosphate supply. In: Harley and Scott Russell (eds.), The soil-root interface, pp. 391–403. London: Academic Press.
10. Lambers, H. 1979 Energy metabolism in higher plants in different environments. Thesis Groningen, The Netherlands.

53. Relation between root respiration and root activity

B. W. VEEN

Centre for Agrobiological Research, Wageningen, The Netherlands

It is generally assumed that the major part of the energy derived from respiration in higher plants is used for growth and maintenance processes [2, 4, 6]. In roots a considerable part of the respiration is used for uptake and transport of ions [3], so that for a quantitative approach to the relation between root activity and root respiration three processes have to be taken into consideration: root growth, uptake and transport of ions and maintenance processes.

Root respiration and other root activities such as ion uptake and root growth mutually influence each other. For instance, a decrease in root respiration under low oxygen conditions has a negative effect on ion uptake, whereas decrease in ion uptake by low ion concentrations has a negative influence on root respiration.

The aim of the experiments is to quantify the respiratory costs of uptake and transport of ions, root growth, and maintenance processes of a maize root system. Maintenance respiration is considered to be linearly related to root fresh weight.

METHODS AND RESULTS

Simultaneous measurement of nitrate and potassium uptake, water uptake, root growth and root respiration [5] were carried out for a period of 7 days. Under our growing conditions as a rule the NO_3^- uptake rate in equivalents was twice the K^+ uptake rate.

Because NO_3^- and K^+ are the main elements taken up, NO_3^- is chosen as a measure of the total ion uptake. Root respiration could be separated into three components by multiple regression analyses, using root growth, ion uptake and root volume as the independent variables. To increase the relative differences between root growth, ion uptake and root volume, two days before the first experimental day part of the root system was cut away. Only the youngest 5 unbranched crown roots were left on the plant. For the same reason different light intensities were used as indicated in Table 2. Table 1 shows the oxygen consumption per unit of NO_3^- uptake, per unit of root growth and per unit of root volume.

R. Brouwer et al. (eds.), Structure and Function of Plant Roots, 277–280. All rights reserved.

Table 1. Oxygen consumption of a maize root in relation to ion uptake, growth and root size. (S_d = standard deviation)

Oxygen consumption per unit of:		S_d
ion uptake $mg\ O_2.meq.^{-1}.NO_3^{-1}$	36.8	2.1
root growth $mg\ O_2.g^{-1}$	24.5	1.2
root weight $mg\ O_2.g^{-1}.h^{-1}$	0.032	0.005

Table 2. Relative energy consumption of uptake, growth and maintenance processes

Day nr	Light int. $W.m^{-2}$	Per cent of total respiration used for		
		ion uptake	growth	maintenance
1	35	13	78	8
2	35	16	75	9
3	70	39	52	9
4	70	49	42	9
5	35	48	38	14
6	70	60	24	16
7	0	41	25	33

The uptake respiration amounts to 36.8 mg O_2 per meq. NO_3^-. The NO_3^- uptake is about 90% of the total anion uptake, so that on basis of unit anion uptake the oxygen consumption is 33.1 mg O_2.

Lundegardh [3] estimated the ratio

$$\frac{\text{equivalents absorbed anions}}{\text{mols consumed oxygen}} = 1$$

under optimal conditions for ion uptake. This means that in his experiments 32 mg O_2 was used per meq. absorbed ions, which is in good agreement with our estimation of 33 mg O_2/meq. anions. Table 2 shows the relative amounts of oxygen used for the three processes during the 7 experimental days. Although the influence of light intensity and root cutting are not strictly separated, the data show a relative large growth respiration and a relative small uptake respiration

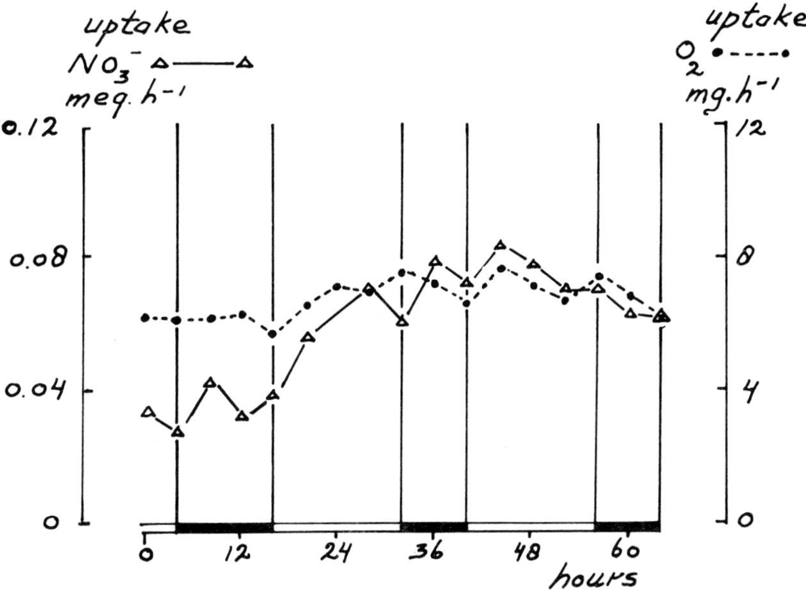

Fig. 1. Effect of changing the light period from 12 h to 16 h on NO_3 uptake and root respiration.

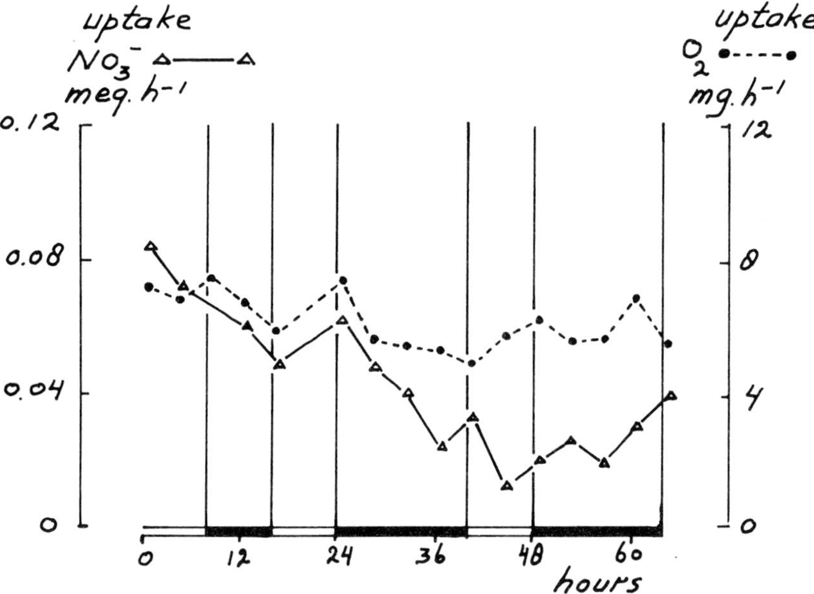

Fig. 2. Effect of changing the light period from 16 h to 8 h on NO_3 uptake and root respiration.

during the first days, because the root size limits the uptake rate [5], while root cutting induces a high relative root growth rate [1].

Uptake respiration is of considerable importance and amounts under normal conditions to 60 per cent of the root respiration.

The last experimental day, when no light was provided, the relative decrease in uptake respiration was greater than the influence on growth- and maintenance respiration.

Although an important part of the root respiration is used for ion uptake, its relative amount is not constant, and is influenced by the amount of light supplied to the shoot. This is demonstrated in the Figs. 1 and 2. Changing the light period from 12 h to 16 h per day, has a strong positive effect on the ion uptake, but only a small effect on root respiration (Fig. 1). Decreasing the light period from 16 h to 8 h per day has a strong negative effect on ion uptake, but only a small effect on root respiration (Fig. 2). A possible explanation for the effects of light on root respiration and ion uptake is, that after decreasing day length, a decreased carbohydrate availability reduces root respiration. Forced to decrease its activities, the uptake of ions by the root is given the lowest priority.

REFERENCES

1. Brouwer, R. and Kleinendorst, A. 1965 Effect of root excision on growth phenomena in perennial ryegrass. Meded. Jaarboek I.B.S. 280, 11–20.
2. Hansen, G. K. and Jensen, C. R. 1977 Growth and maintenance respiration in whole plants, tops and roots of *Lolium multiflorum*. Physiol. Plant 39, 155–164.
3. Lundegardh, H. 1949 Quantitative relations between respiration and salt absorption. Ann. Roy. agric. Coll. Sweden 16, 372–403.
4. Penning de Vries, F. W. T. 1974 Substrate utilization and respiration in relation to growth and maintenance in higher plants. Neth. J. agric. Sci. 22, 40–44.
5. Veen, B. W. 1977 The uptake of potassium, nitrate, water and oxygen by a maize root system in relation to its size. J. exp. Bot. 28, 1389–1398.
6. Wit, C. T. de, Brouwer, R. and Penning de Vries, F. W. T. 1970 The simulation of photosynthetic systems. In: Prediction and measurement of photosynthetic productivity. Pudoc, Wageningen.

54. The efficiency of root respiration in different environments

HANS LAMBERS

School of Agriculture and Forestry, University of Melbourne, Parkville, Vic. 3052, Australia

An interesting development in the field of energy metabolism in plants is the increase in knowledge on the biochemistry of CN-resistant respiration. We now know that CN-resistant respiration resides in the mitochondria and in addition to cytochrome oxidase they contain an 'alternative' oxidase. The alternative pathway branches from the cytochrome pathway; the branchpoint is ubiquinone. The alternative electron transport chain is *not* coupled with energy conservation. The only product of NADH-oxidation *via* this pathway, next to water, is heat. [For reviews on the alternative pathway: 4, 10.]

For an evaluation of the physiological significance of the alternative pathway it is important to know more about the regulation of its activity. It has been shown that the alternative pathway does not compete for electrons with the cytochrome pathway [2] and the present biochemical evidence suggests that the alternative pathway is an 'overflow' rather than a 'leak' for reducing equivalents [10].

Another point of interest is the extent to which the alternative pathway is operative *in vivo*, under physiological conditions. In roots of a large number of species an active alternative pathway has been demonstrated [4]. Its activity depends on the growth conditions, age of the plants and the species under investigation.

It is the aim of this paper to present information on the physiological significance of the alternative pathway and its activity in different environments. The implications of the experimental results in terms of the whole plant's functioning will be discussed.

THE PHYSIOLOGICAL SIGNIFICANCE OF THE ALTERNATIVE PATHWAY

Various hypotheses regarding the physiological significance of the nonphosphorylating pathway have been proposed. Lambers [4] concluded that an important role for the alternative pathway in roots of higher plants is that of an 'energy overflow'. The pathway is of significance in oxidation of sugars which are not required for carbon skeletons, energy production, osmoregulation or

R. Brouwer et al. (eds.), Structure and Function of Plant Roots, 281–285.

storage. Only a few experiments presenting evidence for the 'energy overflow model' will be presented here [for review: 4].

Photosynthate availability

During the ontogeny of *Plantago lanceolata* the rate of photosynthesis (per gram shoot) decreases by approximately 40%. However, carbohydrate consumption for growth and shoot respiration, (also per gram shoot) is constant. The rate of root respiration decreases in parallel with that of photosynthesis and this decrease is largely due to a decrease in activity of the alternative pathway [9].

Carbohydrate storage

In young roots of *Hypochaeris radicata*, which lack the capacity to store carbohydrates, the alternative pathway accounts for ca. 75% of root respiration. Older roots of *H. radicata*, which store sugars in their taproot, respire at a lower rate and the alternative pathway contributes to a small extent in root respiration [6].

The above, and additional, results have led to the conclusion that the alternative pathway is of significance in oxidation of sugars which are translocated to the roots in excess of the roots' requirement. Consequently, an active alternative pathway in the roots, indicates that more carbohydrates are imported than required to satisfy the roots' demand.

ENVIRONMENTAL EFFECTS ON ROOT RESPIRATION

One may ask why an overflow mechanism, which appears to be rather wasteful in terms of energy conservation, did not disappear during phylogeny. Several possibilities can be considered.

1. The roots might have no or only weak control over the rate of sugar import.
2. Root growth might depend on a flux of sucrose from the shoots, without sugar itself being limiting. It has to be realised that all compounds, translocated in the phloem, generally travel with sucrose [3].
3. It might be useful to translocate more carbon to the roots than required under constant conditions, since this might enable a plant to cope with changes in the environment. Possibility 3 has been the subject of experiments to be discussed below.

The N-source

Various legumes, including *Lupinus albus* [11], have an operative alternative pathway when grown non-symbiotically with NO_3. However, when they rely on symbiotic N_2 fixation the alternative pathway is essentially inoperative. Thus, roots of legumes relying on combined N receive an excess of carbon, which could be used for N_2 fixation.

Salinity

When *Plantago coronopus* is grown in either non-saline or saline (50 m*M* NaCl) culture solution, the alternative pathway contributes to ca. 50% in root respiration. However, when *P. coronopus* plants are transferred from non-saline into saline culture solution the activity of the alternative pathway temporarily decreases to almost zero whilst the cytochrome pathway is largely unaffected [7]. The concentration of sorbitol, a compatible osmotic solute in *Plantago* [1], in the roots increases quickly after transfer into saline solution. The amount of carbohydrate used in sorbitol synthesis is the same as that saved in respiration, due to the decreased activity of the alternative pathway [7]. Therefore, the plants were able to adapt to the saline environment without interfering with carbon utilization for growth and energy production.

Nutrient supply

When *Plantago lanceolata* plants are transferred from a nutrient-rich ($\frac{1}{4}$ 'Hoagland') into a nutrient-poor (2 per cent of $\frac{1}{4}$ 'Hoagland') solution, shoot growth rapidly decreases. A rapid increase in the activity of the alternative pathway is observed, presumably due to oxidation of carbohydrates which otherwise would have been used in shoot-growth [9]. Possibly, the alternative pathway allows the plant to cope with a temporary imbalance between production and requirement for carbohydrates.

THE ROOTS AS A SINK FOR CARBOHYDRATES

The above results suggest that roots are generally luxuriously supplied with carbohydrates. However, when the rate of photosynthesis decreases as in *P. lanceolata* [9], the roots receive less carbon, indicating that the roots are in a weak competitive position.

Although the roots do not appear to compete for carbohydrates with the shoot, different parts of a root system do compete with each other. At the end of a night after a 12 h day of low light intensity, the activity of the alternative pathway in *Zea mays* roots decreases. However, the root tips maintain a high activity of the pathway, indicating that they still receive sufficient carbon to even maintain the 'energy over flow' [5].

If NO_3 is added to only one half of a split root system of *Phaseolus vulgaris* the roots supplied with NO_3 are supplied with more carbohydrates than the roots deprived of NO_3. A higher activity of the alternative pathway is observed in the NO_3-supplied roots (H. Lambers and J. S. Pate, unpublished).

DISCUSSION

From the observations presented on the activity of the alternative pathway in roots has come a further understanding of the physiological significance of this pathway. Some of the results suggest that the significance of an 'overflow' in the energy metabolism of roots may be in adaptation to a fluctuating environment. However, more work needs to be done, both on plants grown in the field and on plants grown under carefully controlled fluctuating conditions, before a role of the alternative pathway in adaptation can be accepted. The above information also adds to a further understanding of source-sink relations and the regulation of root growth.

A large proportion of root respiration is often maintained by the nonphosphorylating pathway. However, we should place this into perspective: expressed as a proportion of net photosynthesis it does not appear to be a major sink for carbon in mature plants.

REFERENCES

1. Ahamd, I., Larher, F. and Stewart, G. R. 1979 Sorbitol, a compatible osmotic solute in *Plantago maritima*. New Phytol. 82, 671–678.
2. Bahr, J. T. and Bonner, W. D. 1973 Cyanide – insensitive respiration II. Control of the alternate pathway. J. Biol. Chem. 248, 3446–3450.
3. Canny, M. J. 1973 Phloem transport. Cambridge: University Press.
4. Lambers, H. 1980 The physiological significance of cyanide-resistant respiration. Plant, Cell and Environment, in press.
5. Lambers, H. and Posthumus, F. 1980 The effect of light intensity and relative humidity on growth rate and root respiration of *Plantago lanceolata* and *Zea mays*. J. Exp. Bot., in press.
6. Lambers, H. and Van de Dijk, S. J. 1979 Cyanide resistant root respiration and taproot formation in two subspecies of Hypochaeris radicata L. Physiol. Plant 45, 235–239.
7. Lambers, H., Blacquière, T. and Stuiver, C. E. E. 1980 Interactions between osmoregulation and the alternative respiratory pathway in *Plantago coronopus* as affected by salinity. Physiol. Plant, in press.

8. Lambers, H., Layzell, D. B. and Pate, J. S. 1980 Efficiency and regulation of root respiration in a legume: Effects of the N source. Physiol. Plant, in press.
9. Lambers, H., Posthumus, F., Stulen, I., Lanting, L., Van de Dijk, S. J. and Hofstra, R. 1980 Energy metabolism of *Plantago lanceolata* L. as dependent on the supply of mineral nutrients. Physiol. Plant, in press.
10. Solomos, T. 1977 Cyanide resistant respiration in higher plants. Ann. Rev. Plant Physiol. 20, 279–297.

55. The influence of mechanical resistance and phosphate supply on morphology and function of corn roots

B. W. VEEN

Centre for Agrobiological Research, Wageningen, The Netherlands

and

F. R. BOONE

Soil Tillage Laboratory, Agricultural University, Wageningen, The Netherlands

Mechanization in agriculture increasingly involves the use of heavy machinery, which causes soil compaction under unfavourable moisture conditions. Soil compaction has a significant effect on root development of a crop by reducing soil aeration and by increasing soil mechanical resistance. The influence of mechanical resistance on root development and root function was studied in pot experiments, where a sandy loam soil was compacted to different pore volumes, after fertilization to different phosphate levels. Three maize plants were grown in 28 cm high and 14 cm wide pots, and harvested two and three weeks after emergence. Roots were collected from layers of 3 cm.

RESULTS

Mechanical resistance has no influence on root fresh weight (Fig. 1), but influences root elongation rate (Fig. 2), root distribution (Fig. 3) and root diameter (Fig. 4). Phosphate supply (Pw) has a positive effect on root fresh weight and root diameter (Figs. 3 and 4).

The influence of mechanical resistance on root morphology has consequences for ion uptake (Fig. 5) and shoot growth (Fig. 1). Shoot growth is related to the P-content of the shoot, only at the lower P-levels. At high P-level and high mechanical soil resistance, growth is less than expected on basis of P-content (Fig. 6). This suggests the presence of a second growth-limiting factor in the root environment.

Fig. 7 shows an increasing K-limitation for shoot growth at increasing P-supply. The higher the mechanical soil resistance, the lower the potassium content in the shoot at the different phosphate levels. In layers with high root density, depletion of nutrient elements results in decrease in root growth (Fig. 8).

R. Brouwer et al. (eds.), Structure and Function of Plant Roots, 287–291.

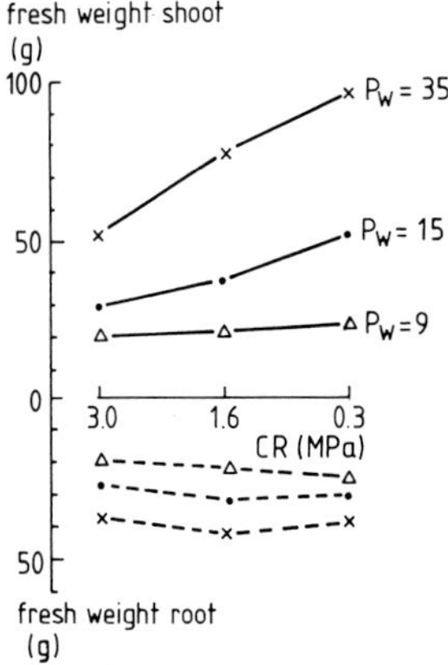

Fig. 1. Influence of soil mechanical resistance on fresh weight of shoot and root, three weeks after emergence at three levels of phosphate availability.

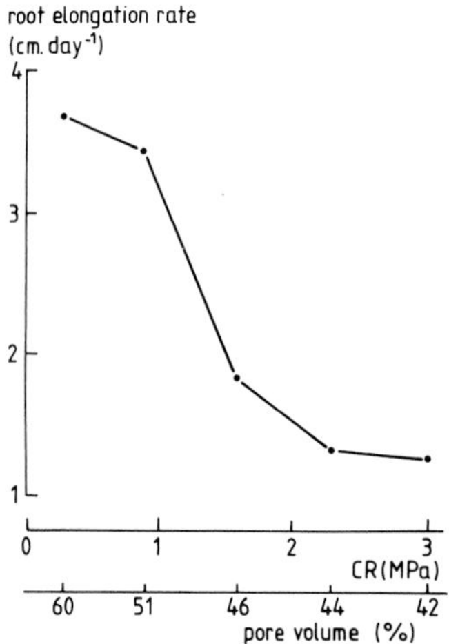

Fig. 2. Relation between the elongation rate of main root axes and the mechanical resistance of the soil.

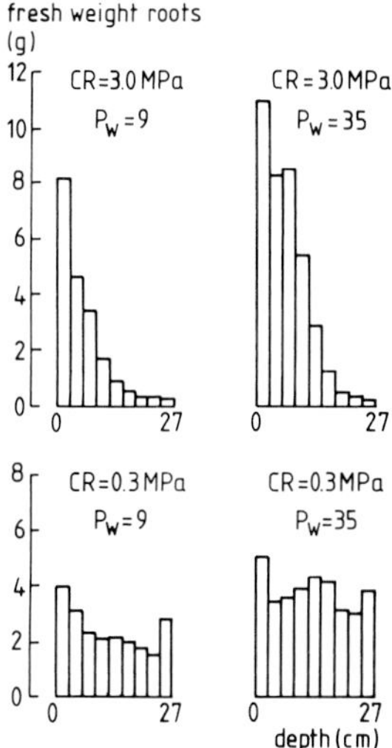

Fig. 3. Root distribution in pots, three weeks after emergence, in relation to soil mechanical resistance (CR) and phosphate availability (Pw).

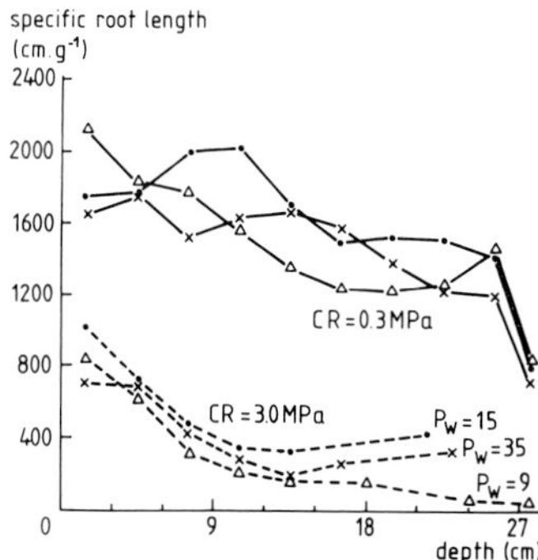

Fig. 4. Influence of mechanical resistance (CR) on root length per unit root weight at different soil depths.

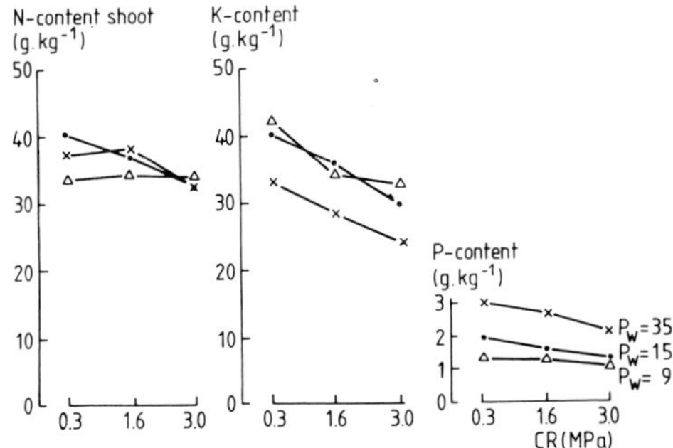

Fig. 5. N-, K- and P-contents of dry shoots, three weeks after emergence of plants grown in soil with different mechanical resistance and different phosphate supply.

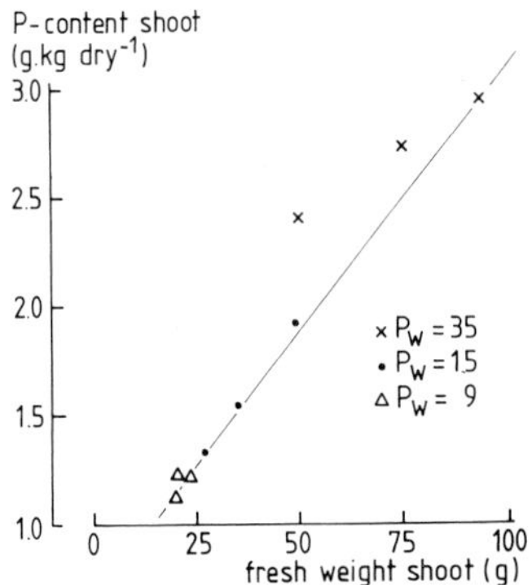

Fig. 6. Relation between P-content and shoot growth.

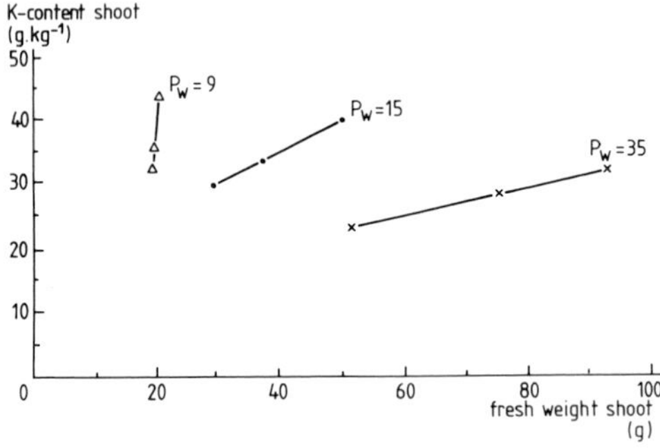

Fig. 7. Relation between shoot growth and K-content of the shoot, at different phosphate supply levels.

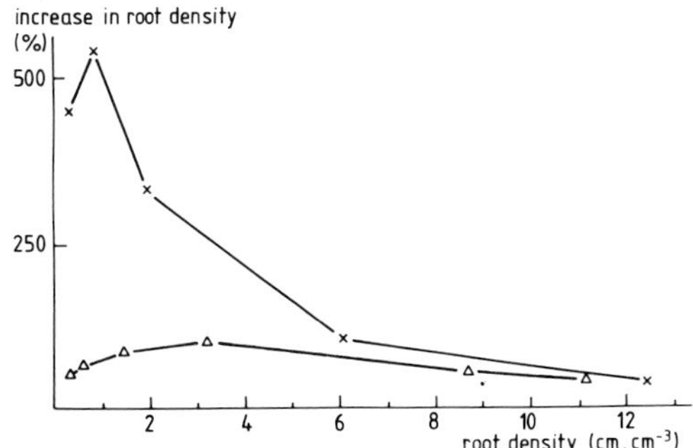

Fig. 8. Influence of root density on the relative increase in root length in the period 2–3 weeks after emergence, at two levels of phosphate supply (x = Pw35, Δ = Pw9).

Uptake of ions is only possible from the deeper soil layers. An important part of the root system, situated in the upper soil layer does not take part in the process of ion uptake.

CONCLUSIONS

The influence of mechanical soil resistance on shoot growth can be explained qualitatively by its influence on root morphology. When relating ion uptake quantitatively to a certain root parameter (length, surface area, weight), availability of nutrients in the different soil layers must be taken into consideration.

56. Genetical variability of the size of the lucerne root system determined by its electrical capacity

OLDŘICH CHLOUPEK

Plant Breeding Station, 664 43 Želešice near Brno, Czechoslovakia

Diallel crosses (F_1) and self-fertilized progeny (I_1) of five lucerne clones were evaluated. The size of the root system of closely spaced plants was measured 10 times during five cuts. One terminal of the measuring bridge was connected with the basal parts of shoots (combined) and the second terminal earthed by an electrode. In this way the electrical capacity of the root system in relation to earth was measured. The values of electrical capacity are relative characteristics of the root system size provided the measurements are performed in the same field with uniform soil moisture [1, 2]. The data obtained were evaluated by means of diallel analysis [3].

The average root system size of self-fertilized variants was only 84.2 per cent of that of diallel crosses, but a significant correlation was found between the size of the root system of self-fertilized progeny and that of diallel crosses. The general combining ability, specific combining ability and reciprocal effects were mostly

Table 1. Results of variance- and diallel-analysis of the root system size of lucerne plants

Average variance (pF)	Periods between cuts					Total values
	I	II	III	IV	V	
for the variability under variants (crosses)	520^{++}	$1\,220^{++}$	$3\,312^{++}$	$3\,502^{++}$	4017^{++}	2066^{++}
under individual plants	134	112	591	541	786	240
for general combining ability	33	158^{++}	638^{++}	279^{++}	2438^{++}	
specific combining ability	45^{+}	57^{++}	117	293^{++}	159	59^{++}
reciprocal effects	36^{+}	42^{++}	146^{+}	117^{+}	172^{+}	72^{++}
error	15	12	65	60	87	27

pF = pikofarad, $^{+}$ significance level 0.05, $^{++}$ significance level 0.01

R. Brouwer et al. (eds.), Structure and Function of Plant Roots, 293–294.

significant in all five periods and the ratio of their average variances was 1:0.38:0.29. Using formulae for predicting yield of synthetic varieties it was found that size of the root system was the most accurate indirect selection criterion for fodder yield. The results will be published later in detail.

REFERENCES

1. Chloupek, O. 1972 The relationship between electric capacitance and some other parameters of plant roots. Biologia Plantarum 14, 227–230.
2. Chloupek, O. 1977 Evaluation of the size of a plant's root system using its electrical capacitance. Plant and Soil 48, 525–532.
3. Griffing, B. 1956 Concept of general and specific combining ability in relation to diallel crossing systems. Aust. J. Biol. Sci. 9, 463–493.

57. Genetical variation in efficiency of plant root systems

NASIR EL BASSAM

Institute of Crop Science and Plant Breeding, Federal Research Centre of Agriculture, D-3300 Braunschweig, F.R.G.

Genes are ultimately responsible for all physiological, developmental and morphological characteristics of plants. About 30 per cent of the plant genome contributes more or less in growth and development of roots [4]. Therefore it is of a great interest for plant breeders to consider specific root features in selection programmes. Aspects of the efficiency and results of the experimental approach will be discussed in this contribution.

EFFICIENCY OF PLANT ROOTS

The efficiency of roots mainly depends on their permeability and extensions [2]. Both features are genetically and environmentally dependent. However, several environmental growth factors and cultivation measures may mask or abolish the

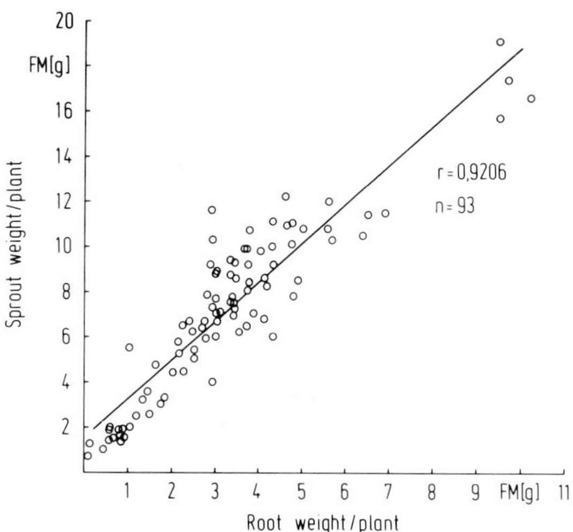

Fig. 1. Relationship between the weight of shoots and roots of spring barley plants after four weeks from germination in solution culture.

R. Brouwer et al. (eds.), Structure and Function of Plant Roots, 295–299. All rights reserved.

genetical variability. A specific root efficiency therefore can be determined under controlled conditions [1].

EXTENSION OF ROOTS

Some morphological and phenological characteristics of the five cultivars of spring barley have been investigated by means of hydroponics. The yield potential of these cultivars has been ascertained for several years by the Federal Office for Cultivar Registration directly in fields. A very close correlation has been found between the weight of the root and shoot systems of all plants (Fig. 1), but no correlation could be established between the root weights, root lengths and yield potential of the investigated cultivars. Plants of cultivars carrying a high yield potential develop more nodal axes than those cultivars of a lower yielding potential (Table 1).

Table 1. Yield potential of five varieties of barley and number of nodal axes in a constant environment after four weeks from germination in solution culture

Variety	Yield potential	Number of the nodal roots
Aramir	high – very high	10.2 ± 3.8
Aura	medium – high	5.9 ± 1.9
Union	low	4.7 ± 2.1
Stankas F. G.	very low	4.3 ± 1.1
Hauters S. G.	old variety	2.7 ± 1.9

WATER PERMEABILITY OF ROOTS

The routes which nutrient and water follow from the soil to shoots – across the cortex to the vascular stele, thence upwards in the xylem – separately controlled and the ratio in which nutrients and water enter plants can differ greatly from that in the solution external to their roots [3]. In this contribution water permeability will be understood and measured as water absorption by the root system.

The technique as described under [1] enables studies on root development and root efficiency under controlled conditions. Effects of soil compaction, nutrient supply, salt concentration and soil water potential can be examined at different root zones. Roots develop in soil columns (total length 120 cm) of four compartments in plexiglass tubes of 30 cm each.

In the root zones 0–30, 30–60, 60–90 cm of the cultivar Aramir (yield potential:

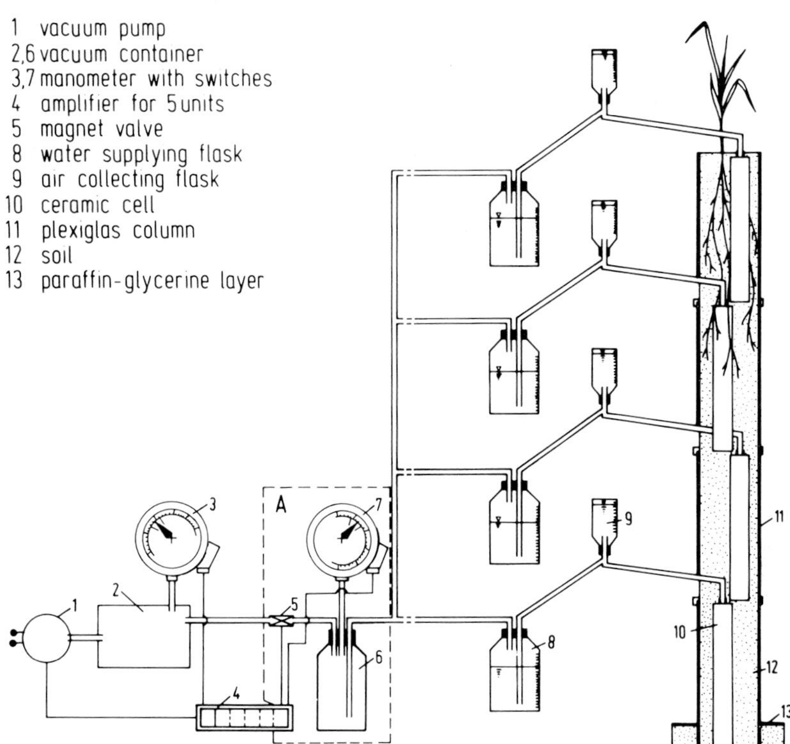

Fig. 2. Installation for root investigations.

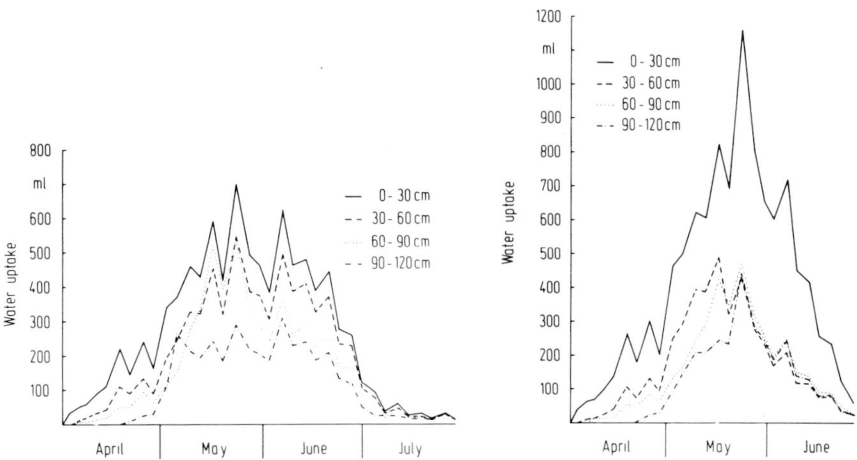

Fig. 3. Daily water absorption of the root zones of two cultivars of barley (left: Aura, right: Stankas F. G.) at soil water potential of −0.2 bar.

very high – high) nearly the same quantity of water has been absorbed during the vegetation period at soil water potential of −0.2 bar. However, roots of the zone 0–30 cm of the cultivar 'Stankas Frühgerste' (low yield potential) absorbed significantly more water than the other deeper zones (Fig. 3).

A comparison between two other cultivars i.e. 'Aura' (modern cultivar, yield potential: medium-high) and 'Hauters Sommergerste' (old variety) showed a similar water uptake of the whole root system at soil water potential of −0.1 bar until the stage of ear emergence. Roots of 'Hauters Sommergerste' after this stage absorbed less water than the first cultivar. Decreasing the soil water potential from −0.1 to −0.35 bar caused an earlier differentiation in water uptake at earlier stage (Fig. 4).

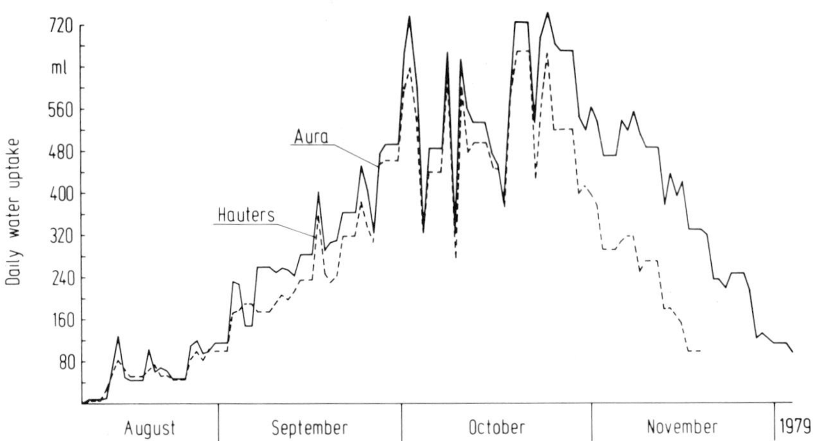

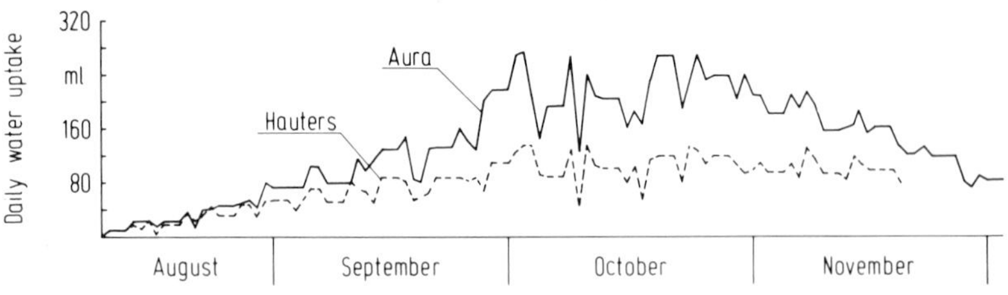

Fig. 4. Water absorption of the whole root system of two cultivars (Aura and Hauters S. G.) at two soil water potentials (above: −0.10 bar and below −0.35 bar).

CONCLUSIONS

The efficiency of the different root zones in soils seems to be determined for the yield potential of cultivar. Differentiation in water uptake could take place under certain conditions at earlier growth stages which is very useful for screening of cultivars. No correlation could be established between the weights of roots on the one hand and the root efficiency as well as the yield potential on the other hand. The determination of the number of nodal root axes may be useful in genotypical identification.

REFERENCES

1. El Bassam, N. and Sommer, C. 1980 Eine Methode zur in situ-Ermittlung der Leistungsfähigkeit des Wurzelnetzes von Genotypen. Z. Acker- und Pflanzenbau 149, 391–397.
2. Kramer, P. J. 1969 Plant and Water Relationships, New York: McGraw-Hill Book Co.
3. Russel, R. S. 1977 Plant root systems: Their function and intraction with the soil. McGraw-Hill Book Company (UK) Limited.
4. Zobel, R. W. 1975 The genetics of root development. In: Torrey, J. G. and Clarkson, D. J. (eds.), The development and function of roots, pp. 261–275. London, New York, San Francisco: Academic Press.

V. RESPONSES OF ROOTS TO STRESS

58. Structure and function in absorption and transport of nutrients

DETLEF KRAMER

Institut für Botanik, Fachbereich Biologie, Technische Hochschule Darmstadt, D-6100 Darmstadt, FRG

The term transfer cell applies to cells that are thought to be particularly active in transport of inorganic or organic solutes across the plasmalemma. Structurally the most significant feature is the differentiation of cell-wall protuberances on the inner surface of the wall. These wall ingrowths – which often form a considerable wall labyrinth – are surrounded by the plasmalemma, which is therefore largely increased. Transfer cells are regarded as analogues of the microvilli in many animal cells, which are specialized in transport processes. It is widely held that in both systems the amplified plasmalemma surface leads to a largely increased number of pumping sites per cell, thus promoting active transport at the symplast/apoplast border [1]. The occurrence of transfer cells has been demonstrated to be common in all plant organs. In roots, transfer cells are common in the xylem parenchyma of leguminous nodules [7] as well as in rhizomes [8]. However, until 1976 no transfer cells had been described that could be directly related to the absorption of nutrients by the root and their transport to the shoot.

In this article two systems of transfer cells will be described that were detected in our laboratory, and it will be shown that both systems function in ion transport under conditions of ecological stress.

TRANSFER CELLS IN THE EPIDERMIS OF ROOTS

The epidermal cells of a primary plant root can be regarded as the most important site of nutrient absorption. It is the site of direct interactions between the symplast of the root and the root environment, with the cell wall and other less structured components like mucilage in between.

Under normal growth conditions these cells are not differentiated as transfer cells. However, we observed that under particular situations of nutritional stress epidermal cells may develop the characteristic features of transfer cells.

In *Helianthus annuus*, epidermal cells differentiate into transfer cells when plants are subjected to iron deficiency. A few days after replacement of a complete nutrient solution by one without any iron, longitudinal growth of roots stops and

R. Brouwer et al. (eds.), Structure and Function of Plant Roots, 303–307.

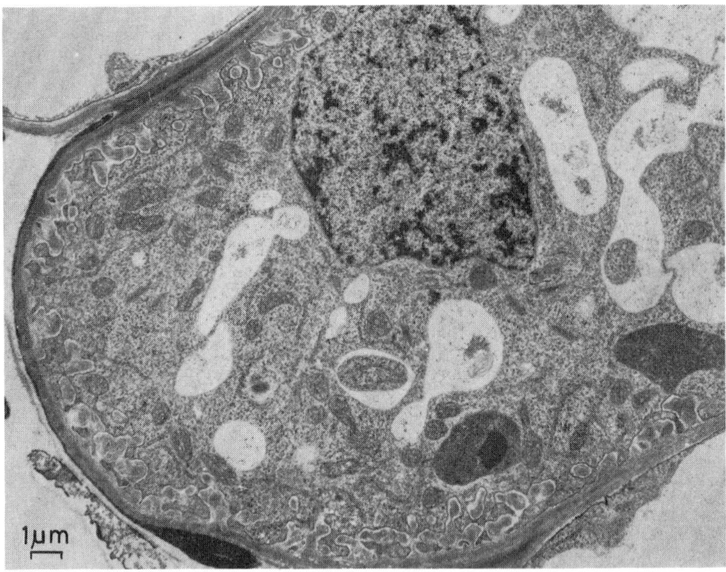

Fig. 1. Epidermal cell of a *Helianthus* root, differentiated as a transfer cell after several days of iron deficiency.

a zone about 5–15 mm in length immediately behind the root apex thickens and forms many short root hairs. On ultrathin sections all living cells facing the external medium epidermal cells or exodermal cells that face dead hair cells appear as typical transfer cells [3]. They have a well-developed wall labyrinth, which is most pronounced on the side facing the external medium. Their cytoplasm occupies most of the cell volume and contains a large number of mitochondria (Fig. 1).

From physiological investigations it is known that *Helianthus*, like other so-called iron-efficient plants, is able to overcome a particular stress, *viz.* the non-availability of iron due to a high pH in the soil. Precipitated iron is dissolved after acidification of the rhizosphere and secretion of chelating substances, and also the capacity for iron absorption is increased by a factor of 100 to 1000 [6]. High pH is a typical feature of calcareous soils, so that this complex is certainly an important aspect of the calcifuge/calcicole-problem. We presume that the transport processes described above may be located in the root epidermal transfer cells.

In roots of two salt-tolerant *Atriplex* species, *A.hastata* and *A.hortensis*, epidermal transfer cells could be induced by high (200–500 mol m^{-3}) salt concen-

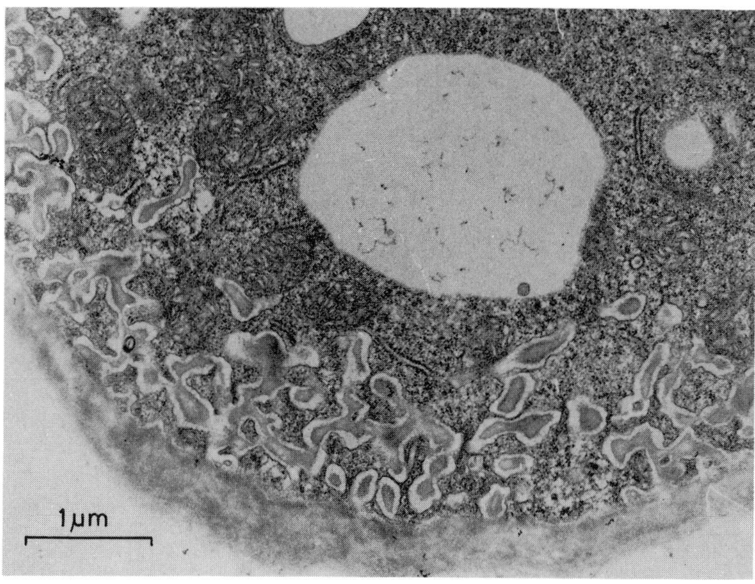

Fig. 2. Epidermal cell of a root of *Atriplex hortensis*. The plant had been exposed to salt stress (300 mol m^{-3} NaCl).

trations in the nutrient solution (Fig. 2). Data obtained by X-ray microanalysis of root epidermal cells suggested a relatively high selectivity in potassium uptake, in view of a K^+/Na^+ ratio 20 to 50 times higher than in the nutrient solution [5]. However, a screening test, in which different NaCl concentrations were combined with different Fe^{3+} concentrations, showed clearly that the transfer-cell features observed as an effect of salt treatment do not appear when the Fe^{3+} concentration in the nutrient solution is increased in direct proportion on to the NaCl concentration. Thus it must be concluded that salinity may be another factor that causes iron deficiency, to which *Atriplex* species react in the same way as *Helianthus*. The mechanism of this interference between NaCl and Fe^{3+} is still unknown.

TRANSFER CELLS IN THE XYLEM PARENCHYMA OF LEGUMES

Several legume species like *Phaseolus coccineus* and *Glycine max*. have the ability to control the composition of the xylem sap secondarily. Although largely salt-sensitive, these plants can resist low and temporary salt stress by reabsorbing Na^+ from the transpiration stream and secreting K^+ into the vessels. This

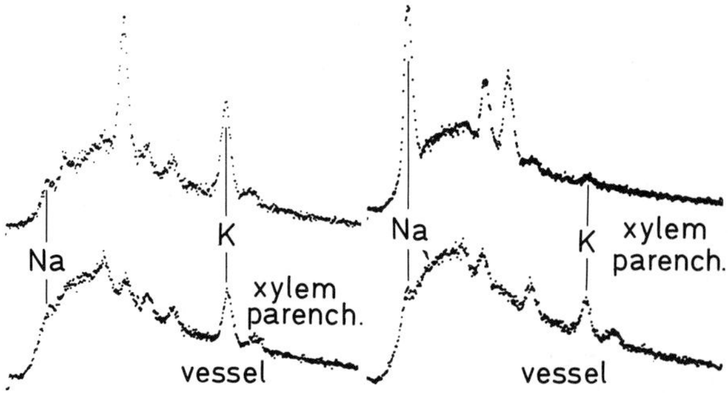

Fig. 3. X-ray spectra of xylem-parenchyma cells and vessels in the proximal part of a root of *Phaseolus coccineus*. The left spectra correspond to a control plant, the right ones were taken from a plant that had grown in the presence of 50 mol m^{-3} NaCl.

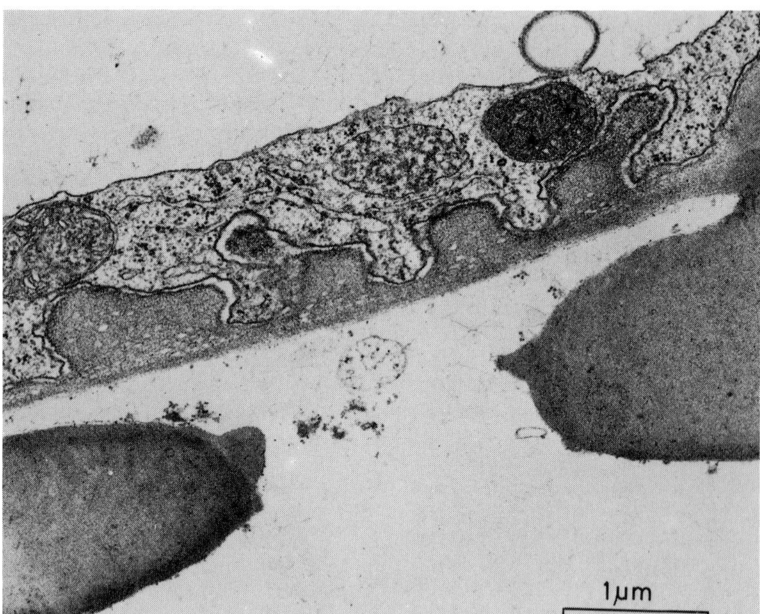

Fig. 4. Electron micrograph of a xylem-parenchyma cell of the proximal part of a root of *Phaseolus coccineus*.

mechanism is located in the upper part of the root and the lower part of the shoot [2]. By X-ray microanalysis we have shown that the actual Na^+/K^+ exchange is a transport process across the plasmalemma of the xylem-parenchyma cells that surround the vessels [4] (Fig. 3). This membrane represents the border between the symplast of the xylem parenchyma and the apoplast of the vessel.

Electron microscopical investigations revealed that in *Phaseolus coccineus* these xylem-parenchyma cells are differentiated as transfer cells with the wall ingrowths restricted to the area of the half-bordered pits (Fig. 4). This might serve as an indication that these half-bordered pits are in fact sites of intensive and controlled transport processes.

The cytoplasm of xylem-parenchyma transfer cells is characterized by a large number of cisternae of rough endoplasmic reticulum, which often surround the plasmalemma invaginations around the wall ingrowths. In many cases fibrillar bridges between both membranes can be seen.

The same features were observed in roots of *Glycine max.* (soy bean). A Na^+/K^+ exchange mechanism was established in proximal parts of roots of the salt resistant soy bean variety 'Lee'.

In conclusion it may be suggested that root transfer cells are found in a variety of plant species at different locations and are perhaps often induced as a response to nutritional stress.

REFERENCES

1. Gunning, B. E. S. 1977 Transfer cells and their roles in transport of solutes in plants. Sci. Progr. Oxf. 64, 539–568.
2. Jacoby, B. 1964 Function of bean roots and stems in sodium retention. Plant Physiol. 39, 445–449.
3. Kramer, D. 1980 Transfer cells in the epidermis of roots. In: Plant Membrane Transport current conceptual issues, pp. 393–394. Amsterdam-New York-Oxford: Elsevier/North Holland Biomedical Press.
4. Kramer, D., Läuchli, A., Yeo, A. R. and Gullasch, J. 1977 Transfer cells in roots of *Phaseolus coccineus*: Ultrastructure and possible function in exclusion of sodium from the shoot. Ann. Bot. 41, 1031–1040.
5. Kramer, D., Anderson, W. P. and Preston, J. 1978 Transfer cells in the root epidermis of *Atriplex hastata* L. as a response to salinity: a comparative cytological and X-ray microprobe investigation. Austr. J. Plant Physiol. 5, 739–747.
6. Kramer, D., Römheld, V., Landsberg, E. and Marschner, H. 1980 Induction of transfer-cell formation by iron deficiency in the root epidermis of *Helianthus annuus* L. Planta 147, 335–339.
7. Pate, J. S., Gunning, B. E. S. and Briarty, L. G. 1968 Ultrastructure and functioning of the transport system of the leguminous root nodule. Planta 85, 11–34.
8. Yeung, E. C. and Peterson, R. L. 1974 Ontogeny of xylem transfer cells in *Hieracium floribundum*. Protoplasma 80, 155–174.

59. Some aspects of structural and functional modifications induced by drought in root systems

NICOLE VARTANIAN

Phytotron, CNRS, 91190 Gif-sur-Yvette, Ecologie végétale, Université de Paris XI, 91405 Orsay, France

According to ecological observations and assumptions on drought resistance, it is usually considered that the root system plays a decisive role in plant adaptation to water deficit. However, few demonstrative experiments and analyses have been performed which could corraborate this consideration. As a matter of fact, the adaptative response to environmental water stress requires a comprehensive research of the mechanisms involved in morphogenetic and metabolic changes induced by progressively increasing water deficit during plant growth. Hence, the aim of the present paper is not only to describe some aspects of structural and functional modifications induced by drought in root systems, but also to show an example of the effect of environmental stress on the morphogenetic program of growth and differenciation within the whole plant. According to Loomis [4] definition: 'Growth differenciation balance, or correlation, is based upon the tendency of nutrient supplies to be used preferentially in growth, and particularly in top growth, whenever conditions are sufficiently favorable. Any factor, such as a deficient supply of water, which checks growth without reducing correspondingly photosynthesis will tend to increase any differentiation response of which the plant is capable'. Thus, plants in which development is shifted towards differentiation, may produce some characters confering better resistance depending on the species. The formation of short tuberized roots in mesophytic species of annual Dicotyledons [7, 8], as an adaptative response to drought resistance at the root level may provide an illustration of these reflections.

Progressive drought stress is induced in plants as described previously [7, 10]. Plants of mesophytic species *Sinapis alba* are grown under controlled environmental conditions at the Phytotron of Gif-sur-Yvette, with a 16 h photoperiod, irradiance approximately 100 $W.m^{-2}$, a constant temperature of 22 °C and 50

R. Brouwer et al. (eds.), Structure and Function of Plant Roots, 309–318.

per cent relative air humidity. The seedlings are cultivated on a sandy soil initially watered at field capacity and protected against water loss from physical evaporation by a 'parafilm' sheet. As no further addition of water takes place during the whole growth period, water disponibility is slowly reduced by transpiration only, and plants will necessarily adapt to a progressively increasing water deficit.

RESULTS

Kinetic of drought adaptative rhizogenesis

With increasing water deficit in soil and plant, growth of the tap root is decreased [10] and a sudden burst of new lateral roots occurs as if rhizogenesis was stimulated at a threshold value of plant water potential: minus ten bars [7]. These new roots remain short.

The kinetic of this particular rhizogenesis (Fig. 1) appears sigmoidal, showing a 3 phase evolution [6]:

- A latent phase during an initial 10 days period where the response is discrete but present.
- An exponential phase leading to a maximum in the number of short roots between the 21th and 27th day of growth.

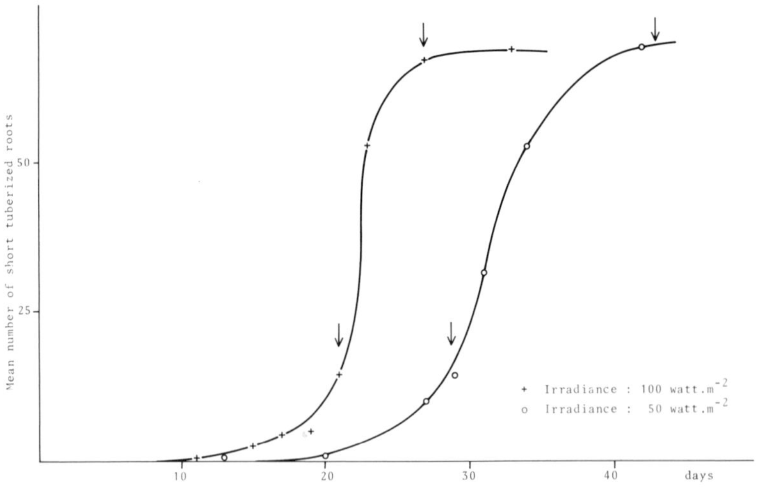

Fig. 1. Kinetic of rhizogenesis during drought in *Sinapis alba*: mean number of short tuberized roots produced as a function of time, under two irradiance conditions.

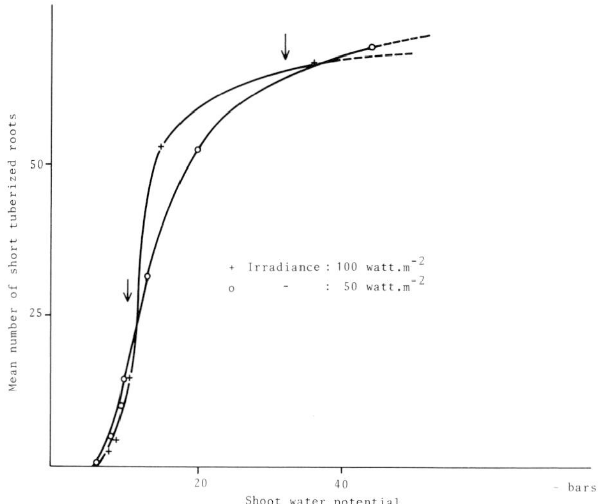

Fig. 2. Rhizogenesis as a function of shoot water potential (arrows indicate the limits of the exponential phase of rhizogenesis).

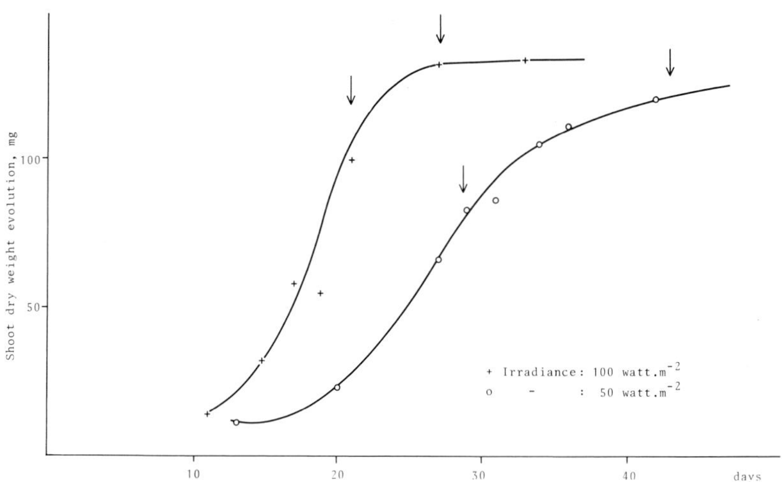

Fig. 3. Shoot dry weight evolution as a function of time.

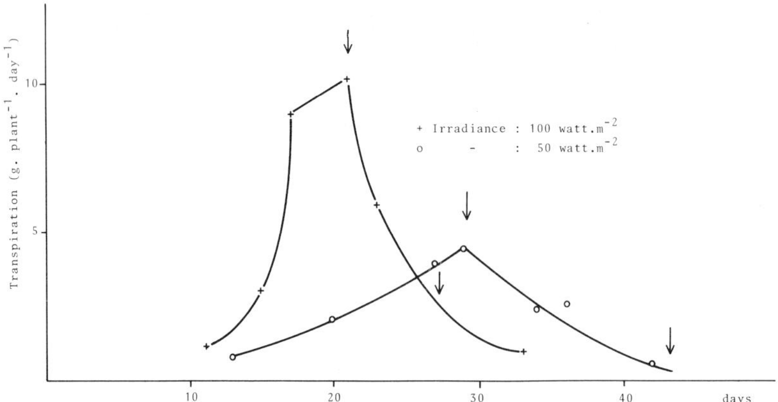

Fig. 4. Transpiration rate as a function of time.

– A third phase corresponding to a 'plateau': the number of new roots remains constant in spite of increasing water stress.

With irradiance reduced to half of its value lowering the evaporative demand, a similar evolution is still observed though initiation of rhizogenesis appears delayed (Fig. 1). However, when the rhizogenesis is plotted as a function of plant water potential (Fig. 2) the delay observed between the 2 conditions of irradiance is eliminated and the curves coincide, showing that the difference was only due to slower dehydration of soil and plant, and that rhizogenesis can directly be related to water deficit.

It is thoroughly interesting to notice that the threshold value for the initiation of these new roots is set at approximately minus ten bars, the same water potential level which reduces shoot growth. Fig. 3 shows that the burst of rhizogenesis is included within the phase of cessation of shoot growth induced by increasing water deficit in shoot tissues. As for the relationship between rhizogenesis and transpiration, it can be seen (Fig. 4) that the occurence of short roots may be associated with the phase of decreasing transpiration due to stomata closure.

Morphology and anatomy of short roots

1. The most distinctive morphological features characterizing these roots are as follows:

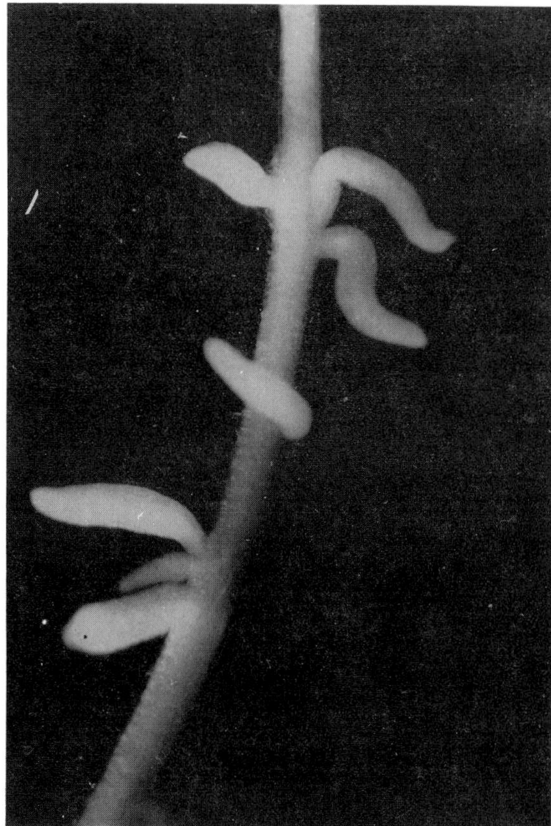

Fig. 5. Short tuberized roots on the tap root (× 15).

- localization, both on tap root (Fig. 5) and lateral roots (Fig. 8), essentially in the upper layers of soil.
- absence of root hairs under drought conditions (Fig. 5).
- size: less than 1 mm in length and $\frac{1}{2}$ mm in width, they appear twice as large as lateral roots bearing them.
- swelling of the basis which increases progressively under aging, giving them a tuberized aspect.

2. Studied by scanning electron microscopy with a technique which preserves the surface texture of the roots, avoiding any distortion (12), three zones can clearly be distinguished on the short roots (Fig. 6):

- an apical zone with regularly ordered cells of cap and meristematic region.

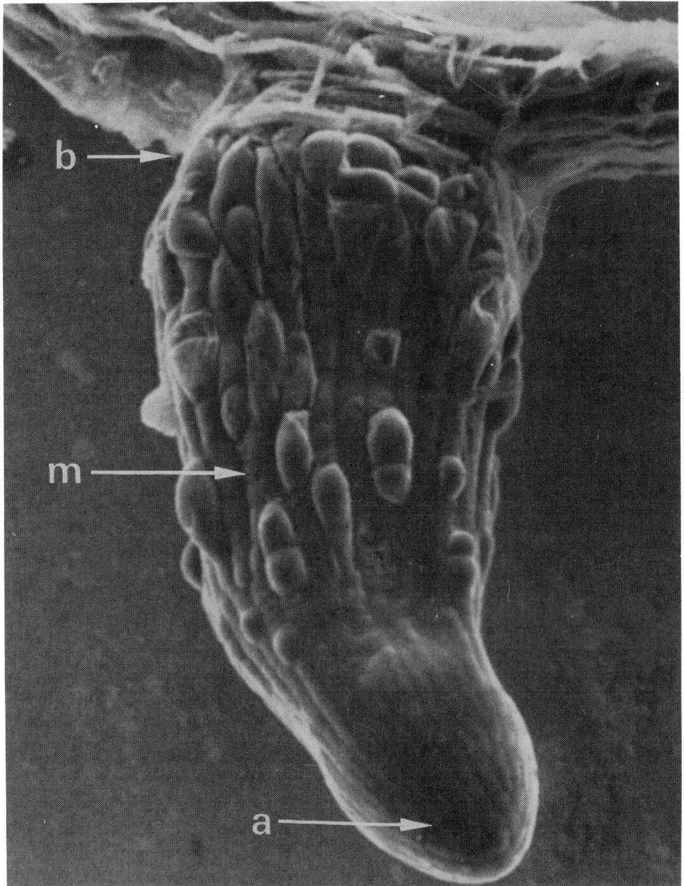

Fig. 6. Scanning electron micrograph of short tuberized roots (× 250). a, apical zone – m, medium zone – b, enlarged basis.

- a medium zone where cell rhizodermis organization shows alternating rows of short and long cells as in normal roots of Sinapis cultivated in well watered soils.
- an enlarged tuberized basis, where all the cells have the same swollen bead-like aspect.

The turgid look of rhizodermal cells in a very dry soil suggests an important mechanism of osmoregulation.

3. Ontogenesis of short tuberized roots, studied in photonic microscopy [9] has shown that they are initiated by pericycle divisions just as normal roots.

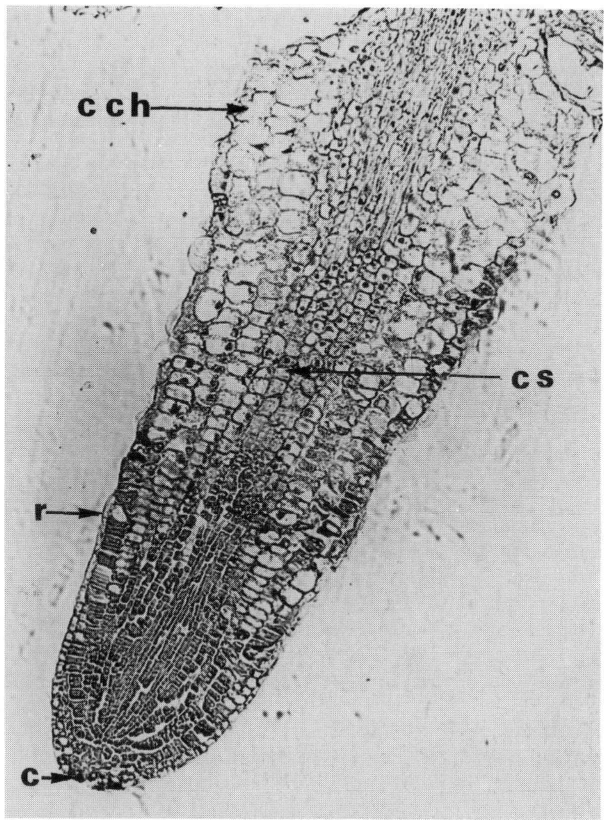

Fig. 7. Longitudinal axial section of short tuberized root (× 100). c, cap – r, rhizodermis – c ch, hypertrophied cortical cells – cs, central stele.

Longitudinal sections of mature differentiated short tuberized roots (Fig. 7) show a structure of primary root, with cap, rhizodermis, sub-apical meristem, cortex, endodermis, pericycle and central stele. No change appears in the number of cells layers as compared with normal root, the most distinctive feature being the radial growth of cortical cells of the outer layer, which enlarges early on the short root. With aging in dry conditions, cortical cells become hypertrophied and accumulate large amount of starch.

Survival and evolutive potentiality of short tuberized roots

Hardening during drought conditions has often been related [3, 5] to high accumulation of carbohydrates. In root systems of *Sinapis alba* subjected to drought, we have observed a high increase in the glucose level [11] and as previously noted, accumulation of starch in the short roots. Those metabolic processes may be partly responsible [3, 5] for the high degree of survival of plants bearing short tuberized roots which can persist a 3 months dehydration.

The role of short tuberized roots in plant growth recovery appears clearly after

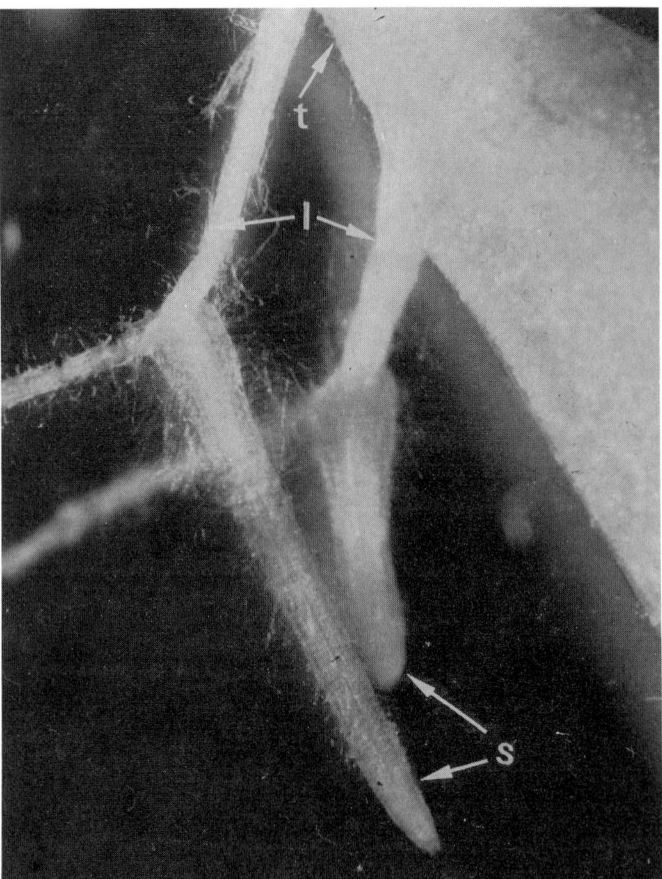

Fig. 8. Morphological aspect of rehydration: formation of root hairs and elongation of short tuberized roots (× 25). t, tap root – l, lateral root – s, short root elongation.

rehydration: due to the intensive development of root hairs and the capacity of short roots to elongate rapidly (Fig. 8) and branch out, they soon give rise to a new functional root system while all other roots have been suberized during the drought period.

When the whole plant has resumed growth, an analysis of the after effects of water stress on growth parameters (S. Atayi and N. Vartanian, unpublished results) indicates that root growth is enhanced in plants having produced short tuberized roots: root dry weight is increased as compared to regularly watered control plants. Elimination of water deficit, and rapid increase in transpiration rate during stress recovery, demonstrate that no after effect of wilting prevents them from returning to their original values, thereby suggesting a high capacity of absorption due to increased absorbing root surfaces.

Occurrence of adaptative response of root system to drought in flowering plants

In Dicotyledons, this response has been found in many different families and seems to be dependent on familial, rather than on specific, characteristics. As until now primitive families only, situated at the basis of the different evolutive phylaea of Dicotyledons, have shown this reaction, it might be suggested that greater meristematic potentialities at the root level could exist in primitive families.

Genotypic variation within species is also investigated. The data obtained up to now on various genotypes of the Cruciferae *Sinapis alba* and *Brassica napus*, indicate large differences in the mean number of short tuberized root observed and in the kinetic of their formation.

CONCLUSIONS

These results are consecutive to a reflexion on an ecological problem and to its descriptive approach. Investigations are now attempted in different fields as biochemistry, genetics, systematics: this seems necessary to understand further how the signal water stress is translated in terms of metabolism changes [1] leading to new morphogenesis.

REFERENCES

1. Hsiao, T. C., Acevedo, E., Fereres, E. and Henderson, D. W. 1976 Stress metabolism. Water stress, growth and osmotic adjustment. Phil. Trans. Roy. Soc. Lond. B 273, 479–500.
2. Iljin, W. S. 1957 Drought resistance in plants and physiological processes. Ann. Rev. Plant Physiol. 8, 257–272.

3. Levitt, J. 1972 Water stress. In: Responses of plants to environmental stresses, pp. 322–446. New York: Academic Press.
4. Loomis, W. E. 1958 Growth correlation. In: W. E. Loomis (ed.), Growth and differentiation in plants, pp. 197–217. The Iowa State College Press.
5. Stocker, O. 1961 Les effets morphologiques et physiologiques du manque d'eau sur les plantes. In: Recherches sur la zone aride XV. Echanges hydriques des plantes en milieu aride ou semi-aride, pp. 69–113. Paris: Unesco.
6. Sabatier, G. and Vartanian, N. 1981 Cinétique de la rhizogenèse adaptative à la sécheresse. I. Relations avec l'évolution des paramètres hydriques et morphologiques chez le *Sinapis alba* L. (in preparation).
7 Vartanian, N. 1971 Action morphogénétique du facteur hydrique sur le système racinaire du *Sinapis alba* L. I. Rhizogenèse et potentiel hydrique racinaire. Rev. Gén. Bot. 78, 19–29.
8. Vartanian, N. 1972a Induction par la sécheresse de racines courtes, tubérisées chez des plantes annuelles: Crucifères et quelques autres familles. C. R. Acad. Sci. 274, série D, 1947–1500.
9. Vartanian, N. 1972b Action morphogénétique du facteur hydrique sur le système racinaire du *Sinapis alba* L. II. Etude histologique des ébauches racinaires produites par la sécheresse. Rev. Gén. Bot. 79, 139–165.
10. Vartanian, N. 1973 Particularités adaptatives de la Moutarde blanche, *Sinapis alba* L., à la sécheresse. In: R. O. Slatyer (ed.), Plant response to climatic factors. Proceedings of the Uppsala symposium, pp. 277–288. Paris: Unesco.
11. Vartanian, N. 1977 Influence des facteurs hydriques de l'environnement sur le système racinaire: aspects morphologiques, histologiques et écophysiologiques. Thèse de Doctorat d'Etat, no. 1937, 150 p. Orsay.
12. Vartanian, N., Wertheimer, D. et Courderc, H. 1981 Scanning electron microscopic aspects of short tuberized roots, with special reference to cell rhizodermis evolution under drought and rehydration (in preparation).

60. Ultrastructure of maize root cells under water stress conditions

MILADA ČIAMPOROVÁ

Institute of Experimental Biology and Ecology, Slovak Academy of Sciences, 88534 Bratislava, Czechoslovakia

The degree of ultrastructure damage caused by polyethylene glycol 4000-induced water deficit depends on the duration of the stress period and on the stage of differentiation of the epidermal and cortical maize root cells at distances of 0.5, 1 and 2 mm from the root cap junction [1–3]. The cells of the endodermis and the central cylinder of maize primary root under conditions of water stress have now been investigated.

In the range of 0.5 to 2 mm from the root cap junction of the intact roots, the xylem vessels were discernible but they still contained all the normal structural components in their cytoplasm at a distance of 2 mm (Fig. 1). The cytoplasm of the sieve elements, however, was almost completely disintegrated at a distance of 0.5 mm (Fig. 2).

A shorter period of stress (24 h) did not cause substantial ultrastructural changes in the cells of the endodermis and the central cylinder (Fig. 3). However, the same water stress (Fig. 4) caused nuclear chromatin aggregation, polyribosome reduction, elongation and parallel arrangement of the endoplasmic reticulum (ER) elements, reduction of the mitochondrial cristae and structural changes of the dictyosomes in the epidermal cells (1 and 2). Similar structural alterations were also observed in the tissues of the central cylinder following longer periods of water stress (48 and 72 h). The effects of water deficit could be seen in all cells apart from sieve elements that had already had their cytoplasmic components disintegrated as a result of their differentiation. Fellows and Boyer [4] consider sieve elements in sunflower leaves to be protected against water deficit by their internal solutes.

The structural responses of maize root cells differed however in the degree of their ultrastructure damage. For example, while the structure of epidermal cells had already been severely damaged after 48 h stress (Fig. 6), the structural integrity of the cells of endodermis and the central cylinder was quite well preserved even after 72 h of water stress (Fig. 5). The greatest range of structural changes occurred always in the epidermal cells (Fig. 7). The hypodermal cells not only occupy the peripheral position within the root but they are more vacuolated

R. Brouwer et al. (eds.), Structure and Function of Plant Roots, 319–322. All rights reserved.

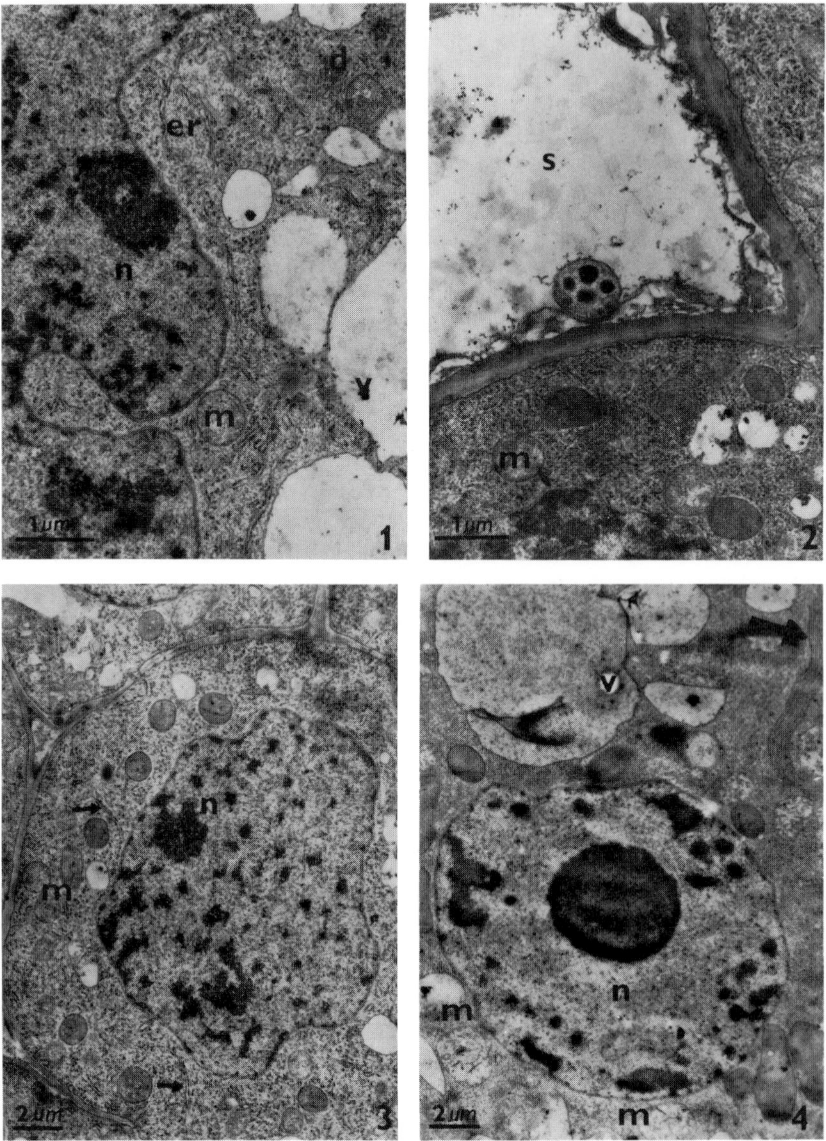

Figs. 1–4. s = sieve element, n = nucleus, m = mitochondria, er = endoplasmic reticulum, d = dictyosome, v = vacuole.

Fig. 1. Xylem vessel of an intact root. × 10 500.

Fig. 2. Part of the sieve element from an intact root. × 9900.

Fig. 3. Vascular parenchyma cell stressed for 24 h. The arrows indicate short, unchanged ER elements. × 6675.

Fig. 4. Epidermal cell stressed for 24 h. The arrow indicates parallel arrangement of ER elements. × 3277,5.

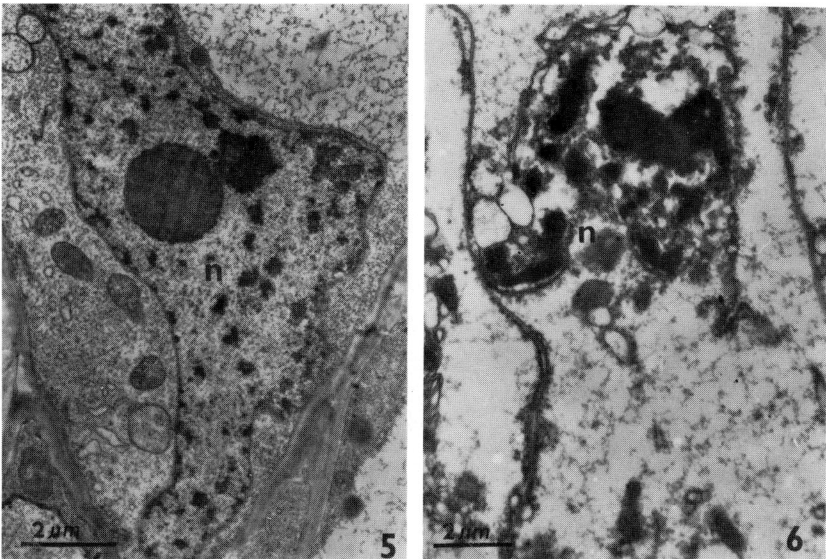

Fig. 5. Vascular parenchyma cell with quite well preserved structural integrity following 72 h water stress. n = nucleus. × 6375.

Fig. 6. Severely damaged nuclear structure in the epidermal cell stressed for 48 h. n = nucleus. × 5242,5.

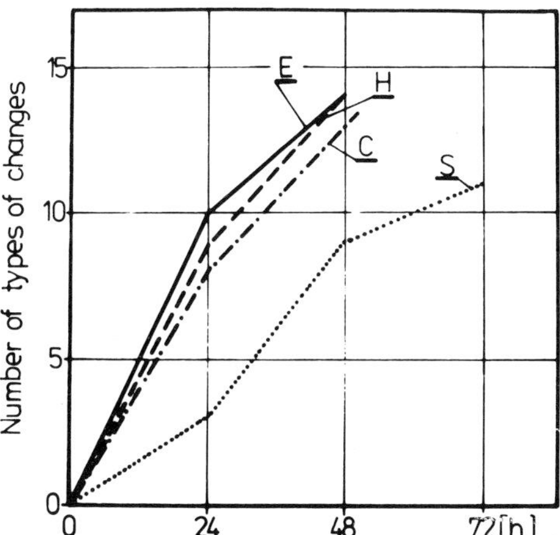

Fig. 7. Different degree of the cell structure damage following the increasing length of water stress duration in the cells of epidermis (E), hypodermis (H), cortex (C), endodermis and central cylinder (S).

than the adjacent cortical layers in maize roots [5]. Therefore the degree of their ultrastructure damage was almost as high as in the epidermal cells and higher than in the cells of the other cortical layers. At the three stages of differentiation observed, the cell ultrastructure was best preserved in the cells of the endodermis and the central cylinder. The degree of the ultrastructural damage caused by water stress follows a gradient that decreases centripetally. Apart from the differentiation stage of the stressed cells, the effects of water deficit on the maize root cell ultrastructure depend also on the site of the stressed cells within the root and, thereby on their distance from the unfavourable environment.

REFERENCES

1. Čiamporová, M. 1976 Ultrastructure of meristematic epidermal cells of maize root under water deficit conditions. Protoplasma 87, 1–15.
2. Čiamporová, M. 1977 Ultrastructure of differentiating epidermal cells of maize root under water deficit conditions. Biol. Plant. 19, 107–112.
3. Čiamporová, M. 1980 Ultrastructure of cortical cells of maize root under water stress conditions. Biol. Plant, in press.
4. Fellows, R. J. and Boyer, J. S. 1978 Altered ultrastructure of cells of sunflower leaves having low water potentials. Protoplasma 93, 381–395.
5. Luxová, M. 1975 Some aspects of the differentiation of primary root tissues. In: Torrey, J. G. and Clarkson, D. T. (eds.), The development and function of roots, pp. 73–90. London: Academic Press.

61. The effect of flooding on metabolism and structure of maize roots

GALINA M. GRINIEVA

K.A. Timiriazev Institute of Plant Physiology, USSR Academy of Sciences, Moscow 127296, U.S.S.R.

This paper is an attempt to analyse the adaptive mechanism in maize plants for maintaining tolerance to hypoxia occurring in flooded conditions. A specific feature of flooding, however, is that increasing hypoxia is associated with permanent watering [2–5, 8–12].

Maize plants are relatively tolerant to hypoxia and well adapted to flooding conditions. In our experiments the seeds (*Zea mays* L. cv. Voronezskaja 76) were germinated and after 10 days the plants were flooded with Knop culture solution diluted twice, so that the lower part of the stem was under water. The control plants were grown under conditions of normal humidity [3, 5, 11]. In young flooded plants an increased growth of adventitious roots and of the lower part of the stem was accompanied by morphological changes. The growth of the main root, however, was retarded (Table 1), without significant ultrastructural damage [3].

The specific morphological adaptive changes observed under flooding were related to the development of aerenchyma or intercellular gas spaces [6, 7]. It was found that the expansion of the primary cortex of the main root promoted the development of aerenchyma on the one hand and the ability to retain water on

Table 1. Effect of 8-day-flooding on some morphological and anatomical changes in young maize plants

Parameters	Control	Flooding
Plant height, cm	29.93 ± 0.72	25.13 ± 0.4
Main root length, cm	23.62 ± 0.37	18.36 ± 0.95
Number of adventitious roots	8.06 ± 0.22	11.19 ± 0.37
Leaf width, cm	1.06 ± 0.09	1.39 ± 0.11
Diameter of the stem base, cm	0.37 ± 0.06	0.47 ± 0.08
Cortex area of the main root, μm	13.7 ± 0.1	19.9 ± 0.1
Diameters (Σ) of the metaxylem vessels of the main root (μm)	15.5 ± 0.7	28.3 ± 0.4
Number of bundles in the cross section of the stem base	108.0 ± 2.4	116.0 ± 3.2
Water content in the stem base tissues, per cent	91.3 ± 0.3	93.7 ± 0.2

R. Brouwer et al. (eds.), Structure and Function of Plant Roots, 323–326.

the other. In the basal part of the stem the xylem vessels increased in size and number. The water was stored in the xylem vessels (Table 1), but the rate of water loss in this zone increased after flooding, because mechanical elements remained undeveloped. Thus, under flooding the rate of water loss by root tissues and by the basal part of the stem is not identical. In the light the growth of the flooded main root was inhibited, whereas the number of lateral roots, the growth of the stem base, the number of root vascular bundles, the area of root cortex increased and root aerenchyma developed. It should be stressed that these structural features of the flooded root system were not observed in plants grown in the dark.

The distribution of ^{14}C in the organs in conditions of flooding differed from that of control plants. The highest radioactivity during flooding was found in the tissues of the stem base, when the leaf was supplied with sucrose-U-^{14}C. The distribution of the label in the fractions of sugar, organic and amino acids in the main roots remained virtually unchanged during flooding. In the adventitious roots the radioactivity of organic acids decreased by a factor of almost 3, while that of amino acids increased by 4–5. Radioactivity in the fraction of organic acids in the lower part of the stem increased. The high radioactivity found in the sugar fraction was probably due to reduced metabolism of sucrose or accumulation of reserve sugars [5, 11].

The enhanced basipetal transport of assimilates caused active respiration in the basal part of the stem. Measurement of O_2 tension in tissues of the basal part of the stem showed that at longer flooding periods saturation with O_2 of tissues in this zone increased more than four fold, compared with the control possibly due to the spaces formed in leaf sheats. The tissues in the lower stem were enriched by oxygen transported from the leaves [7–10]. In young maize plants – as they were flooded during 3–7 days – respiration in the main roots gradually diminished whereas respiration in the adventitious roots appeared to have increased when compared with the control [3]. Intensive consumption of oxygen by adventitious roots was clearly associated with activation of glycolysis as appeared from the high CO_2 evolution and ethanol accumulation (Table 2). It was also shown that the content of the components in the respiratory chain underwent substantial change. The content of cytochromes *b* and *c* in the mitochondrial fraction from maize roots was proven to increase markedly after flooding when compared with the control [3]. It was also found – by means of fluorescent microscopy – [1] that the number of reduced pyridine nucleotides increased in the adventitious roots after flooding compared to that in the main roots (Table 2). Thus, on the flooding of young maize plants, respiratory metabolism is activated in adventitious roots and in the basal part of the stem. In the main root on the contrary, the activity of the respiratory system is reduced.

Table 2. The rate of respiration, ethanol and glucose content and the fluorescence intensity ratio of reduced pyridine nucleotides (Fpn)/oxidized flavoproteins (Ffp) after 8-day-flooding

Part of root	O_2 uptake	CO_2 evolution	Ethanol	Glucose	Fpn/Ffp
	$\mu l.h^{-1}$ mg protein		$mg.mg^{-1}$ protein		
Main root	102.6	119.0	0.56	2.1	10.17
Adventitious roots	181.7	203.1	0.94	2.6	21.70

In maize plants the adaptive responses to flooding were quite clearly manifested in tissues of the stem basal part and of adventitious roots by an enhanced growth while growth of the main root was retarded.

In the basal part of the stem the xylem vessels increased in size and number thus increasing the plant's ability to accumulate water. The content of oxygen increased due to the spaces formed in leaf sheaths. The enhanced basipetal transport of assimilates caused active respiration in the basal part of the stem with a rise in cytochrome content in the respiratory chain.

In the main root on the contrary, respiratory activity decreased. The expansion of the primary cortex of the main root promoted the ability to retain water on the one hand and the development of aerenchyma on the other. No essential ultrastructural damages were observed in the cells of the main root.

Under natural conditions, plant properties characteristic of flooding began to appear in response to the concurrent action of hypoxia, excess water and light. All our evidence confirms an important regulatory role of the base of the stem in the conditions of excess moisture in the corn plant.

REFERENCES

1. Chirkova, T. V., Dragunova, E. V. and Burgova, M. P. 1977 Redox reactions of flavoproteins and pyridine nucleotides from roots of plants different in their resistance to oxygen deficiency studied in vivo. Fiziol. rast. 24, 126–131.
2. Crawford, R. M. M. 1978 Metabolic adaptation to anoxia. In: Plant life in anaerobic environments, pp. 119–136. Mich.: Ann. Arbor Sci.
3. Grinieva, G. M. 1975 Control of plant metabolism under oxygen deficiency. Moscow: Nauka Publishers (in Russian).
4. Grinieva, G. M. 1980 The effect of flooding on corn root metabolism and structure. In: Structure and function of roots, 2nd Internat. Symposium, Abstracts, p. 26, Bratislava, Czechoslovakia.
5. Grinieva, G. M. and Nechiporenko, G. A. 1977 Distribution and transformation of sucrose-U-^{14}C in maize plants in conditions of flooding. Fiziol. rast. 74, 44–50.

6. Hook, D. D. and Scholtens, J. R. 1978 Adaptations and flood tolerance of tree species. In: Plant life in anaerobic environments, pp. 299–332. Mich.: Ann. Arbor Sci.
7. Kawase, M. 1979 Role of cellulase in aerenchyma development in sunflower. Amer. J. Bot. 66, 183–190.
8. Lambers, H. 1979 Energy metabolism in higher plants in different environments. Thesis, Groningen.
9. Luxmoore, R. T. and Stolzy, L. H. 1972 Oxygen diffusion in the soil plant system. VI. A synopsis with commentary. Agron. J. 63, 725–729.
10. McManmon, M. and Crawford, R. M. M. 1971 A metabolic theory of flooding tolerance: the significance of enzyme distribution and behaviour. New Phytol. 70, 299–306.
11. Nechiporenko, G. A. and Grineva, G. M. 1976 Effect of duration of flooding on distribution of ^{14}C-sucrose-U-in maize. Fiziol. rast. 23, 984–989.
12. Vartapetian, B. B., Andreeva, I. N., Nuritdinov, N. 1978 Plant cells under oxygen stress. In: Plant life in anaerobic environments, pp. 13–88. Mich.: Ann. Arbor Sci.

62. The response of *Zea mays* roots to chilling

JOZEF KOLEK, IGOR MISTRÍK, MARGITA HOLOBRADÁ, TATIANA PŠENÁKOVÁ and OTÍLIA GAŠPARÍKOVÁ

Institute of Experimental Biology and Ecology SAV, 885 34 Bratislava, Czechoslovakia

The hardiness of plants in withstanding chilling temperatures has captivated the interest not only of those concerned in theoretical research but of others involved in agronomical practice. While in plant breeding considerable attention has been paid to the temperate zone in order to obtain resistant cultivars, investigation into the biological principles of cold and frost resistance has been on the rise for only the last 10–15 years.

Investigations are generally focussed upon problems of frost influence observed in the aerial parts of plants. The plant root is a very rare object of research into cool temperature effects.

The maize roots response to temperatures down to 5 °C was examined with diversely resistant hybrids and with different near-isogenic lines of maize with different resistances to low temperature.

CHANGES IN THE ULTRASTRUCTURE OF ROOT CELLS

Numerous lipid drops show up after 3–4 days exposure to a temperature of 5 °C in the root cells of the non-hardy line (TVA 708) of maize. These drops concentrate near the membranes of cell organelles, the vacuole, the nucleus and the mitochondria [3]. The visualisation of lipid drops is obviously due to the breakdown of lipoprotein complexes of membranes.

LEAKAGE OF SUBSTANCES

Efflux of electrolytes

Under normal physiological conditions, a portion of the ions taken up by the roots is released into the root environment as a consequence of the dynamic maintenance of the ionic status of the plant.

The resistant line of maize (TVA 877) leaked more K^+ than the non-resistant line (W 64 A) during 5 days of exposure to a temperature of 5 °C (Fig. 1). This is in agreement with the course of the development of the effusate conductivity (Fig.

R. Brouwer et al. (eds.), Structure and Function of Plant Roots, 327–335.

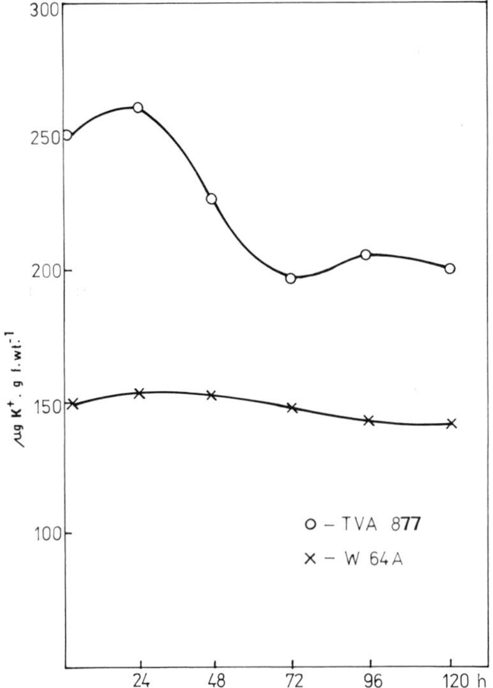

Fig. 1. Leakage of K^+ from the roots of resistant (TVA 877) and non-resistant (W 64 A) line of maize after exposure to chilling temperature (5 °C).

2). The leakage of Ca^{2+} also increased (Fig. 3) although its absolute amount was lower in relation to K^+. The resistant line leaked more electrolytes than the non-resistant one under the impact of chilling temperature. This difference, however, is only relative, since the root tissue of the resistant line contained more electrolytes than the tissue of the non-resistant line. When the amount of the K^+ leakage was subsequently converted as a proportion of overall content, the ratio was reversed. In comparing the values of net K^+ uptake in both lines, the amount of net uptake was found to be considerably higher in the non-resistant line at temperatures of 25 and 5 °C (Table 1).

From these experiments the conclusion may be drawn that a simple estimation of the amount of electrolytes leakage at a low temperature may not be as reliable a test for the resistance to cold as is often maintained. The amount of electrolyte leakage depends on their content in the tissue. The overall cation content and the

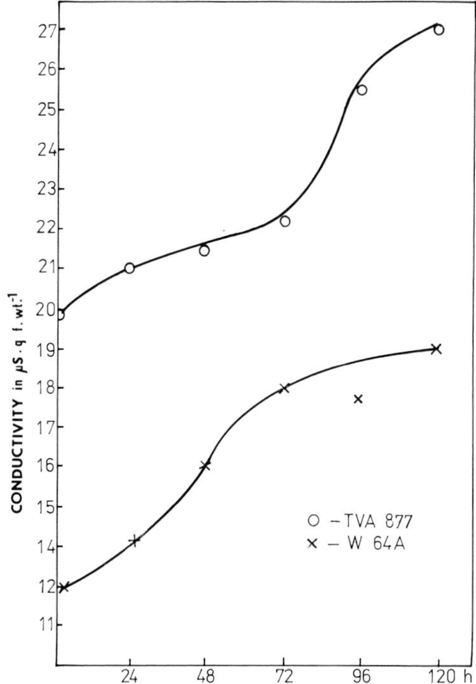

Fig. 2. Conductivity of effusate from the roots of resistant (TVA 877) and non-resistant (W 64 A) line of maize seedling exposed to the temperature of 5 °C.

Table 1. The net uptake values of maize TVA 877 (resistant) and W 64 A (non-resistant) at the two temperatures (in $\mu g.g^{-1}$ $F.W.H^{-1}$)

Line	Temperature	
	25 °C	5 °C
TVA 877	0.767	0.255
W 64 A	1.023	0.358

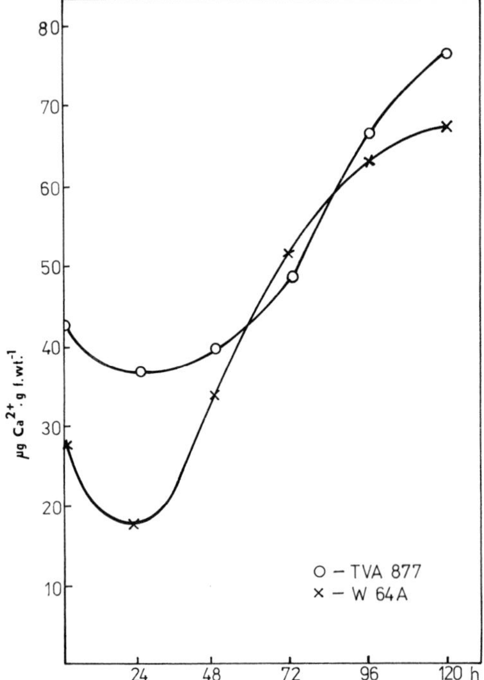

Fig. 3. Ca^{2+} content in the effusate from the roots of two lines of maize exposed to the temperature of 5 °C.

level of metabolic activity define the amount of the net uptake of that particular cation.

The leakage of sugars and substances with absorption at 256 and 280 nm

It is known [5] that approximately 80% of the organic substances leaked during the cold hardening process are sugars. For 5 days the leakage of sugars from the roots of maize seedlings was followed at a temperature of 5 °C with two lines of maize. As seen from Fig. 4, the non-resistant line (W 64 A) released more sugars than the resistant line (TVA 877). Sugar leakage decreased as did the leakage of K^+, which was traced at the same time.

The content of substances absorbed at 256 and 280 nm dropped in both lines (Table 2). Leakage of these substances was higher in the non-resistant line. At the

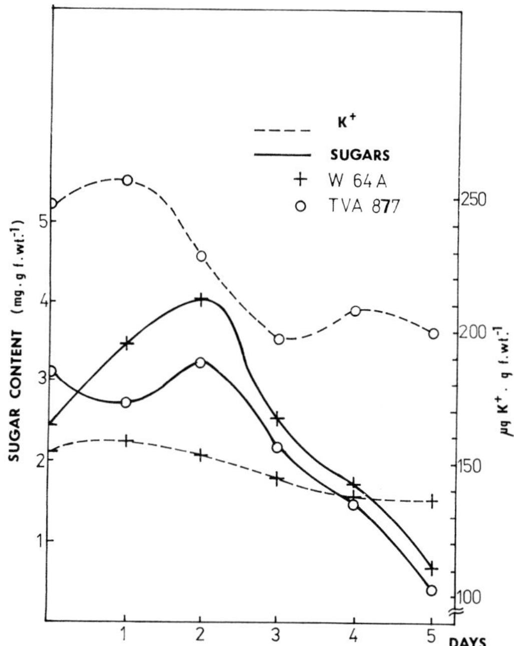

Fig. 4. Leakage of sugars from the roots of resistant (TVA 877) and non-resistant lines (W 65 A) of maize seedlings exposed to the chilling temperature of 5 °C.

Table 2. The leakage of organic substances with absorbancy at 265 and 280 nm

Temperature in °C	Line	Absorbancy, nm	
		265	280
25	TVA 677	0.208	0.170
	W 64 A	0.234	0.185
5	TVA 877	0.198	0.154
	W 64 A	0.209	0.182
frozen tissue	TVA 877	0.518	0.471
	W 64 A	0.710	0.642

end of the experiments, the tissue was frozen with liquid nitrogen. After thawing, substances were estimated in the effusate. The non-resistant line appeared to contain unproportionally more of these substances than the resistant line. As was true in the case of cation leakage, the amount of leakage is defined, to an essential degree, by the absolute content in the tissue.

UPTAKE AND LEAKAGE OF PHOSPHATE AND SULPHATE

The uptake of anions was markedly influenced by the chilling temperature [2]. While sulphate uptake dropped steadily during the 7 days of experiments, phosphate uptake seemed to adapt itself to the stress situation. The absorption rate as well as the metabolism of phosphate was nearly the same as it was at the start of the experiments. Differences were observed with various types of maize

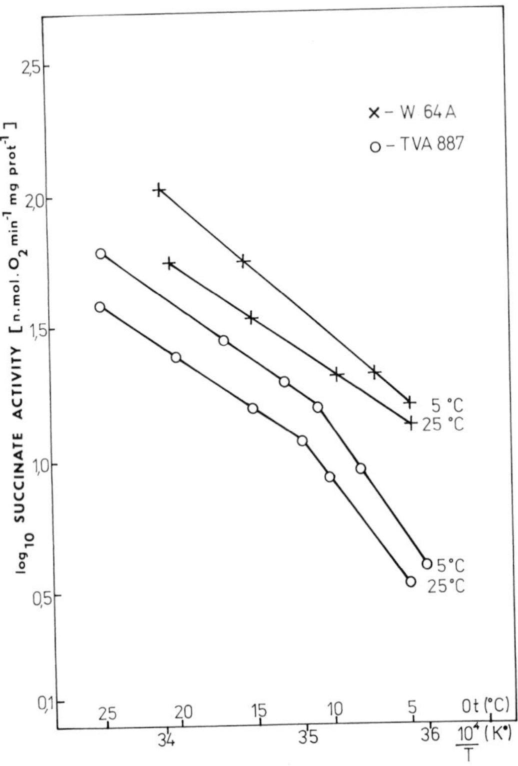

Fig. 5. Arrhenius plot of isolated mitochondria of the maize roots (succinate oxidation in the stage 3): the comparison of two lines of maize with different resistance.

roots in the uptake experiments. The nodal roots showed the greatest ability for anion absorption and differences were observed between two lines of maize with different levels of resistance against low temperature. This temperature, however, brought about an increase in anion leakage.

MEMBRANE SYSTEMS

The sensitivity and reactivity of membrane systems to stress is generally known. In the sensitive line, chilling produced a visualization of lipids which gathered along the membrane systems. Oxygen uptake was measured to verify the transition stage of mitochondrial membranes isolated from the roots of two different resistant lines. Fig. 5 shows the activity of succinate oxidation and Fig. 6 the activity of NADH oxidation. A typical break in the Arrhenius plot was as-

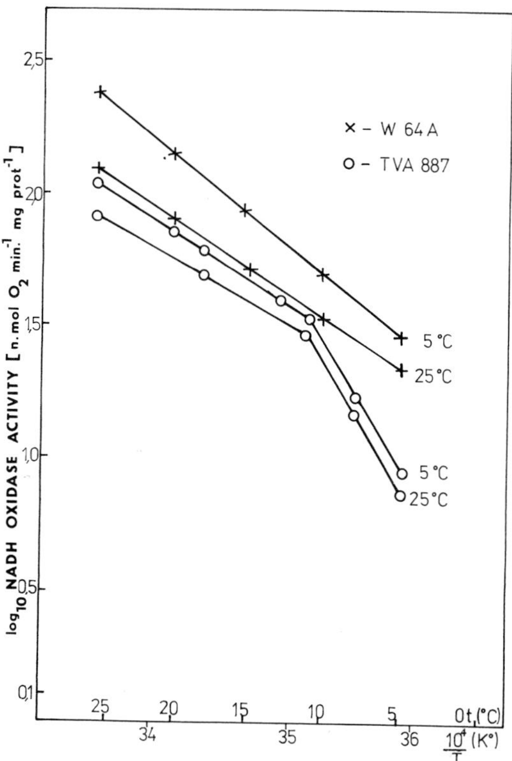

Fig. 6. Arrhenius plot of NADH oxidase activity within isolated maize mitochondria.

certained for the lines designated as resistant to low temperature. Conversely, the non-resistant line did not exhibit this break. Simultaneously, it may be seen from the graphs that it is not important whether or not mitochondria are isolated from plants grown at 25 or 5 °C. The important factor is the temperature at which oxygen uptake by the isolated mitochondria is measured.

In connection with the behaviour of isolated mitochondria under various temperatures and the level of lipids in the root tissue of both lines was also examined. The total lipid content remained at the same level after 5 days exposure to low temperature. The content of polar lipids increased. The dominant phospholipids are phosphatidylcholine and phosphatidylethanolamine. The level of both phospholipids rose with the length of exposure to low temperature, this rise being more conspicuous in the resistant line. The GLC analyses of free and bound fatty acids showed an increase in the content of unsaturated fatty acids of linoleic and linolenic acid in both lines. The double-bound index also increased. The resistance rate on isolated mitochondria yielded results opposite to those expected. The hardy line exhibited a marked change in the transition stage of the mitochondrial membrane within the temperature range of 10–12 °C. The degree of fatty acid desaturation expressed by the double-bond index was lower in the non-resistant line (1.59) than in the resistant one (1.61). The content of the bound linolenic acid in the 'resistant' line, however, doubled during its exposure to low temperature.

NUCLEIC ACIDS

Qualitative and quantitative changes in protein content are determined by mechanisms of their synthesis. Numerous experimental data like ours bear out the rise in soluble proteins during plant hardening. These changes are due more to the breakdown of large protein molecules than to their *de novo* synthesis. Our observation of nucleic acid metabolism under the influence of low temperature (5 °C) underscored two experimentally verified facts: first, that the synthetic apparatus of nucleic acids is relatively resistant to the chilling temperature, secondly, that even after exposure of one week this apparatus readapts itself very quickly to normal conditions. This seems to be due to the fact that the phosphate uptake and metabolism adapt to low temperature. In our experiments [1] low temperature produced a decline in RNA synthesis, but no essential differences between maize lines were found. With the lengthening exposure to low temperature a higher adaptability to chilling was noted in the resistant line. Low temperature causes an inhibition of the rRNA maturation process in the root and the inhibition of RNA synthesis is shown by slower incorporation of ^{3}H-uridine and by disturbances in transcription and maturation of rRNA.

REFERENCES

1. Gašpariková, O. 1981 The effect of low temperature on the RNA synthesis in Zea mays roots. Biológia (Bratislava) 36.
2. Holobradá, M., Mistrík, I. and Kolek, J. 1980 The effect of root temperature on the uptake of anions by the root system of *Zea mays* L. Biológia (Bratislava) 35, 259–265.
3. Klasová, A. 1980 Response of roots to reduced environmental temperature. Biológia (Bratislava) 35, 687–689.
4. Li, P. H. and Palta, J. P. 1978 Frost hardening and freezing stress in tuber-bearing Solanum species. In: Li, P. H. and Sakai, A. (eds.), Plant cold hardiness and freezing stress, pp. 49–71. New York: Academic Press.
5. Palta, J. P. and Li, P. H. 1978 Cell membrane properties in relation to freezing injury. In: Li, P. H. and Sakai, A. (eds.), Plant cold hardiness and freezing stress, pp. 93–116. New York: Academic Press.

63. The study of root growth under low temperatures at the cell level

OKTYABRINA PAVLOVNA RODCHENKO

Siberian Institute of Plant Physiology and Biochemistry, Siberian Branch, USSR Academy of Sciences, Irkutsk 33, U.S.S.R. 664033

The study of root growth at low temperatures was carried out simultaneously with the examination of the changes in the main metabolic processes: respiration, protein, nucleic acid and phosphorus metabolism. This enabled us to distinguish the important steps responsible for productivity.

Assuming the growth response of the roots to be a main criterion of resistance, we thought it necessary to determine the maximal productivity of meristem in unfavourable environments. Recent studies in the mechanism of the hormonal effect of the roots on above ground parts, stimulated an examination of the specific role of root growth in the function of its synthetic activity [1].

a) The root cells of the same age of three maize varieties with different resistance (Omskaya [2] – cold resistant, Hybrid Bukovinsky 3 – less resistant and Uzbekskaya zubovidnaya – inresistant) have been used in the work [2, 5].

b) The initial material was selected for growth rate in optimal environments. Plants with roots 3–4 cm long (after two days germination) were used for the experiment.

c) The response to low temperatures was determined under wide temperature changes.

d) The recovery rate of growth disturbances in the cells affected by the unfavourable factor was defined.

The rates of cell formation and cell change to elongation specify the variety response to low temperatures. The rate of cell formation decreases 13 times in Uzbekskaya zubovidnaya, 9 times in Hybrid Bukovinsky 3 and 6 times in Omskaya 2. The resistant variety continues growth in a stable regime, that is the rate of division is reduced in the same manner as the rate of cell changing to elongation which results in meristem retention and the high rate of recovery of the growth processes.

The detailed study of cell formation – kinetics of cell proliferation and the structure of mitotic cycle – allows us to conclude that the duration of the cycle depends on degree of cold resistance of the varieties. An increase in the mitotic

R. Brouwer et al. (eds.), Structure and Function of Plant Roots, 337–338.

cycle duration as well as a decrease in the number of multiplicating cells – the proliferation pool, cause the rate of cell multiplication to be reduced.

The duration of the mitotic cycle and its periods was determined graphically by the method of Quastler and Sherman and by Quastler equation taking into account the period of saturation of cell population with H^3-thymidine [3, 4].

Wide temperature reduction inhibited the passage of G_2 cells through the mitotic cycle much more than through all other periods including mitosis. The determination of the duration of mitosis by a radiation method just after low temperature exposure showed no significant variation in the duration of mitosis among the varieties. Varietal differences in mitosis duration occurred when the processes of mitotic cycle preceding mitosis were affected by low temperatures.

The state of walls in the elongating cells at low temperatures may estimate the resistance of the varieties. Varietal characteristics are revealed in the process of cell wall formation (the rate of formation, its composition). It reflects the effect of regulatory mechanisms which govern the chronological sequence of genetic programme realization in ontogenesis. The resistant variety shows the enhanced differentiation as evidenced by a number of factors: the ratio of the components of carbohydrate fraction in cell walls, increased content of dry matter in cell walls (due to hemicellulose and cellulose) and shorter time period necessary to reach the final sizes.

The susceptible variety shows a sharp decrease in the synthesis of all the components of the cell wall. The volume of the cells increases due to water saturation; the surface of the cell wall increases but its dry matter drops greatly and a large amount of pectin substances especially soluble pectins is accumulated.

The investigation shows that cell lignification begins long before the completion of cell growth. Under temperature reduction, the content of lignin increases, and the less the rate of its accumulation is reduced, the greater is the rate of cell growth retarded [1].

REFERENCES

1. Ivanov, V. B. 1979 Peculiarities of spatial and temporal cellular organization of root growth in relation to root function. Plant Physiol. 26, 888–898.
2. Maricheva, E. A., Rodchenko, O. P. and Nechaeva, L. V. 1975 Respiration of the cells of maize roots during deceleration of their growth by lowered temperature. Fiziol Rastenii 22, 591–597.
3. Quastler, H. and Sherman, F. G. 1959 Cell population kinetics in the intestinal epithelium of the mouse. Exptl. Cell Res. 17, 420–438.
4. Quastler, H. 1960 Cell population kinetics. Ann. N. Y. Acad. Sci. 90, 580–591.
5. Rodchenko, O. P. and Maricheva, E. A. 1973 Effect of low positive temperature on growth and content of protein nitrogen in the cells of maize root. Fiziol Rastenii 20, 597–602.
6. Rodchenko, O. P., Skvortsova, R. G. and Akimova, G. P. 1979 Varietal responses of maize to low temperatures. Fiziol. Biokhim. Kult. Rastenii 3, 229–234.

64. Nitrogen assimilation in wheat in relation to the root zone temperature

N. D. ALJOCHINA, A. I. KLUIKOVA and S. S. KENZHEBAEVA

Plant Physiology Department, Biology Faculty, Moscow State University, U.S.S.R. 117234

Temperature is one of the most important environmental factors controlling growth and development of plants. Low temperature produces changes in metabolic processes and plants mobilize various mechanisms to neutralise the negative action of unfavourable temperature. One of the mechanisms of adaptation is that of counteracting 'disbalance' of the rates of processes and retention of the required intensity of metabolism. This mechanism of adaptation is based on the changes in enzyme activity [1, 5].

Labeled nitrogen absorption and the activity of enzymes (nitrate reductase-NR, glutamate dehydrogenase-GDH, glutamine synthetase-GS, glutamate synthase-GOGAT) in cold-hardened winter wheat and maize requiring warmth, were investigated. The seedlings were grown in a nutrient solution (NH_4NO_3 concentration – 1.5 mM) at 25 °C (treatment 1 or 12 °C (treatment 2). The air temperature was 25 °C in both cases. Extraction and enzyme assays were the same as described [2, 3].

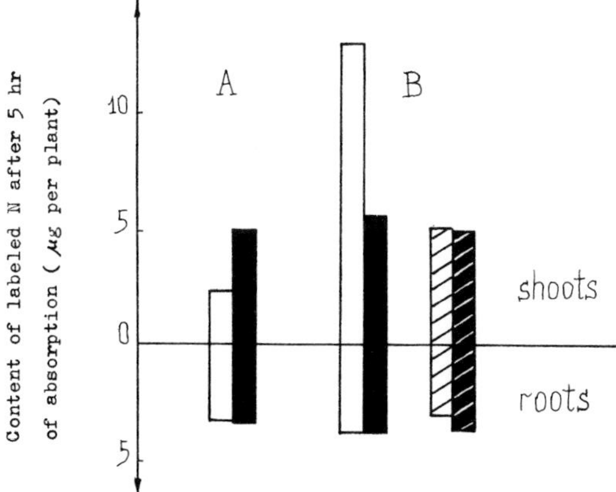

Fig. 1. Influence of the temperature on the absorption and translocation of ammonium (A) and nitrate (B) by 12-day old wheat seedlings. White area: treatment 1; black area: treatment 2; hatched area: reduced nitrate of treatment 1 and 2.

R. Brouwer et al. (eds.), Structure and Function of Plant Roots, 339–341.

Table 1. Total activity of the enzymes of the wheat seedlings (μmol product per plant in hour)

Treatment		NR	GS	GDH	GOGAT
1	Shoots	10.73	6.62	1.80	0.20
	Roots	0.95	0.39	0.59	0.02
2	Shoots	13.23	1.15	0.95	0.16
	Roots	2.01	0.85	1.16	0.06

A low temperature is required at early stages of winter wheat development. Under these conditions the metabolic processes, and the activities of enzymes which regulate them, reflect the genetic adaptation of a winter wheat to lower temperature.

There was a marked decrease in nitrate uptake and a slight increase in ammonium uptake at 12 °C (see Fig. 1). All nitrates were reduced in the roots and the shoots of treatment 2. These results correlated with data on NR activity (see Table 1).

The NR, GDH, GS, and GOGAT activities were higher in the roots of the wheat seedlings, grown at the low temperature (Table 1). The high level of catalytic activity displayed by the winter wheat enzymes is a result of the adaptation to the growth in cold seasons and provides an effective nitrogen assimilation at low temperature.

Low temperature is unfavourable for development of maize. Using labeled nitrogen, it was demonstrated that there is an increase of glutamine synthesis in roots of 15-day old maize grown at 12 °C [4]. GDH, GS and GOGAT levels were influenced differently by temperature. The NADH- and NADPH-dependent GDH activity as well as that of GOGAT were lower in the roots at 12 °C. The GS activity was higher [2, 3]. The increase of the GS level is associated with the adaptive changes of its properties. The energies of activation for enzymes of treatments 1 and 2 were 18.3 and 10.1 kkal/mol, respectively. Michaelis constants of GS from maize roots grown at 25 °C were 8.6 mM for glutamate and 1.0 mM for NH_4^+. For treatment 2 GS half-saturation concentrations of glutamate and NH_4^+ were 6.8 mM and 0.2 mM, respectively. The enzymes from roots of treatment 1 and 2 had different optimum temperatures (50 °C and 47 °C) and were distinguished by their thermostability.

Changes in all these characteristics indicate that adaptive rearrangement in the GS system in roots of maize seedlings occurs due to temperature of growth.

REFERENCES

1. Roberts, D. W. A. 1969 Some possible role for isozymic substitution during cold hardening in plants. Int. Rev. Cyt. 26, 303–325.
2. Aljokchina, N. D. and Kenzhebaeva, S. S. 1977 The activity of glutamate synthetase in roots in relation to temperature conditions during the growth. Fiziol. rast. 24, 1129–1134 (in Russian).
3. Aljokchina, N. D. and Sokolova, S. A. 1975 Changes in glutamate synthetase activity during the growth in the different temperature conditions. Fiziol. rast. 22, 97–102 (in Russian).
4. Aljokchina, N. D. and Shirshova, E. D. 1970 Nitrogen assimilation with plants. Proc. of Univ. Biol. Sci. I, pp. 5–18 (in Russian).
5. Kchotchachka, P. and Somero, J. 1977 The strategy of biochemical adaptation. Moskva: Mir, pp. 205–308 (in Russian).

65. Effect of aluminium toxicity on root morphology of barley

CH. HECHT-BUCHHOLZ and C. D. FOY

Institut für Nutzpflanzenforschung, Pflanzenernährung, Technische Universität Berlin (West), Germany

Light- and electronmicroscopic investigations were carried out in order to find the differences, in order to explain the differential Al tolerance of two barley varieties [2], Kearney (Al-sensitive) and Dayton (Al-tolerant). The plants were exposed to 9 ppm Al in an acidified nutrient solution (pH 4,8).

RESULTS AND DISCUSSION

Typical symptoms

Typical symptoms of Al toxicity [6] characterized by stunted brown root tips and inhibition of the newly emerging lateral roots occurred in the more tolerant Dayton some days later than in the Kearney.

Incipient symptoms

Incipient symptoms of Al toxicity in the root tips of both varieties observed with the light microscope also in wheat [5] and in maize [1] were: destruction of root cap cells, swelling and destruction of epidermis and cortex cells resulting in a disintegrated outer shape.

Electronmicroscopic investigations revealed rapid autolysis of the Al-affected cells, beginning with the disorganization of the plasmalemma. This was very similar to changes described in Ca-deficient tissue [3, 4]. The Al-induced changes in cell structure occurred in Dayton later (2–4 d) than in Kearney (24 h).

R. Brouwer et al. (eds.), Structure and Function of Plant Roots, 343–345.

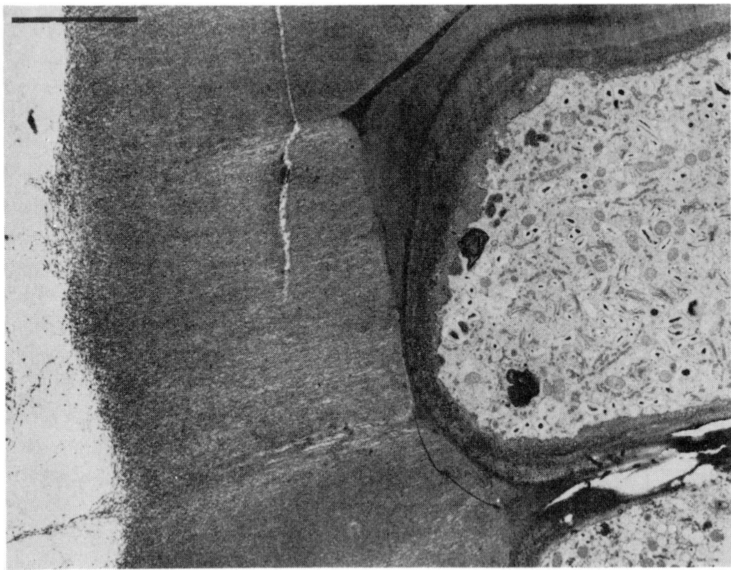

Fig. 1. Part of root cap cell of Kearney (control plant) with ruthenium stainable mucigel outside and between cell wall and plasmalemma. The ultrastructure of the cytoplasm (mitochondria, secret vesicles of the golgi apparatus) is intact. Scale line = 5 μm.

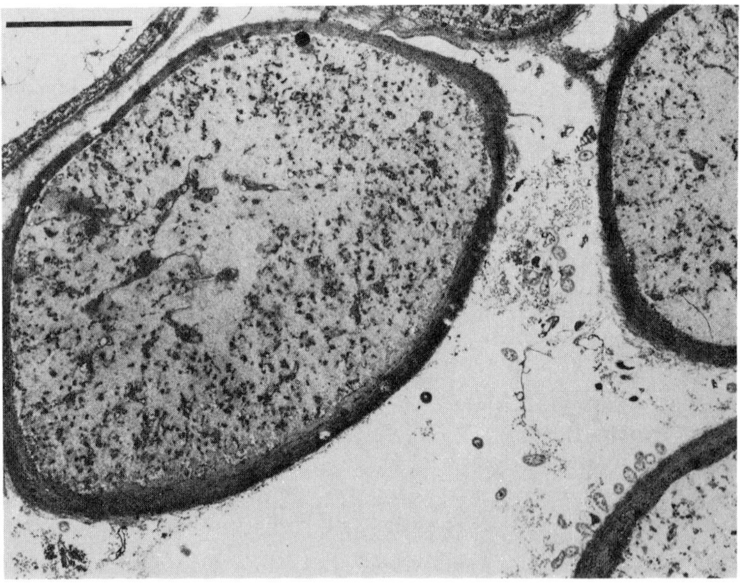

Fig. 2. Root cap cell of Kearney after exposure to Al. The ultrastructure of the cytoplasm is destroyed. No mucigel is visible outside the cell. Scale line = 5 μm.

MUCIGEL PRODUCTION

A greater production of slime (mucigel) could be responsible for the longer resistance of Dayton to Al stress. When Al was not added (control plant) the difference between Kearney and Dayton regarding the mucigel layer around the root cap was not apparent. The electronmicroscopic investigation revealed more mucigel inside the slime secreting root cap cells after 6 and 24 h exposure to Al in Dayton than in Kearney. This is probably due to a secondary effect, as the root cap cells of Dayton are resistant to Al toxicity for a longer period of time than are those of Kearney. After 4 days of exposure to Al the mucigel covering the root cap was destroyed. Furthermore, the ultrastructure of the mucigel secreting root cap cells were impaired both in Kearney (Figs. 1, 2) and in Dayton.

CONCLUSION

The greater tolerance of Dayton to Al can be explained by a longer resistance of the plasmalemma towards Al stress. Whether a stronger mucigel production could be resonsible for increased Al tolerance has to be examined in more detail.

ACKNOWLEDGMENT

The author would like to thank Carmen Wolfram for skilful assistance.

REFERENCES

1. Baier, R., Münnich, H., Heinke, F. und Göring, H. 1976 Zytologische Untersuchungen zur Wirkung von Aluminium auf Maiswurzeln. Wiss. Z. d. Humboldt-Univ. zu Berlin, Math. Nat. R XXV, 840–844.
2. Foy, C. D., Armiger, W. H., Briggle, L. W. and Reid, D. A. 1965 Different aluminium tolerance of wheat and barley varieties in acid soils. Agron. J. 57, 413–417.
3. Hecht-Buchholz, Ch. 1979 Calcium deficiency and plant ultrastructure. Commun. in Soil Science and Plant Analysis 10, 67–81.
4. Hecht-Buchholz, Ch. 1978 Aluminium toxicity in barley – a light and electronmicroscopic study. Proc. of the 8th Intern. Colloquium on Plant Analysis and Fertilizer Problems Auckland, New Zealand, N.Z. DSIR Inform. 134, 179.
5. Henning, S. J. 1975 Aluminium toxicity in the primary meristem of wheat roots. Ph. D. Thesis, Oregon State Univ., Corvallis Oregon.
6. McLean, F. T. and Gilbert, B. E. 1927 The relative aluminium tolerance of crop plants. Soil Sci. 24, 163–175.

66. Effect of heavy metals and chelating agents on potassium uptake of cereal roots

CLAIRE BUJTÁS and EDITH CSEH

Department of Plant Physiology, Eötvös University, H-1445, Budapest, Hungary

Heavy metals are being deposited on both soil and plants in undesirable ratios and quantity as a result of industrial effluent. Some of these metals (e.g. Fe, Cu, Zn, Mn, Co) are essential in low and inhibitory in high concentration; others (e.g. Cd, Pb) have an overall toxicity. These ambiguous aspects of the content of trace elements in the soil raise genetical, physiological and ecological problems.

It has been suggested and in some cases proved that metals inhibit the transport ATPases of the plasmalemma of roots [1, 3]. Therefore – investigating the primary effect of Cu^{2+} and Cd^{2+} on plant metabolism –, we studied the effect of these ions on K^+ uptake in excised cereal roots. We have measured the quantity of K^+ taken up by radioactive tracer methods using $^{86}Rb^+$.

Cu^{2+} and Cd^{2+} strongly inhibit the uptake of K^+ in wheat (Fig. 1), but they have little if any effect on K^+ absorption of boiled roots or on that of living roots at temperatures near to 0 °C, and they do not affect the electrolyte efflux from previously K^+-saturated roots. We have concluded that Cu^{2+} and Cd^{2+} affect only the metabolism-dependent absorption of K^+. They do not alter the passive *ad*sorption, the value of which is estimated from these experiments to be about 0.1–0.3 μmol K^+/g fresh weight.

K^+ uptake of winter wheat varieties shows a minimum in late winter (Fig. 2). The absolute value of the minimum is very near to the value of the passive adsorption. Cu^{2+}-inhibited K^+ uptake is always nearly equal to the passive adsorption. So, the inhibitory effect of Cu^{2+} is far greater in the natural period of active growth than in late winter in the slowly growing plants. (These plants were always grown in controlled environment in darkness.) A seasonal change can

R. Brouwer et al. (eds.), Structure and Function of Plant Roots, 347–350.

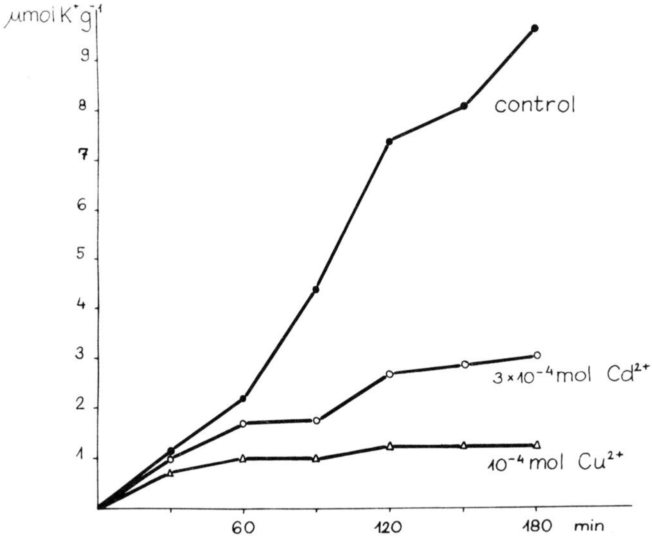

Fig. 1. Time-course of Cd^{2+}- and Cu^{2+}-inhibition of K^+ uptake in excised wheat roots (uptake solution 5×10^{-4} mol $CaSO_4$ + 10^{-4} mol KCl).

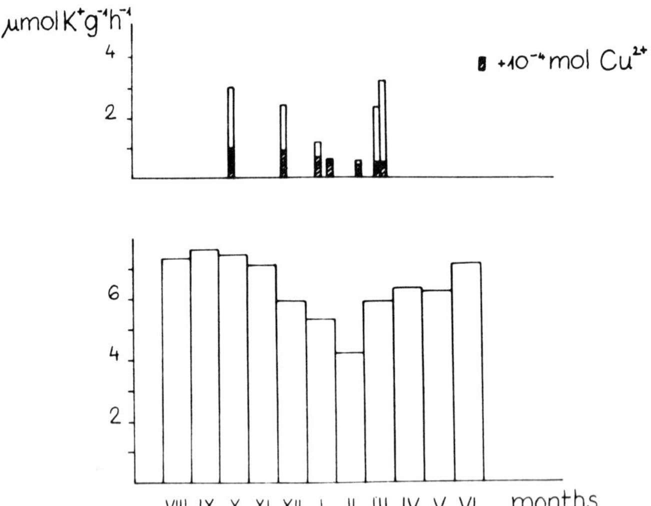

Fig. 2. Seasonal change in K^+ uptake of two winter wheat varieties (upper diagram: GK-Szeged, lower diagram: F481).

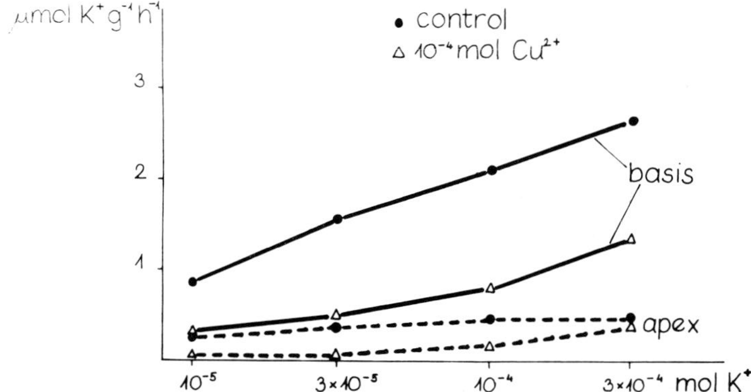

Fig. 3. Effect of Cu^{2+} on K^+ uptake of maize roots (uptake solution 5×10^{-4} mol $CaSO_4$ + 10^{-4} mol KCl).

also be observed in the K^+-absorption of apical and basal parts of maize seminal roots.

In maize roots Cu^{2+} inhibits the uptake of K^+ both in basal and apical parts (Fig. 3). The value of the Cu^{2+}-inhibited K^+ uptake of the apical parts is nearly equal to the passive K^+ adsorption of boiled maize roots. Cd^{2+} strongly inhibits

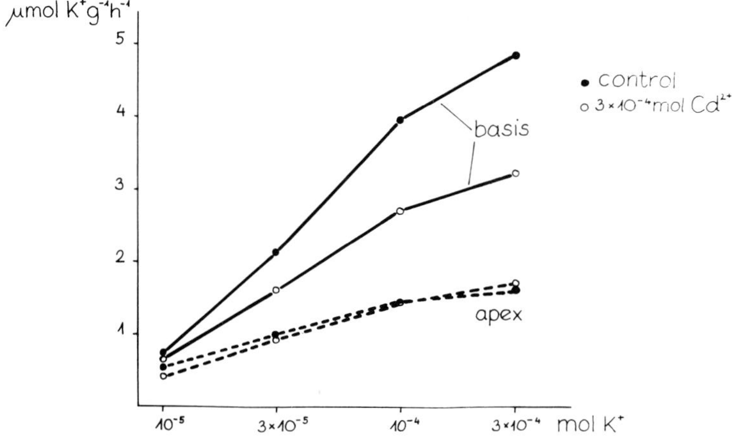

Fig. 4. Effect of Cd^{2+} on K^+ uptake of maize roots (uptake solution 5×10^{-4} mol $CaSO_4$ + 10^{-4} mol KCl).

the K^+ influx in basical parts but it has not any marked effect in the apical parts (Fig. 4).

Chelators added to the nutrient solution in appropriate proportion to the heavy metals might protect cereal plants growing in water culture from the damage caused by the metals, especially from Cu^{2+} [2]. In excised wheat roots the inhibition of K^+ uptake by Cu^{2+} can also be prevented by adding chelators simultaneously with Cu^{2+}. If the inhibition has already been developed, it can be reversed to some extent by washing the roots for more than 1 h with chelators.

REFERENCES

1. Cocucci, M., Ballarin-Denti, A. and Marré, M. T. 1980 Effects of orthovanadate on H^+ secretion, K^+ uptake, electrical potential difference and membrane ATPase activities of higher plant tissues. Plant Sci. Lett. 17, 391–400.
2. Cseh, E., Bujtás, C. and Szabados, M. 1979 Preliminary notes about the effects of heavy metals on plant nutrition. Proc. 1st Internat. Symp. Plant Nutrition, Varna, Bulgaria. Vol. II, pp. 315–319.
3. Keck, R. W. 1978 Cadmium alteration of root physiology and potassium ion fluxes. Plant Physiol. 62, 94–96.

67. Transport and metabolism of sulphate under salt stress

NINA I. SHEVYAKOVA

K.A. Timiriazev Institute of Plant Physiology, Academy of Sciences, Moscow 127276, U.S.S.R.

The incorporation of sulphur into important cell components such as sulphur-containing amino acids, vitamins, proteins, lipids and so on begins with the uptake of sulphate followed by some initial stages of its conversion. Under the normal concentration of sulphate ions in the soil sulphur metabolism in plants proceeds at the optimal level. But under soil salinity a surplus of sulphate and chloride in plants leads to drastic changes in sulphur metabolism.

The excess of sulphate induces the synthesis of secondary-type S-metabolites such as 3′-phosphoadenosine 5′-phosphosulphate (PAPS), taurine, methionine sulphoxide and sulphone, some of which, we believe, are responsible for the survival of plants under sulphate salinity. For instance, the binding of the sulphate group from PAPS to the cell wall or membrane lipids, polysaccharides and proteins results in an increase in hydrophilic properties of these compounds and an enhancement of their capability of exchanging ions. The accumulation of other metabolites (methionine sulphoxide) can be seen in the plants of extremely high salt status under a severe irreversible disturbance of the sulphur metabolism. The high concentration of methionine sulphoxide leads to toxic effects in plant tissues. The oxidative degradation of the toxic metabolites to inorganic sulphate is of importance for the adaptive changes in metabolic processes and is typical of conditions of sulphate salinity. It prevents the accumulation of toxic S-metabolites in plant tissues and it takes place in plants only under moderately high sulphate ion concentrations in the environment (50–100 mM Na_2SO_4).

Another type of sulphur metabolism occurs in plants cultivated under chloride salinity. Disturbances in sulphur metabolism in this type of salinity are induced mainly through chloride inhibition of oxidative reactions, which in turn leads to the restriction of formation of metabolites. As a result, the sulphur oxidative cycle in plants is inhibited and sulphur starvation occurs. The content of sulphur in plants drops almost to the level of sulphur deficit. Thus there are two different types of adaptive changes and drastic disturbances of the sulphur metabolism in plants under chloride and sulphate salinity.

These metabolic changes are closely connected with the transport of sulphate. Experiments show that [1] the main feature of sulphate transport is its complete

R. Brouwer et al. (eds.), Structure and Function of Plant Roots, 351–353.

dependence upon metabolic processes. We believe that low-molecular-weight products of sulphur metabolism – such as sulphur-containing amino acids – active sulphate and others are the modifiers of allosteric or genetic types in carrier-mediated transport of sulphate to the roots. It should, however, be taken into account that under natural salinity all these mechanisms are influenced and complicated by the environment. So, in the presence of Na_2SO_4 the functional activity of these mechanisms in plant cells is diminished as a result of the activation of methionine and methionine sulphoxide oxidative degradation to inorganic sulphate. That is why the methionine sulphoxide repression of ^{35}S uptake by excised cotton roots is observed only in the culture solutions free of Na_2SO_4. Moreover, Na_2SO_4 (42 mM) represses methionine conversion into cysteine [2]. Some *Aspergillus* mutants are unable to metabolize methionine to homocysteine. The methionine added to the media fails to regulate biosynthesis of S-amino acids in these mutants too. This suggests that under sulphate salinity neither methionine nor methionine sulphoxide are able to control ^{35}S uptake. But the transport inhibition or repression may be carried out by metabolic sulphate (Fig. 1). As a result, a differential uptake of ^{35}S from salt-free solution by plants raised in sulphate or chloride salinity is observed. In the case of sulphate salinity, ^{35}S uptake decreases. In chloride salinity its uptake increases. The stimulation of ^{35}S uptake is closely connected with sulphur deficit in plants under chloride salinity; it can be recovered by certain inhibitors (puromycin, putrescine).

Thus, in plants growing under conditions of sulphate and chloride salinity the specific changes of sulphate uptake systems are observed [3]. These changes are

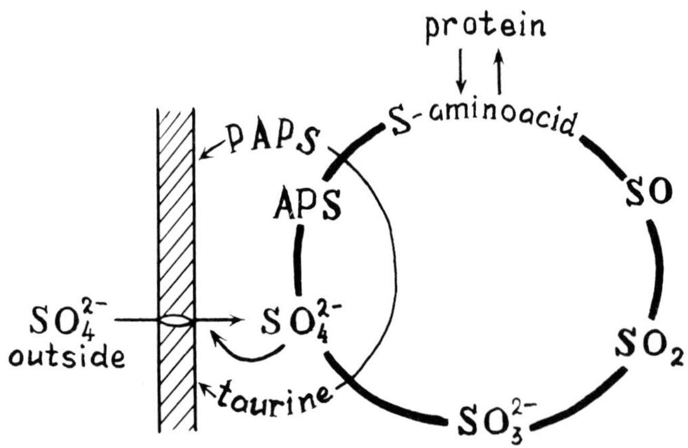

Fig. 1. Sulphur oxidative cycle in plants under Na_2SO_4 excess.

dependent in a high degree upon the peculiarities of adaptive reactions in sulphur metabolism.

REFERENCES

1. Bradfield, G., Somerfield, P., Mein, T., Holby, M., Babcock, D., Bradley, D. and Segel, J. H. 1970 Regulation of sulfate transport in filamentous fungi. Plant Physiol. 46, 720–727.
2. Paszewski, A. and Grabski, J. 1974 Regulation of S-amino acids biosynthesis in Aspergillus nidulans. Role of cysteine and or homocysteine as regulatory effects. Mol. Gen. Genet. 132, 307–320.
3. Shevyakova, N. I. 1979 Sulphur metabolism in plants, pp. 167. Moscow: Nauka.

VI. INTERACTION BETWEEN ROOTS AND SHOOTS

68. Shoot: root functional equilibria Thermodynamic stability of the plant system

RYSZARD K. SZANIAWSKI*

Department of Botany, Technical University, D-8 München 2, Germany

Considerable interest is shown in the thermodynamic description of biological systems. Plants like other living organisms are thermodynamic open systems. Such systems tend to maintain steady state (or homeostasis) keeping an orderly balance among subsystems that process matter-energy or information.

There are several examples available from the literature that strongly support the idea of homeostatic equilibrium between shoot- and root-subsystems. Thus, light of one-third to full intensity had little effect on the relative distribution of dry matter among shoots and roots of Loblolly pine seedlings [6]. Under stable nutrient stress, after some time for adaptation, growth rates of the organs of birch seedlings returned to normal [5]; Irrespective of root temperature during plant growth, the constant ratio between photosynthetic activity of the shoots and salt uptake activity of the roots was reported for several plant species [2].

Our experiments illustrate the existence of a functional equilibrium between shoots and roots. When sunflower plants were grown in nutrient solution at different root temperatures (10 to 30 °C) and one common shoot ambient temperature (25 °C) – given a short time for adaptation – no significant effect on relative growth pattern of shoots and roots was noted. This was shown by the constancy of the regression coefficients of the allometric relationship between shoots and roots growth (Table 1). Such stability (i.e. when a growing organism does not attempt to maintain a constant size, but rather attempts to maintain a constant growth pattern) is often called homeohesis [1]. The result is healthy and well-balanced individuals almost independent of root temperature conditions.

Equilibrium between shoots and roots is also seen from the results presented in Fig. 1: the maintenance respiratory activity of the shoots and of the roots remained in a constant proportion, irrespective of root temperature conditions applied during plant growth. Maintenance respiration is a function of protein turnover and may also reflect metabolic activity of living organs. In other words,

* Permanent address: R. K. Szaniawski, 02-541 Warsaw, Narbutta 20, Poland.

R. Brouwer et al. (eds.), Structure and Function of Plant Roots, 357–360.

Table 1. The allometric relationship: $\lg(y) = a + k \lg(x)$, where x is dry weight of roots and y is dry weight of shoots of sunflower plants after adaptation to different root temperatures

Root temperature (°C)	Regression coefficient k	Correlation coefficient r^2
10	1.13	0.93
20	1.19	0.95
30	1.15	0.94
mean for all treatments	1.15	0.95

energy from maintenance respiration is used mainly to combat the spontaneous entropy production which characterize the nonequilibrium systems. Such a balance as presented in Fig. 1 indicates that shoot and root subsystems are not only in functional equilibrium, but also that their basal metabolism processes are integrated enough to become actively self-regulating.

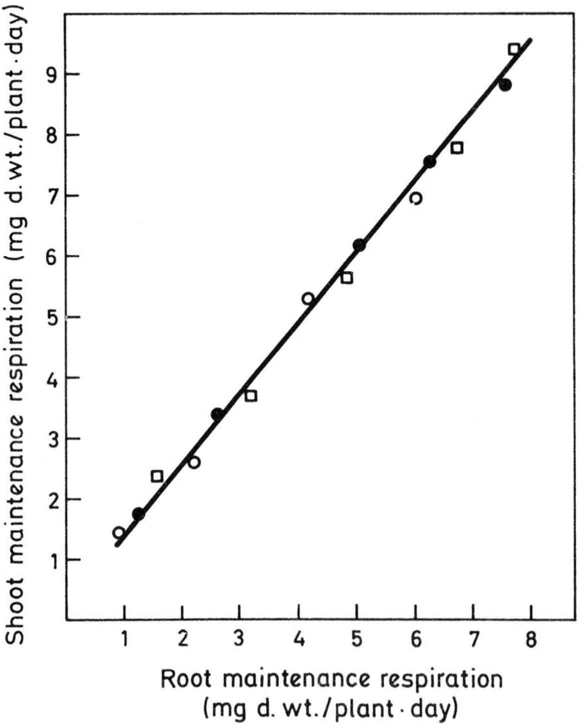

Fig. 1. The balance between shoots and roots maintenance respiratory activity. Root temperature: ●, 10 °C; ○, 20 °C; □, 30 °C.

Table 2. Respiratory losses from roots of sunflower plants after adaptation to different root temperatures

Root temperature (°C)	Per cent of gross photosynthesis respired by roots
10	14.7
20	14.4
30	12.7

Another aspect of the energetic balance between shoots and roots of sunflower plants is presented in Table 2. At all temperatures applied during plant growth the respiratory losses from the roots were relatively constant if expressed as the percentage of gross photosynthesis activity. This may again suggest that under a given set of environmental conditions, metabolic activity of shoots and roots efficiently adjust to each other.

Classical thermodynamics deals with static concepts and its conclusions are independent of the mechanisms that lead to the equilibrium end stages. Steady state stages in all living systems are controlled by negative feed-backs. The kind of informational channel involved in the regulation of described equilibria – whether or not it is a hormonal regulation and/or a simple negative feed-back which operates through the changes in substrate concentration, or those in water balance – is still difficult to evaluate. It seems that within a wide, but still physiological range of environmental conditions, the plant system is able to achieve homeostatic balance between shoot and root growth and functioning activity. Such balances can only be achieved through very efficient adjustment processes, which should show auto-oscillating attainment of equilibrium [4]. This last fact was experimentally confirmed [3]. It was not found however, in the experiments presented here, probably because of the fast growth and the relatively short time of adaptation after transferring the plants to different root temperature regimes.

ACKNOWLEDGMENT

The work was supported by a grant from the Alexander von Humboldt Foundation.

REFERENCES

1. Callow, P. 1976 Biological machines – a cybernetical approach to life, 1th Edn. London: Arnold (Publishers) Limited.

2. Davidson, R. L. 1969 Effect of root/leaf temperature differential on shoot/root ratios in some pasture grasses and clover. Ann. Bot. 33, 561–569.
3. Drew, A. P. and Ledig, F. T. 1980 Episodic growth and relative shoot : root balance in Loblolly pine seedlings. Ann. Bot. 45, 143–148.
4. Gladyshev, G. P., Ershov, Y. A. and Loshchilov, V. I. 1980 Thermodynamic principles of the behavior of biological systems. J. Theor. Biol. 83, 17–42.
5. Ingestad, T. 1980 Growth and nutrient allocation at stable nutrient stress. In: FESPP II Congress Abstracts, pp. 79, University of Santiago de Compostela, Spain.
6. Ledig, F. T., Bormann, F. H. and Wenger, K. F. 1970 The distribution of dry matter growth between shoot and roots in Loblolly pine. Bot. Gaz. 131, 349–359.

69. Root-shoot relationships in mature grass plants

ARTHUR TROUGHTON

Welsh Plant Breeding Station, Aberystwyth, SY23 2EB, U.K.

Under suitable conditions a perennial grass plant produces new root axes from the base of its tillers almost continuously, whilst old ones senesce and decay. So at any time the root system is composed of axes (main roots originating directly from tillers) of varying ages. If the tiller bases are kept dry, production of new axes ceases and the plants are forced to live with ageing ones. This results in a slow decline in shoot growth, followed eventually by death of the plant.

The cause of death is presumably linked to the ageing of the root axes, young ones evidently performing some function not carried out by older ones. This is not the ability to absorb and translocate water to the shoot as the rate of transpiration per unit leaf weight or area and leaf water deficit have not been found to differ significantly between plants with only old roots and with axes of varying ages. Likewise the ability to absorb and translocate nitrogen, phosphorus and potassium to the shoot system has been found to be unrelated to the age of the root axes. The ability to absorb and translocate calcium has, however, been found to decline with the age of the root axes, and the evidence strongly suggests that the plants eventually die of a deficiency of this element [6].

The relationship between longevity of plants not producing new axes and the ability of the ageing axes to absorb and translocate calcium was studied. This was done by comparing the percentage of calcium in young shoot growth, removed by periodic defoliation, with the length of time taken for the plants to die. The percentage of calcium was determined in a sample of tissue derived from all the living plants in a treatment and so reflects the ability of plants which were still growing, not those on the point of death, to absorb and translocate calcium.

Plants of different species varied considerably in their length of life when not producing additional root axes and this was related to the percentage of calcium in the shoot tissue of growing plants (Fig. 1). A similar though not such an

R. Brouwer et al. (eds.), Structure and Function of Plant Roots, 361–365.

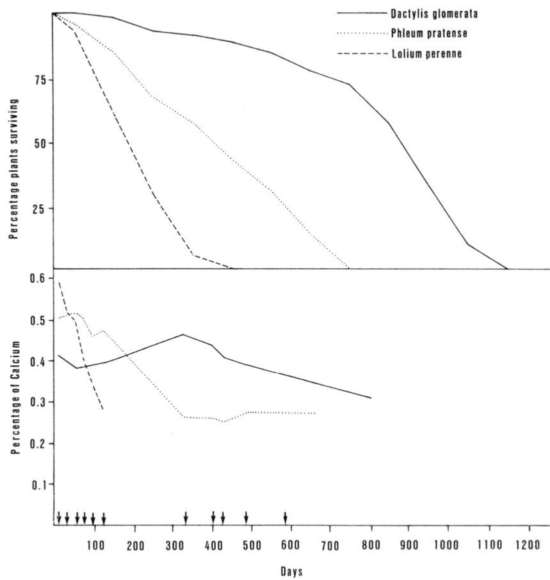

Fig. 1. Comparison of plants of different species not producing additional root axes. Percentage of plants surviving and percentage of calcium in the shoot tissue removed by defoliation.

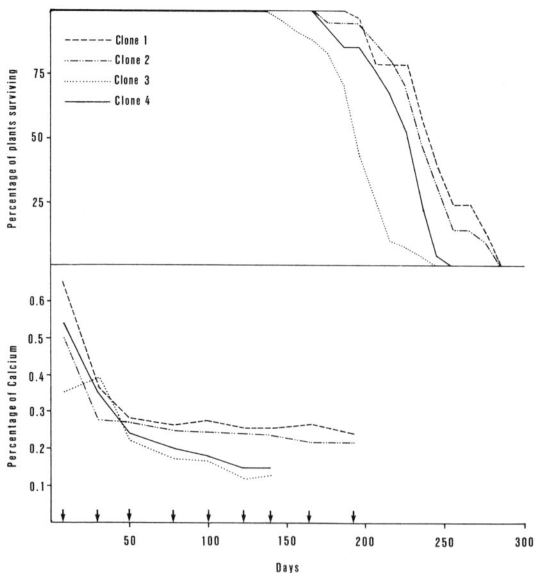

Fig. 2. Comparison of clones of *Lolium perenne* not producing additional root axes. Percentage of plants surviving and percentage of calcium in the shoot tissue removed by defoliation.

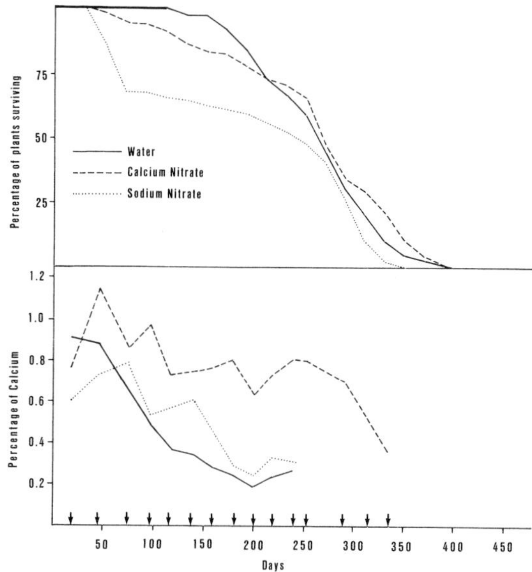

Fig. 3. Effect of mineral nutrition upon plants of *Lolium perenne* not producing additional root axes. Percentage of plants surviving and percentage of calcium in the shoot tissue removed by defoliation.

extreme variation occurred within clones of *Lolium perenne*; length of life again being related to the percentage of calcium (Fig. 2). Increasing the level of calcium nutrition, though it increased the calcium content of the shoot tissue, had little effect upon the plant's length of life (Fig. 3).

These results suggest that the decline in the ability of ageing root axes to absorb calcium is a function of an internal factor which is only slightly influenced by the concentration of calcium at the root surface.

Calcium is principally absorbed by the apical portion of the root where suberization of the endodermis has not taken place, while phosphorus and potassium can be absorbed over a much greater length [4]. The decline in the uptake of calcium could be due to a decrease in the length of root without suberized endodermis, i.e. suberization of the endodermis taking place closer to the apex as the axes aged. There does not seem to be any direct evidence for this hypothesis available but Heimsch [2] found that vascular tissue was differentiated nearer to the apex in slow growing barley roots than in more rapidly elongating ones. As an axis ages it is highly probable that its rate of elongation decreases and that suberization of the endodermis follows the pattern of differentiation of vascular tissue. The plant would presumably die when the length of root with unsuberized endodermis became less than the minimum capable of absorbing sufficient calcium to enable the plant to continue growing. The minimum length is probably very small since the rate of shoot growth before a plant dies is usually very slow, and the requirement for calcium would be minute.

On the basis of this hypothesis, genetic variation in length of life and its relationship with the uptake of calcium could be explained as being due to differences in the initial length of unsuberized root and in the rate at which suberization approached the apex. An increase in the level of calcium nutrition would cause an increase in its rate of absorption [3] and so a decrease in the length necessary to absorb the minimum required for growth, so resulting in an increase in the longevity of the plant. This is not entirely in accordance with the experimental results (Fig. 3). If however, the minimum length was small relative to the rate at which suberization approached the apex, any decrease in the length would only have a small effect upon the plant's longevity, as in fact, occurred. Laterals are assumed to behave in a manner similar to the axes.

The magnesium content of the young shoot tissue varied, in general, in a manner similar to that of calcium. Magnesium resembles calcium in that it is absorbed principally by the apical portion of the root where suberization of the endodermis has not taken place [1] but differs from calcium in that it remains mobile after being incorporated into a plant's tissue whereas calcium becomes immobile [3]. This may account for the lack of precision in its correlation with the plant's length of life.

The relationship between the growth of the root and shoot systems is complex in young grass plants [5] but from the experiments discussed in this paper and the results of others published elsewhere [7] it is evident that it is even more complex in older ones. The results of experiments with young plants cannot therefore be extrapolated to more aged ones with any guarantee of correctness. This opens up a field of research, the limits of which are not yet known.

REFERENCES

1. Ferguson, I. B. and Clarkson, D. T. 1976 Simultaneous uptake and translocation of magnesium and calcium in barley (*Hordeum vulgare* L.) roots. Planta (Berlin) 128, 267–269.
2. Heimsch, C. 1951 Development of vascular tissue in barley roots. Am. J. Bot. 38, 523–537.
3. Mengel, K. and Kirby, E. A. 1979 Principles of plant nutrition. 2nd Ed. pp. 593. Berne: International Potash Institute.
4. Scott Russell, R. and Clarkson, D. T. 1976 Ion transport in root systems. In:N. Sunderland, (ed.), Perspectives in experimental biology. Vol. 2. Botany, pp. 401–411. Oxford: Pergamon Press.
5. Troughton, A. 1977 Relationships between the root and shoot systems of grasses. In: Marshall, J. K. (ed.), The belowground ecosystem. A synthesis of plant-associated processes, pp. 39–51. Range Sci. Dep. Sci. Ser. No. 26, Fort Collins U.S.A.: Colorado State Univ.

6. Troughton, A. 1978 The effect of the prevention of the production of additional root axes upon the growth of plants of *Lolium perenne*. Ann. Bot. 42, 269–276.
7. Troughton, A. 1981 Production of root axes and leaf elongation in perennial ryegrass in relation to dryness of the upper soil layer. J. agric. Sci., Camb., in press.

70. Inter-organ control of photosynthesis mediated by emerging nodal roots in young maize plants

TIMOTEJ JEŠKO

Institute of Experimental Biology and Ecology of the Slovak Academy of Sciences, 885 34 Bratislava, Czechoslovakia

Root systems participate, as a whole, in maintaining the functional equilibrium between the root and shoot of developing plants [1, for review: 8]. This equilibrium was studied from different angles and some of them were discussed during the symposium by R. K. Szaniawski, Z. Starck and E. Czajkowska, D. Richards, E. M. Wiedenroth, and others (Part VI, this volume).

During the ontogenesis of the plant the above mentioned equilibrium between

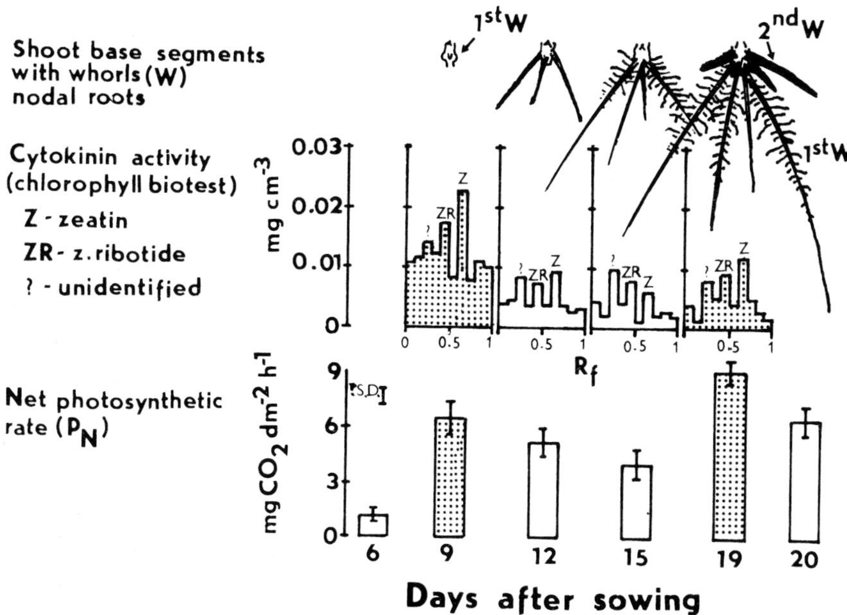

Fig. 1. Emergence of the first and then the second whorl of nodal roots in a young maize plant (top) is characterized by the temporarily increased activity of the free endogenous cytokinins, mainly of zeatin, locally at the base of shoot in region of leaf bases (middle) and, simultaneously, by the temporary increase of the net photosynthetic rate of entire shoot assimilation surface (bottom). Dotted: the events occurring in interval of emergence of the first and then the second whorl of nodal roots.

R. Brouwer et al. (eds.), Structure and Function of Plant Roots, 367–371.

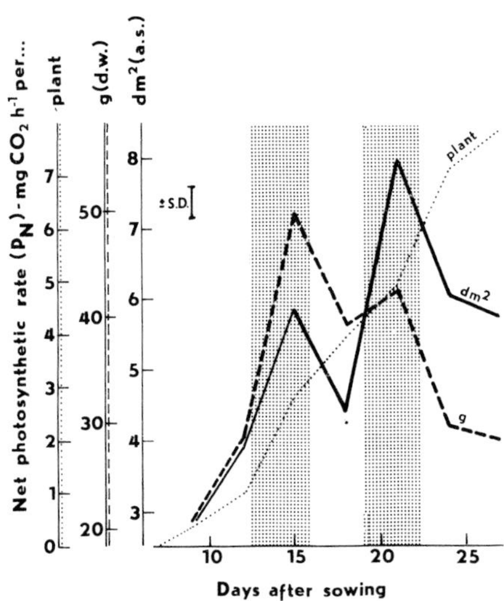

Fig. 2. Part of the ontogenetic course of photosynthetic activity of the young maize plant counted on three different units (plant, dry matter of shoot and assimilation surface of shoot) with temporary increase of the net photosynthetic rate during emergence of the first and then the second whorl of nodal roots. Dotted: the interval of emergence of the first and then the second whorl of nodal roots.

the root and shoot may be affected and transitorily disturbed by the emergence and development of new organs on the plant body. Those are, for example, the nodal adventitious roots growing in whorls close to leaf bases in young sweet sorghum [2, 3] and maize [6] plants.

We consider emerging nodal roots of the first whorl and the second whorl as locally acting new sinks of assimilates, which by their activities cause the short term increases in photosynthetic activity (Fig. 1, 2, 3). These results were obtained in plants in controlled conditions (Conviron E8H growth chamber, Canada), growing in Knop's solution, using an infra-red CO_2 analyzer (Infralyt III, DDR) for net photosynthetic rate (P_N) and dark respiration (R_D) estimation of shoots of plants.

In experiments where all nodal roots initiating extension growth were removed [4, 5], and where the plants were kept with their shoot bases producing the nodal roots in air just above the nutrient solution level (unpublished), we found that the transient increases of photosynthetic activity of shoots are causally connected with the endogenous functions of emerging nodal roots, and not with their eventual exogenous functions of water and mineral absorption.

Because the capacity of emerging nodal roots for assimilates can be neglected, we assume that the hormones produced by new sinks are the main factor

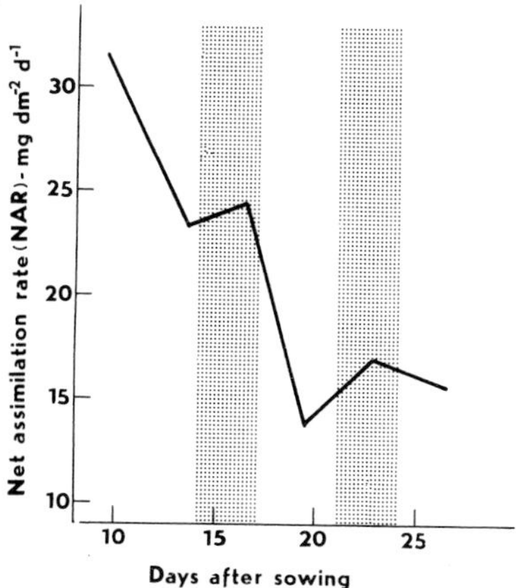

Fig. 3. Temporary increase of the net assimilation rate of the young maize plant during emergence of the first and then the second whorl of nodal roots. Dotted: as Fig. 2.

controlling photosynthetic activity during its transient increase. Fig. 1 [adopted from 6] shows that these hormones are likely to be cytokinins and mainly zeatin, where the rise in activity corresponds with transient increase in photosynthesis, while the relative leaf growth rate (Fig. 4) transitorily decreases. This retardation in leaf area growth and in leaf elongation (unpublished) may be an inhibitory effect of cytokinins on longitudinal growth of leaf cells [7], situated in leaf bases of maize close to the site of the emerging nodal roots.

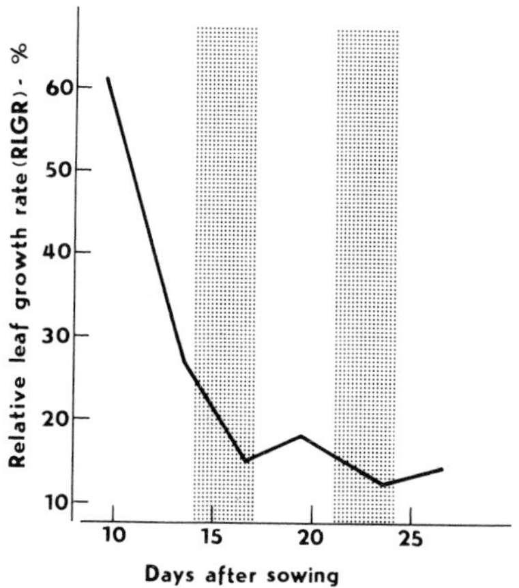

Fig. 4. Temporary decrease of the relative leaf growth rate of the young maize plant during emergence of the first and then the second whorl of nodal roots. Dotted: as Fig. 2.

The events described suggest that photosynthetic activity of the developing plant, which is to a large extent controlled by or dependent on intrinsic factors in each leaf, may also be controlled by events in other parts of the plant. We call this last control the inter-organ control of photosynthesis.

The phenomenon of transient increases in photosynthetic activity of the plant, observed in young sweet sorghum and maize plants during the emergence of their nodal roots in both controlled environments and also in field conditions, was studied later under different experimental situations (cutting treatments, growth of plants in absence of endosperm, different light intensity and temperature, etc.). The results obtained allow us to put forward a hypothesis concerning one of the possible principles of inter-organ control of leaf photosynthesis by roots within the plant, namely that in the emerging nodal roots of the first and second whorls of maize type plants the local activity of free endogenous cytokinins – mainly of zeatin – is increased in the shoot bases (in region of leaf bases). As expected, this increase stimulates photosynthetic activity (P_N, NAR) of the leaves, but retards their growth. Consequently, free sugars and some amino acids accumulate in leaf blades, forming a pool of assimilates, which are then mainly translocated into the emerging nodal roots. The redistribution of assimilates from the leaves to the roots is accompanied by a transient increase in dark respiration (R_D) of the shoot and this increase begins 1 to 2 days later followed by a rise in P_N.

As the nodal roots continue to grow and their meristematic tips, the centres of cytokinin synthesis, gradually recede from the stem and leaf bases, the local maximum of cytokinin activity diminishes and the P_N and NAR of the shoot decreases, while leaf growth is released from partial inhibition.

REFERENCES

1. Brouwer, R. 1963 Some aspects of the equilibrium between overground and underground plant parts. Inst. Biol. Scheikd. Onderz. Landbouwgewassen Wageningen Jaarb., 31–39.
2. Ješko, T. 1968 Growth and developmental changes of the photosynthetic activity of *Sorghum saccharatum* (L.) Moench. Thesis. Inst. Chem., Slovak. Acad. Sci., Bratislava. (Slovak).
3. Ješko, T., Heinrichová, K. and Lukačovič, A. 1971 Increase in photosynthetic activity during the formation of the first node roots and first tiller in *Sorghum saccharatum* (L.) Moench. Photosynthetica 5, 233–240.
4. Ješko, T. 1972 Removal of all nodal roots initiating the extension growth in *Sorghum saccharatum* (L.) Moench. I. Effect on photosynthetic rate and dark respiration. Photosynthetica 6, 51–56.
5. Ješko, T. 1972 Removal of all nodal roots initiating the extension growth in *Sorghum saccharatum* (L.) Moench. Effect on growth analysis data. Photosynthetica 6, 282–290.
6. Ješko, T. and Vizárová, G. 1980 Changes of free endogenous cytokinins during transitorily increased photosynthetic rate initiated by formation of the first two whorls of nodal roots in *Zea mays* L. Photosynthetica 14, 83–85.

7. Maksymowych, R. 1973 Analysis of leaf development, pp. 109. Cambridge University Press.
8. Troughton, A. 1977 Relationships between the root and shoot systems of grasses. In: Marshall, J. K. (ed.), The belowground ecosystem. A synthesis of plant-associated processes, pp. 39–51. Range Sci. Dep. Sci. Ser. No. 26. Fort Collins U.S.A.: Colorado State Univ.

71. Root-shoot interactions in fruiting tomato plants

DENNIS RICHARDS

Horticultural Research Institute, Department of Agriculture, Knoxfield, Victoria 3180 Australia

The inclusion of fruit in studies on root: shoot interactions has, so far, received little attention. In fruiting plants the growth of fruit places an added demand on the root system for mineral nutrients and on the shoot system for photosynthates. This demand becomes increasingly competitive with vegetative growth [3, 6] and may involve both a redirection of assimilates and minerals into the developing fruit and a redistribution of nutrients from other plant tissues [1].

In the present experiment, tomato (*Lycopersicon esculentum* Mill.) plants grown throughout their vegetative and fruiting phases were used to examine the co-ordination between root and shoot growth with reference to the possible controlling mechanisms. Further, the effect of fruit growth on the relationship between root and shoot function was examined.

RESULTS AND DISCUSSION

Tomato plants (c.v KIO) having a *determinate* growth habit were grown in 0.10 L (S = small), 0.26 L (M = medium) and 0.85 L (L = large) containers and, in order that nutrients and water were non-limiting, were fed continuously with nutrient solution. Container size is an ideal way of manipulating the growth and size of the root system [13]. Two further treatments involved transferring plants from the small to the large containers (de-restriction) either when the fruit first appeared (SR_1) or 3 weeks later when fruit were growing rapidly (SR_2).

Root and shoot growth

Fig. 1(a–c) shows the effect of treatment on the growth of the various components of the plant throughout the experimental period. Root growth is represented by root number, shoot growth by leaf area and fruit growth by fruit fresh weight. Reducing container size quickly resulted in reduced growth of roots and shoots, and subsequently, fruit. De-restriction, at either date, resulted in a rapid and almost immediate response in root growth which was quickly followed by rapid growth of shoots. A response in fruit growth was delayed by 5 weeks in either de-

R. Brouwer et al. (eds.), Structure and Function of Plant Roots, 373–380.

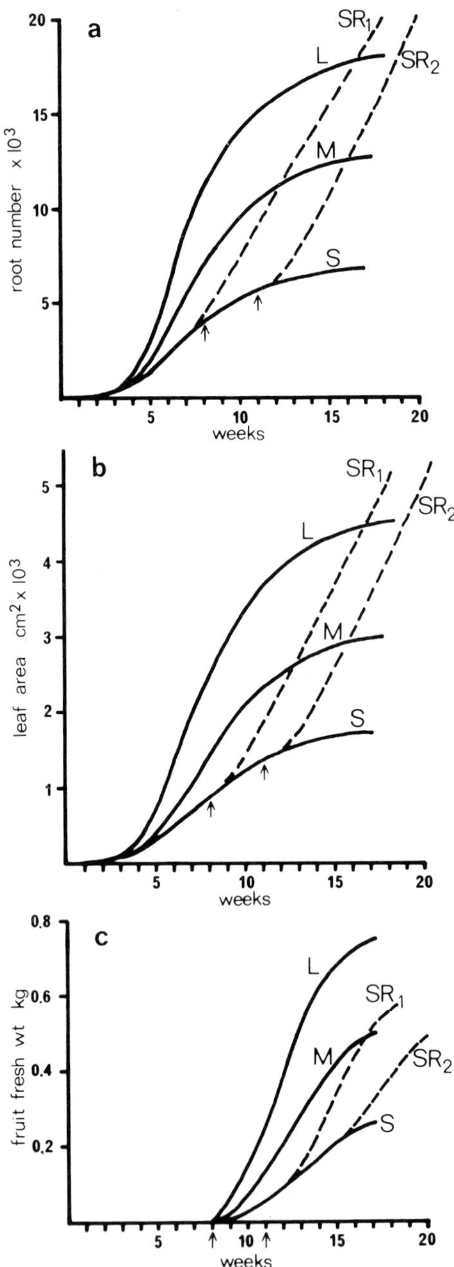

Figs. 1a–c. 1a. The effect of treatment on a) root; b) shoot and c) fruit growth. Arrows indicate the two de-restriction dates.

restriction treatment and was associated with the production of new flowers, and fruit on trusses elevated above the original fruiting canopy. It therefore appears that a cultural treatment which encouraged rapid root growth could breakdown the so-called determinated nature of this tomato variety. Fruit yield for de-restricted treatments was lower than the control (L) because the additional fruit did not have time to size before the experiment was terminated. Despite the varied treatment effects on growth the distribution of dry weight in terms of either the vegetative shoot : root ratio or the top : ratio (which includes shoot and fruit components together) was similar for all pot sizes. Top : root ratio increased dramatically with the onset of fruit growth, reaching values around 12 : 1, while the vegetative ratio remained relatively constant at around 5 : 1.

With water and nutrients supplied in adequate amounts, it appears that the size of the root system exerted considerable control over the size of the shoot and fruit system it supported. When root growth was encouraged, at any stage, the shoot system rapidly responded and subsequently so did fruit growth. Examina-

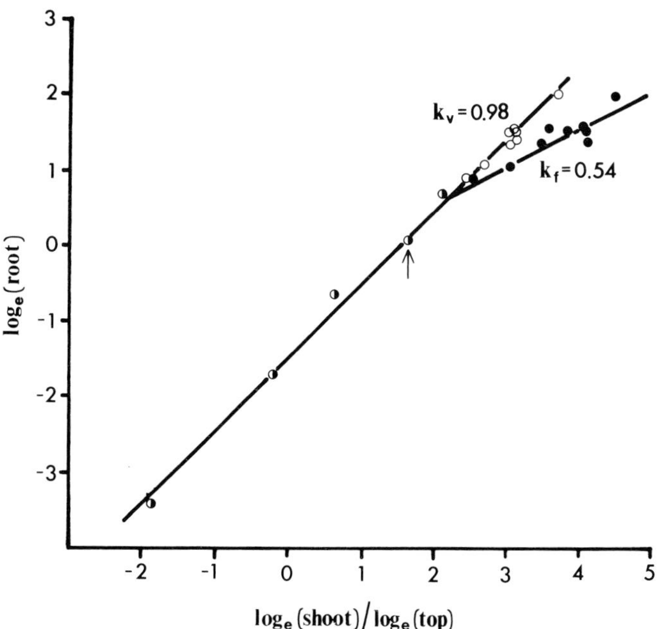

Fig. 2. Allometric relationship between root dry weight and either shoot dry weight (open symbols) or top (i.e. shoot + fruit) dry weight (closed symbols) for plants grown in large pots.

k_v is the slope of root against shoot: k_f is the slope of root against top. Arrow indicates time of first flowers.

Table 1. The allometric relationship between root and shoot dry weight. Linear regressions of the form $\log_e$ (root d wt.) = a + k_v. $\log_e$ (shoot d wt.) where k_v ($\pm$ 95 per cent confidence limits) is the slope and r the regression coefficient ($P < 0.01$).

Treatment	k_v	r
L	0.979 ($\pm$ 0.021)	0.998
M	0.945 ($\pm$ 0.016)	0.997
S	0.926 ($\pm$ 0.034)	0.992
SR_1	0.930 ($\pm$ 0.028)	0.994
SR_2	0.914 ($\pm$ 0.025)	0.993

tion of the allometric growth curves confirms this close co-ordination between root and shoot growth.

A plot of the natural logarithms of root dry weight against those of either shoot dry weight alone or top (i.e. shoot + fruit) dry weight is shown in Fig. 2. In *Lolium perenne* L. the slope of such allometric growth curves decreases dramatically at flowering indicating a physiological change in the relationship between root and shoot growth in favour of the shoot [16]. Here, unlike grasses, a change in slope occurred some time after flowering and only when fruit began to be a sizeable component of the plant. In fact, without fruit in the comparison (i.e. $\log_e$ root against $\log_e$ shoot) there was no alternation in the relationship between root and shoot growth associated with either flowering or fruit development.

Allometric plots for all other treatments were similar in form to that depicted for large pots (Fig. 2). However, there were treatment differences in the values of k_v, the slope of $\log_e$ root against $\log_e$ shoot (Table 1). As pot size decreased the value of k_v decreased. Thus, in dry weight terms, plants in large pots apportion more growth into their roots than those in small pots. For plants released from the small to the large containers there was no change in the value of k_v compared to small pots even though de-restriction greatly stimulated vegetative growth.

The most interesting result of the allometric data is that the co-ordination of growth between the root and the vegetative shoot was unchanged following flowering or fruiting. Although roots and shoots can interact in many conceivable ways the present results and those of others [2, 9] suggest that a major site of control resides in the root system and is hormonal in nature.

Richards and Rowe [13] reached the same conclusion for peach trees, but went further and suggested that the parameter of the root system most closely related to top growth was the number of root tips. They suggested that a relationship between root number and some measure of top growth indicates a controlling mechanism in the root mediated by its cytokinin production [14]. Fig. 3 shows

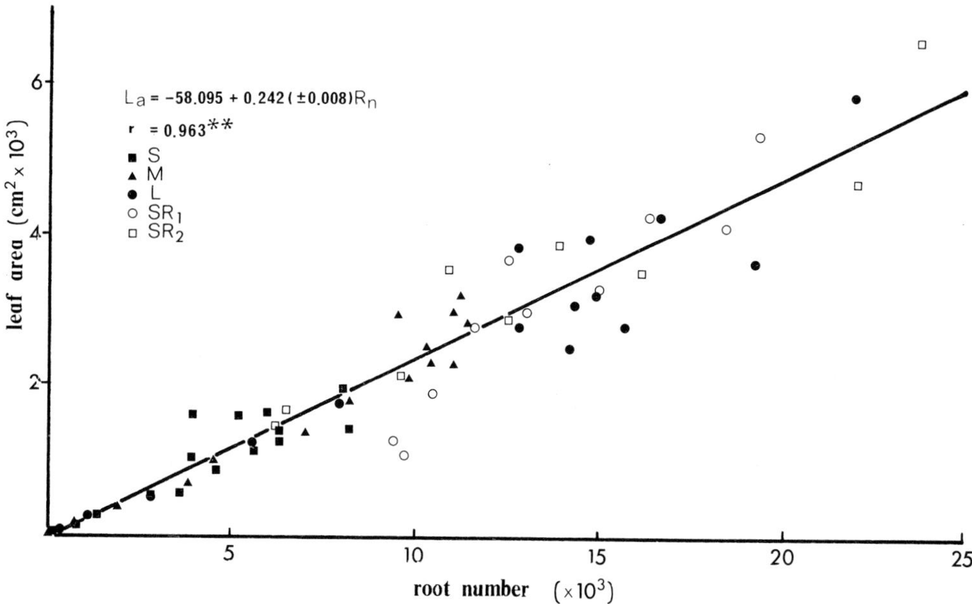

Fig. 3. The relationship between leaf area and root number in vegetative and fruiting plants as measured at 20 consecutive weekly harvests.

such a relationship, root number against leaf area for all of the treatments at all of the different harvest times. Despite large treatment differences in growth (Fig. 1), and a transition from the vegetative to the reproductive phase, there was a very close relationship (accounting for 93 per cent of the variation) between the leaf area and the root number of individual plants.

Taken together the results indicate that the root tip (by its production of cytokinin) exerts considerable control over the growth of the shoot even in the presence of fruit.

Functional equilibria

It has been suggested that in vegetative plants the root and the shoot maintain a certain equilibrium in relation to their outputs that is independent of varying external conditions and root or shoot treatment. Davidson [4] represented this functional relationship as:

$$\text{root mass} \times \text{rate (absorption)} \propto \text{leaf mass} \times \text{rate (photosynthesis)}. \qquad (1)$$

To test this hypothesis Hunt [7] presented data in the form:

$$W_r/W_s \propto 1/(SAR_M/USR), \tag{2}$$

where W_r and W_s are the dry weights of the root and shoot respectively. SAR_M the specific absorption rate of a defined nutrient(s) M, and USR the unit shoot rate.

Thornley [15] suggested that a simpler way of presenting equation (2) was:

$$\Delta W \propto \Delta M \tag{3}$$

where ΔM is the increment in weight of nutrient(s) M and ΔW the increment in total plant dry weight over a given time period.

Whichever equation is chosen functional relationships have been demonstrated in respect to: carbohydrate partitioning in *Lolium perenne* L. and *Trifolium repens* L. [5]; K, Ca and Mg uptake in *Lolium perenne* L.; N uptake in

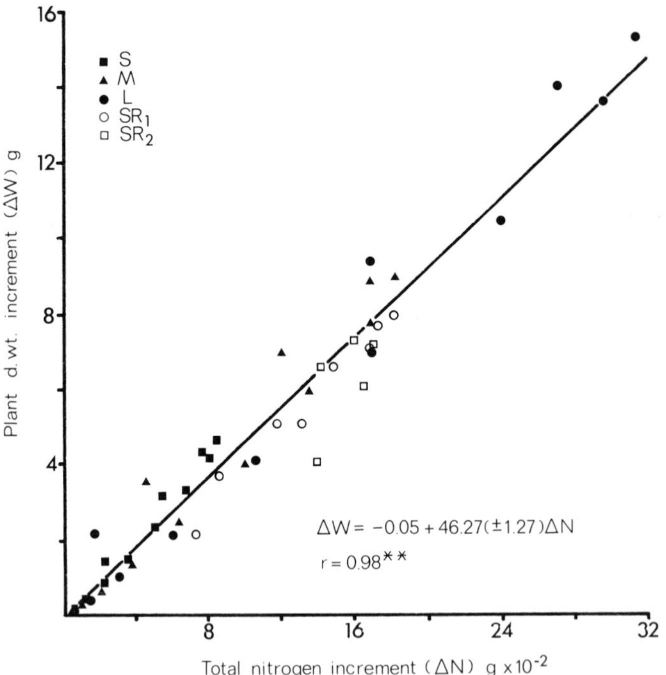

Fig. 4. The relationship between plant dry weight increment (ΔW) and whole plant nitrogen increment (ΔN) between adjacent harvests.

Vaccinium macrocarpon Ait. [8]; N, P, K, Ca and Mg uptake, and water uptake, in *Prunus persica* L. Batsch. [10, 11]. In each case it was postulated that alterations in the shoot : root ratio and/or the activities of the root and shoot system led to the maintenance of a constant internal nutrient status (or water use [10]).

Due to the substantial ontogenetic drift in top: root ratios and to the large treatment effects this experiment produced a diverse population of plants suitable to test for the existence of functional equilibrium conditions in the manner of equation (3). Furthermore, unlike previous work, this relationship could be tested using plants that progressed from the vegetative to the fruiting phase.

Fig. 4 shows the relationship between ΔW, the increment in total plant dry weight and ΔN the increment in total tissue nitrogen between adjacent harvests. Regression analysis produced a single straight line accounting for 96 per cent of the variation suggesting that, independently of treatment effects or a transition to reproductive growth, there was an equilibrium between the dry weight produced mainly by the shoot and the amount of nitrogen taken up by the root. The presence of fruits on the plant, which are considerable sinks for carbohydrates and minerals, may have resulted in a redirection of growth, as shown by the increasing top : root ratios, and of nitrogen [12], but accompanying changes in the activities of the root and shoot systems were such that a balance between shoot and root function was maintained.

REFERENCES

1. Bollard, E. G. 1970 The physiology and nutrition of developing fruits. In: Hulme, A. C. (ed.), The biochemistry of fruits and their products, pp. 387–425. London and New York: Academic Press.
2. Buttrose, M. S. and Mullins, M. G. 1968 Proportional reduction in shoot growth of grapevines with root systems maintained at constant relative volumes by repeated pruning. Aust. J. biol. Sci. 21, 1095–1101.
3. Cooper, A. J. 1972 Partitioning of dry matter by the tomato. J. hort. Sci. 47, 137–140.
4. Davidson, R. L. 1969*a* Effect of root/leaf temperature differentials on root/shoot ratios in some pasture grasses and clover. Ann. Bot. 33, 561–569.
5. Davidson, R. L. 1969*b* Effects of edaphic factors on the soluble carbohydrate content of roots of *Lolium perenne* L. and *Trifolium repens* L. Ann. Bot. 33, 579–589.
6. Fisher, K. J. 1975 Effect of the amount and position of leaf tissue on the yield of single-truss tomatoes. Sci. Hort. 3, 303–308.
7. Hunt, R. 1975 Further observations on root-shoot equilibria in perennial ryegrass (*Lolium perenne* L.). Ann. Bot. 39, 745–755.
8. Hunt, R., Stribley, D. P. and Read, D. L. 1975 Root/shoot equilibria in cranberry (*Vaccinium macrocarpon* Ait.). Ann. Bot. 39, 807–810.
9. McDavid, C. R., Sagar, G. R. and Marshall, C. 1973 The effect of root pruning and 6-benzylaminopurine on the chlorophyll content, $^{14}CO_2$ fixation and the shoot/root ratio in seedlings of *Pisum sativum* L. New Phytol. 72, 465–470.

10. Richards, D. 1977 Root-shoot interactions: a functional equilibrium for water uptake in peach (*Prunus persica* L. Batsch.). Ann. Bot. 41, 278–281.
11. Richards, D. 1978 Root-shoot interactions: functional equilibria for nutrient uptake in peach (*Prunus persica* L. Batsch.). Ann. Bot. 42, 1039–1043.
12. Richards, D., Goubran, F. H. and Collins, K. E. 1979 Root-shoot equilibria in fruiting tomato plants. Ann. Bot. 43, 401–404.
13. Richards, D. and Rowe, R. N. 1977*a* Effects of root restriction, root pruning and 6-benzylaminopurine on the growth of peach seedlings. Ann. Bot. 41, 729–740.
14. Richards, D. and Rowe, R. N. 1977*b* Root-shoot interactions in peach; the function of the root. Ann. Bot. 41, 1211–1216.
15. Thornley, J. H. M. 1975 Comment on a recent paper by Hunt on shoot : root ratios. Ann. Bot. 39, 1149–1150.
16. Throughton, A. 1960 Further studies on the relationship between shoot and root systems of grasses. J. Brit. Grasslands Soc. 15, 41–47.

72. Function of roots in NaCl-stressed bean plants

ZOFIA STARCK and ELŻBIETA CZAJKOWSKA

Dept. of Plant Biology, Warsaw Agricultural University 02-528 Warsaw, Poland

Higher plants have adopted two physiological strategies in response to saline environments; either the uptake of toxic ions is reduced (e.g. Na^+, Cl^- and others), or accumulation of toxic ions in the aerial part may be prevented by their retention in the roots (or stem) and in consequence – reduction of their transport to the youngest, developing part of shoot [2].

Recent reports suggest that phytohormones contribute to the regulation of ion absorption and their distribution within the whole plant [11]. On the other hand ions affect the response of tissues to growth regulators. In NaCl-stressed plants drastic changes in hormonal balance take place: ABA accumulates in plant tissue while some reduction in the levels of auxins, GA-s and cytokinins is found to take place. Exogenously introduced growth regulators partially reversed the negative effect of stress conditions both on photosynthesis and on translocation processes [7, 8, 9]. All these facts indicate the important role played by roots in the regulation of shoot metabolism through the selective supply to the aerial part of both ions and specific phytohormons.

RESULTS

All experiments were done on seedlings of *Phaseolus vulgaris* with developed primary leaves and the first trifoliate leaf just beginning to expand. The plants were grown in water culture, under natural light and air conditions. In most cases plants were treated with NaCl successively during 3 days, decreasing the water potential of the nutrient solution (N.S.) by 4.5×10^3 hPa.

Over a one day period the effect of GA_3, (sprayed on the leaves at a concentration of 3×10^{-4} M) and zeatin in the nutrient solution at 5×10^{-4} M on the absorption rate (AR) was estimated. The absorption rate was calculated accord-

R. Brouwer et al. (eds.), Structure and Function of Plant Roots, 381–387.

ing to the formula [12]:

$$AR = \frac{\log_e R_2 - \log_e R_1}{R_2 - R_1} \cdot \frac{M_2 - M_1}{t_2 - t_1}$$

t – time
R – roots dry matter
M – amount of ions

Over a period of 5 days both phytohormones stimulated K-absorption in NaCl-stressed bean plants; Ga_3 reduced the AR of Na, thus doubling the K/Na ratio. Both GA_3 and zeatin prevented Na-accumulation in the apical part and blades, as well as improved P-translocation to the youngest part of shoot: Fig. 1, 2, adapted from [10]. Similarly kinetin reduced Na and Cl – transport to the shoots of cotton and bean plants [1].

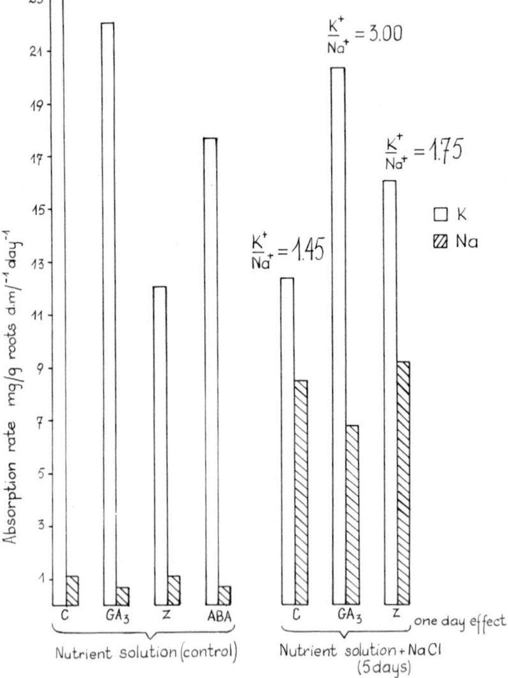

Fig. 1. Effect of GA_3 and zeatin on the absorption rate (AR) of K^+ and Na^+ in 5 days NaCl-stressed bean plants.

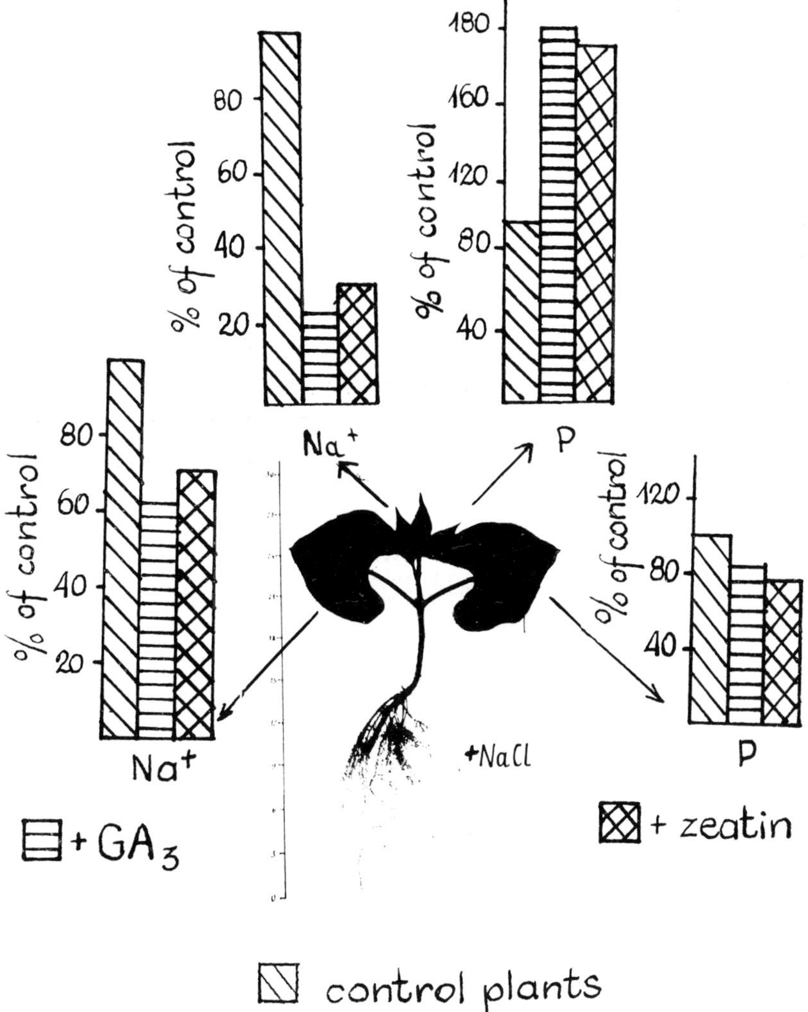

Fig. 2. Effect of GA_3 and zeatin on the distribution of sodium and phosphorus.

Diminished Na-content in the blades of plants treated with GA_3 and zeatin may be the reason for the recovery of the rate of photosynthesis in plants which were stressed but treated with phytohormones.

The results suggest that roots, as organs where biosynthesis of GA-s and cytokinins take place, may indirectly affect photosynthesis and translocation of assimilates. In agreement with this supposition are the results of experiments, in

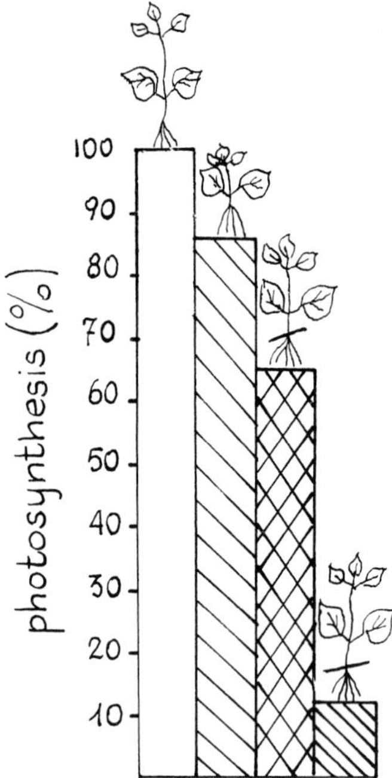

Fig. 3. Photosynthetic activity in NaCl-stressed and grafted bean plants. Assimilation of CO_2, calculated per plant, is assumed as 100 per cent. From the left: control plants, NaCl-stressed – 4 days + 8 days nutrient solution; plants grafted: control aerial part on control roots; plants grafted: control aerial part on 4-day, NaCl-stressed roots.

which plants grown on control nutrient solution were grafted on the roots detached from the 4-day, NaCl-stressed plants, their photosynthesis being estimated after 8 days. (Assimilation of CO_2 was measured by the use of Infralyt IV Juncalor DDR, in the open circuit gas analysis system). Photosynthetic activity was drastically reduced (Fig. 3). Grafting treatment of a control shoot onto another root also diminished photosynthetic activity, but to a smaller degree. All stressed plants grafted on stressed roots died. Plants which were subjected to 4-day NaCl-stress and then transferred for 8 days into nutrient solution, recovered to a great extent their capacity for photosynthesis.

In another experiment (Fig. 4) photosynthetic activity and export of photosynthates from the blades were estimated through the use of $^{14}CO_2$ in 4-day and 12-day stressed bean plants. Both processes were drastically retarded by the prolonged NaCl-treatment. However, grafting of 10-day-NaCl-stressed plants on the control roots, 2 days before the $^{14}CO_2$-exposure improved slightly both $^{14}CO_2$-assimilation and ^{14}C-export from the stressed, primary blades (Fig. 4).

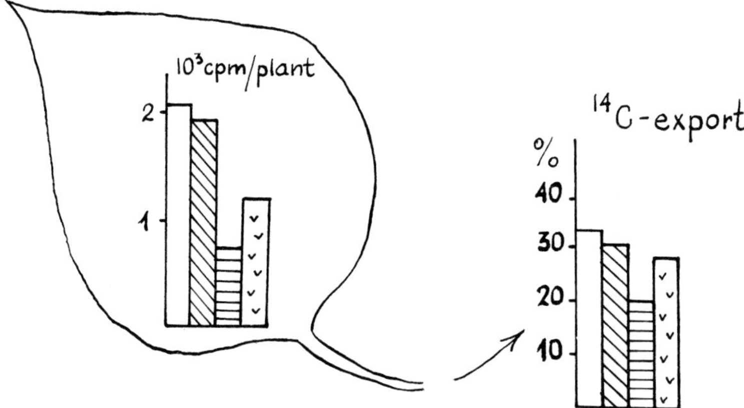

Fig. 4. Photosynthetic activity and export of ^{14}C-assimilates as the effect of NaCl-stress and grafting of control roots. Bars from the left: □ control plants ▧ 4-day, NaCl-stress + 8 day nutrient solution ▤ 12-day, NaCl-stress; ☑ 10-day, NaCl-stress + 2-day inoculation of control roots.

Grafting of control roots 3 days before CO_2-assimilation measurement on the 5-day NaCl-stressed plants also improved photosynthetic rate. Stressed roots inhibited photosynthesis of control blades (Fig. 5). Stressed shoots on stressed roots all died, as in previous experiments.

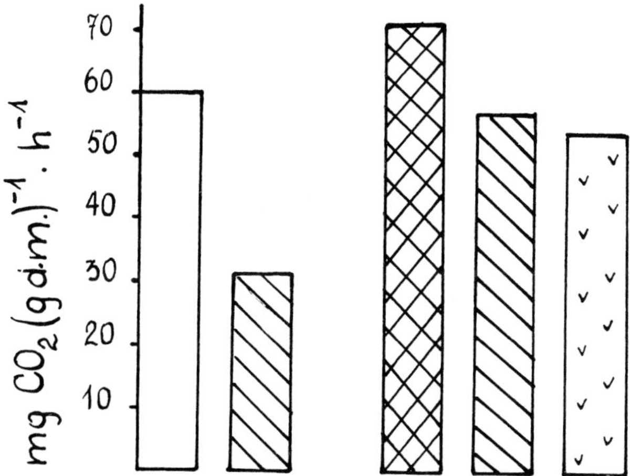

Fig. 5. Effect of root inoculation on the rate of photosynthesis. From the left: □ control plants (c), ▧ 5-day, NaCl-stress + 3 day nutrient solution (s), ⊠ grafting of control aerial part on control roots (c/c) (3 day effect), ▧ grafting of control aerial part on 5 day, NaCl-stressed roots — 3 days effect, ☑ grafting of 5-day, NaCl-stressed aerial part on control roots — 3 days effect, (s/c).

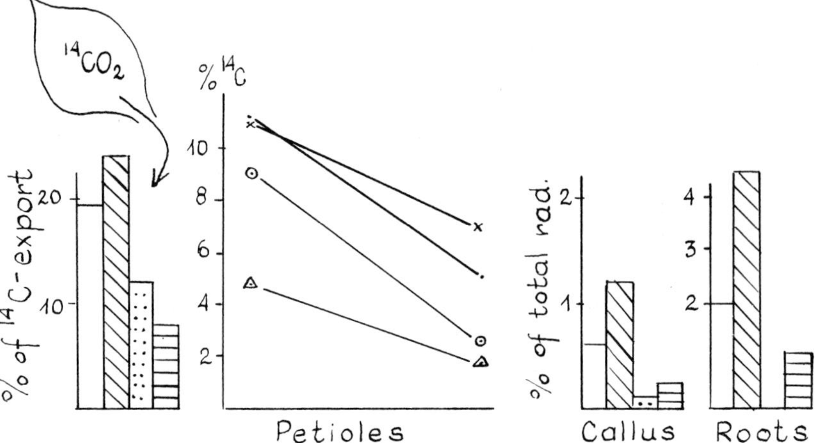

Fig. 6. $^{14}CO_2$-assimilation and ^{14}C-distribution in detached, NaCl-stressed leaves, • control leaves, leaves, × leaves detached from 5-day, NaCl-stressed plants, rooted ($S_p + R$), ⊙ the same series but leaves without roots ($S_p - R$), △ leaves 7-day, NaCl-stressed after detachment and rooted $S_L + R$.

In another series of experiments, detached primary leaves of bean plants were used (Fig. 6). Some leaves, detached from the 5-day, NaCl-stressed plants formed roots (Sp + R). They exported ^{14}C-assimilates to the petioles, roots and callus more efficiently than did the control leaves. Leaves which did not form roots (Sp − R) exported ^{14}C-substances to a very small extent. The gradient of labelled compounds in the petioles was very steep, suggesting the inhibition of phloem transport.

Leaves stressed with NaCl for 7 days after detachment also formed roots even (S_L + R) but it did not prevent the inhibition of export of ^{14}C-photosynthates from the blades and transport through the petioles to the roots and callus.

DISCUSSION

The results of our experiments, as well as others also discussed during the symposium, [3, 6] suggest that the root system takes part in maintenance of the functional balance in the whole plant. This coordination between the root and shoot metabolism may be mediated by the contribution of the root system to hormonal regulation. Cytokinins, and GA-s produced in roots affect ion absorption in roots and their distribution in the whole organism.

Much experimental data has indicated that growth substances may affect membrane properties, contributing to the control of the selective absorption of ions [5, 11 and others].

In salt-stressed plants such changes in membrane properties as composition of lipids [4], ATP-ase activity [4] and conformation of enzyme proteins cause disturbances in the selectivity of ion uptake as well as in the distribution of ions in the cells and among particular organs. Disturbance of membranes by NaCl-stress both in the root and shoot influences photosynthesis and transport of photosynthates, probably as a secondary effect as was suggested earlier [8]. A limited supply of photosynthates to roots enhanced inhibition of their activity as donors of ions and producers of some hormones, through a negative feedback mechanism. As a consequence all these changes increase the inhibition of photosynthesis and translocation of assimilates. A limited supply of ions with changes in their distribution, as well as a deficit of giberellins and cytokinins in the shoot, also affects the growth of aerial parts which again diminishes production of photosynthates.

REFERENCES

1. El Saidi, M. T. and Kuiper, P. J. C. 1972 Mededelingen Landbouwhogeschool Wageningen, Nederland 75, 15, 1–6.
2. Erdei, L. and Kuiper, P. J. C. 1979 Physiol. Plant. 47, 95–99.
3. Jesko, T. 1981 2nd International Symposium, Structure and function of roots, Bratislava, p. 367.
4. Kuiper, P. J. C. 1980 Physiol. Veg. 18, 83–88.
5. Nelles, A. 1977 Planta 137, 293–298.
6. Richards, D. 1981 Structure and function of roots, Bratislava, p. 373.
7. Starck, Z., Karwowska, R. and Kraszewska, E. 1975 Acta Soc. Bot. Pol. 44, 565–587.
8. Starck, Z. and Karwowska, R. 1977 Proc. of the Second Int. Symp. of plant growth regulators, Sofia, Publ. House of the Bulgarian Ac. of Sci. pp. 183–186.
9. Starck, Z. and Karwowska, R. 1978 Acta Soc. Bot. Pol. 47, 245–267.
10. Starck, Z. and Kozińska, R. 1980 Acta Soc. Bot. Pol. 49, 111–125.
11. Steveninck van, R. F. M. 1976 In: Lüttge, U. and Pitman, M. G. (eds.), Transport in Plants II B, pp. 307–342, Berlin, Heidelberg: Springer Verlag.
12. White, R. E. 1973 Studies on mineral ion absorption by plants. Plant and Soil, 38, 509–523.

73. The distribution of ^{14}C-photoassimilates in wheat seedlings under root anaerobiosis and DNP application to the roots

ERNST-MANFRED WIEDENROTH

Humboldt-University Berlin, Section of Biology, 1040 Berlin, G.D.R.

Growth and development of plants are highly dependent on an equalized metabolic balance between root and shoot and factors which retard the metabolic activity of one of them also influence the other. In cereals oxygen supply to the subterranean parts is often insufficient and becomes a limiting factor of root metabolism leading to lower yield or even to a complete cessation of growth. Therefore the reaction of wheat seedlings to hypoxia or anoxia in the rhizosphere is not only of scientific, but also of practical interest.

In previous papers [9, 10] it was shown that in wheat seedlings grown in

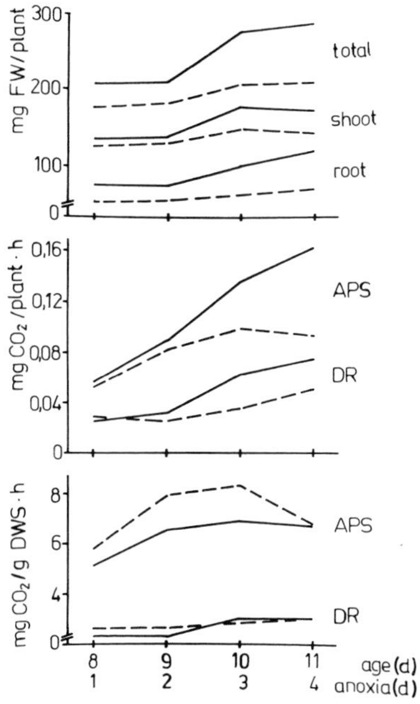

Fig. 1. Time course of fresh weight (FW), apparent photosynthesis (APS), and dark respiration of shoots (DR) in wheat seedlings (*Triticum aestivum* L. cv. Caspar) under root anaerobiosis. solid line = control – broken line = O_2 deficiency variant.

R. Brouwer et al. (eds.), Structure and Function of Plant Roots, 389–393.

nutrient solution under continuous light of low intensity (30 W m^{-2}) no measurable influence on apparent photosynthesis or dark respiration of the shoots was observed 10 h after onset of root anaerobiosis. On the other hand it is known that longer periods of root anaerobiosis cause drastic changes in the metabolic equilibrium (Fig. 1). One to five days of oxygen deficiency in the root medium effects an increasing retardation of growth not only of the roots but also of the shoots. Whilst dark respiration of the shoots is not changed, the values of apparent photosynthesis are higher during the first days of root anaerobiosis than in the control, compensating to some degree for the retardation of growth. The appearance of such an adaptive increase of photosynthesis in spite of light intensity being far below the compensation level requires further investigation, especially because it is known that the root is a rather less effective competetive sink [6].

One to five days of anoxia in the rhizosphere also causes remarkable changes in the distribution of assimilates within the plant. This was investigated by application of $14CO_2$ during photosynthesis. After feeding for 10 min in light, the plants were allowed to respire 24 h in darkness, extracted with boiling ethanol and radioactivity measured with a liquid scintillation counter. The stimulation of root exudation into the root medium is of particular interest (Fig. 2). The increased exudation is related to a decrease in the amount of label in the non-soluble (starch) fraction, whilst the ethanol soluble fraction is unaffected. After 4 days of root hypoxia the relative activity of the starch fraction rises again indicating the onset of a new equilibrium between root and shoot.

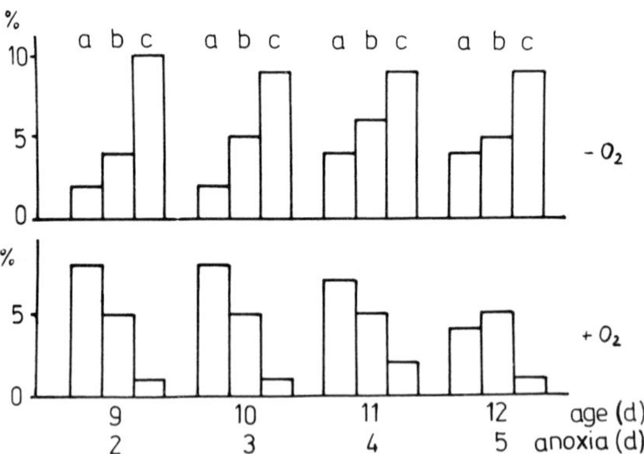

Fig. 2. Distribution of label in percentage of total radioactivity 24 h after $^{14}CO_2$ application to the shoots within the roots (ethanol insoluble fraction a, ethanol soluble fraction b) and the medium of roots (c) of wheat seedlings (*Triticum aestivum* L. cv. Caspar) under root anaerobiosis.

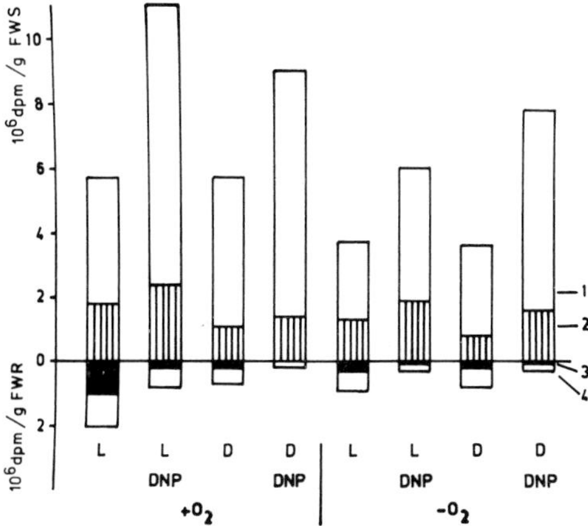

Fig. 3. Distribution of label within shoots (ethanole soluble fraction 1, ethanole insoluble fraction 2) and roots (ethanole insoluble fraction 3, ethanole soluble fraction 4) of 8 days old wheat seedlings (*Triticum aestivum* L. cv. Caspar) 24 h after application of $^{14}CO_2$ to the shoots as influenced by root anaerobiosis and DNP application to the root medium. (Pretreatment: Continuous light (L) or darkness (D) 24 h before $^{14}CO_2$ application).

If the metabolic interactions mentioned above are mainly regulated by the energy balance of the system, application of 2,4-DNP to the roots would be expected to have a similar effect on the distribution of photoassimilates as root anaerobiosis [7]. In our experiments (Poskuta, J. and Wiedenroth, E.-M., unpublished data) application of 10^{-4} M DNP to the root medium of 7-day old wheat seedlings immediately after exposure of the shoots to $^{14}CO_2$ strongly increases the radioactivity of shoots, mainly in the ethanol soluble fraction, and decreases those of the roots in all experimental combinations checked (Fig. 3). Oxygen deficiency in the rhizosphere in the comparable variants always causes a lower amount of label both in root and shoot due to the general retardation of metabolic processes in the whole plant. DNP as well as oxygen deficiency additionally stimulates the exudation from the roots into the surrounding medium. The substances exudated consist of a volatile compound (partially CO_2) and, to a lesser degree, of a non volatile one.

On the basis of our working hypothesis concerning the root shoot relations the results may be explained as follows. Wheat seedlings have a certain pool of carbohydrates in the root system allowing the production of energy equivalents by fermentation to maintain their normal metabolism and probably growth during short periods of anoxia in the rhizosphere without shortening their export of metabolites to the shoot and therefore without influencing the shoot's gas

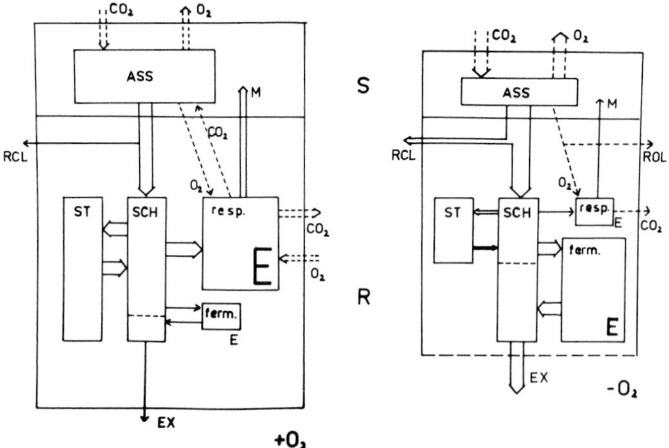

Fig. 4. Scheme of correlations between root and shoot of wheat seedlings under normal aeration and O_2-deficiency in the rhizosphere.
E – metabolic energy, EX – exudate, ferm – fermentation, M – metabolits, R – root, RCL – radial carbohydrate loss, ROL – radial oxygen loss, S – shoot, SCH – soluble carbohydrates, ST – insoluble carbohydrates (starch).

exchange. Blocking the three-carbon-acid cycle fermentation pathways generally leads to the accumulation of toxic compounds, e.g. ethanol and lactate. Even if these could be exuded partially by the roots [1, 2, 9] or if their origin could be avoided e.g. by reducing pyruvate or acetaldehyde to amino acids [4, 5], fermentation processes always have a very low energy yield leading after some days of root anaerobiosis to adaptive changes in the metabolic relations between root and shoot in such a way that the energy balance remains in equilibrium (Fig. 4).

Photoassimilates are used primarily within the shoot and only the surplus is exported to the roots. Under normal conditions they are used for energy production by respiration as a prerequisite for growth, maintenance and intermediary storage [6, 8]. Additionally in the apical meristem some fermentation takes place as a result of a permanent O_2 deficiency [4].

Generally, fermentation instead of respiration (caused by the lack of oxygen in the rhizosphere or by uncoupling the oxidative electron transport from ATP production by DNP) always leads, by ways of energy deficiency in the roots, to a retardation of growth and decreased export of metabolites into the aboveground parts of the plants, causing retardation of growth also in the shoot. On the other hand, the less effective fermentation increases the consumption of the starch pool and thereby after some time the demand for photoassimilates. This turns out to be important following DNP application, that is when the oxidative degradation of starch persists in addition to fermentation necessary to overcome the stress period. Furthermore there may be an increase in the radial movement of carbohydrates and thereby even of losses by exudation due to reduction of vertical translocation. In the experiments mentioned above, the result is sometimes an increase in apparent photosynthesis and always a higher exudation rate to avoid toxification [2, 3], additionally favoured by a rising permeability of the root cortex caused directly by oxygen deficiency [1]. The exudates contain soluble final products of fermentation still rich in energy and therefore represent additional losses in the total balance.

The scheme presented above supplements the hypothesis of an overflow mechanism by Lambers [5] with the view of light intensities far below the saturating level.

REFERENCES

1. Benjamin, L. R. and Greenway, H. 1979 Effects of a range of CO_2 concentrations on porosity of barley roots and on their sugar and protein concentrations. Ann. Bot. 43, 383–391.
2. Crawford, R. M. M. 1978 Metabolic adaptation to anoxia. In: Hook, D. D. and Crawford, R. M. M. (eds.), Plant life in anaerobic environments, pp. 119–136. Ann. Arbor, Mich.: Ann. Arbor Science Publ.
3. John, C. D. and Greenway, H. 1976 Alcoholic fermentation and activity of some enzymes under anaerobiosis. Aust. J. Plant. Physiol. 3, 325.
4. Kohl, J. G. 1978 Anpassung der Pflanzen an Sauerstoffmangel in der Rhizosphäre. Coll. Pflanzenphys. 1, 202.
5. Kohl, J. G., Baierova, J., Radke, G. and Ramshorn, K. 1978 Regulative interactions between anaerobic catabolism and nitrogen assimilation as related to oxygen deficiency in maize roots. In: Hook, D. D. and Crawford, R. M. M. (eds.), Plant life in anaerobic environments, pp. 473–496. Ann. Arbor, Mich.: Ann Arbor Science Publ.
6. Lambers, H. 1979 Energy metabolism in higher plants in different environments. Proefschrift Universiteit Groningen.
7. McDougall, B. 1970 Movement of ^{14}C-photosynthate into the roots of wheat seedlings and exudation of ^{14}C from intact roots. New Phytol. 69, 37–46.
8. Penning de Vries, F. W. T. 1975 The cost of maintenance respiration in plant cells. Ann. Bot. 39, 77–92.
9. Wiedenroth, E. M. and Poskuta, J. 1978 Photosynthesis, photorespiration, respiration of shoots, and respiration of roots of wheat seedlings as influenced by oxygen concentration. Z. Pflanzenphys. 89, 217–225.
10. Wiedenroth, E. M. and Poskuta, J. CO_2 exchange rates of shoots and distribution of ^{14}C-photoassimilates of growing wheat seedlings as influenced by oxygen deficiency of roots. Acta. Soc. Bot. Pol. (in press).

74. Influence of nitrate and aeration on growth and chemical composition of *Zea mays* L.

LOUISE Y. SPEK

Botanical Laboratory, State University of Utrecht, Utrecht, The Netherlands

Both nitrate supply and aeration of the root environment affect plant growth. Enhanced nitrogen fertilization has often been shown to reduce the effect of insufficient aeration. It is an old and still open question, whether nitrate in the root environment can serve as an electron acceptor and alleviate the stress under anaerobic conditions. The present work attempts to gain more insight into these questions by following plant responses to interruptions of nitrate and/or oxygen supply to the roots. This has been done as follows: Pretreatment either $+NO_3^- + A$ or $-NO_3^- + A$. At time 0, plants were transferred to: plus $NO_3^- + A$ (aerated with normal air); plus $NO_3^- + N_2$ (aerated with N_2-gas); minus $NO_3^- + A$ and minus $NO_3^- + N_2$.

PREGROWN WITH NITRATE

Omission of nitrate from the culture solution caused a decrease in shoot growth but not in root growth (Fig. 1A), leading to a decrease in shoot/root ratio [1]. Under nitrogen ($+A + N_2$), root growth was strongly inhibited in the presence and absence of nitrate; the inhibition of shoot growth was relatively weaker. Under nitrogen the nitrate content of the plants decreased, indicating that they were not capable of accumulating nitrate (Fig. 2A, B). However nitrate reductase activity (NRA) stayed high (Fig. 2C, D). The nitrate concentration in the exudate was low and decreasing. In the root, the NRA seems to be stimulated under nitrogen [2, 3]. Although there was no accumulation of nitrate in the roots of the $-NO_3^- + A$ plants, root growth was not decreased. An increase in the accumulation of the total soluble carbohydrates (TSC) (Fig. 2E, F) with a decrease in the nitrate content, seems to be caused by a reduction in the usage of photosynthetic products. This is not in agreement with other experiences [2].

R. Brouwer et al. (eds.), Structure and Function of Plant Roots, 395–398.

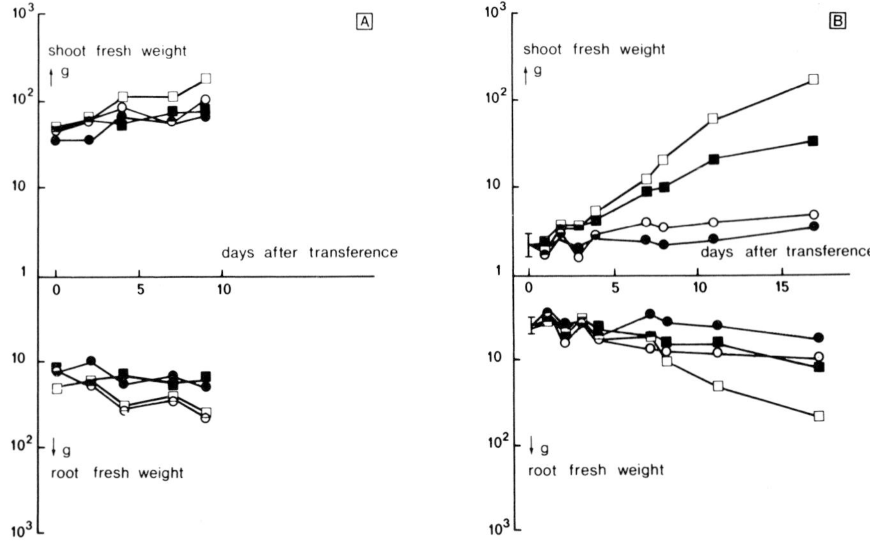

Fig. 1. Time course of the weight after transference from $+NO_3^-$ + A (A) and $-NO_3^-$ + A (B) to: $+NO_3^-$ + A (□); $+NO_3^-$ + N_2 (■); $-NO_3^-$ + A (○) and $-NO_3^-$ + N_2 (●).

PREGROWN WITHOUT NITRATE

Shoot growth is predominantly stimulated by nitrate; root growth by oxygen (Fig. 1B). Root growth with the $+NO_3^-$ + A plants was stimulated after a certain shoot/root ratio was reached [1]. Accumulation of nitrate is only possible with aeration (Fig. 3A, B). The initial nitrate accumulation with the $+NO_3^-$ + N_2 plants can be explained by the relatively large root system in the early stages which provides the shoot adequately with nitrate as a consequence of air channels. As a result of the relatively faster growth rate of the shoot, a shortage in the supply of nitrate by roots occurs rather quickly. Transport of nitrate through the xylem was less than under aerated conditions but the NRA in the shoots of the $+N_2$ and +A plants, was unaltered during a certain period (Fig. 3C). The NRA in the roots of the $+NO_3^-$ + N_2 plants increased remarkably (Fig. 3D) [2, 3]. This may indicate (though not yet proven) that nitrate is used as an electron acceptor [3]. A strong increase in the NO_2^- concentration in the solution was found under these conditions. Obviously (Fig. 3E, F), a reasonably well functioning shoot ($+NO_3^-$ + N_2) must be present to cause an increase in the content of TSC. With the increased accumulation of nitrate in the $+NO_3$ + N_2 plants, at the outset, a decrease in TSC can be seen (Fig. 3A, B; Fig. 3E, F).

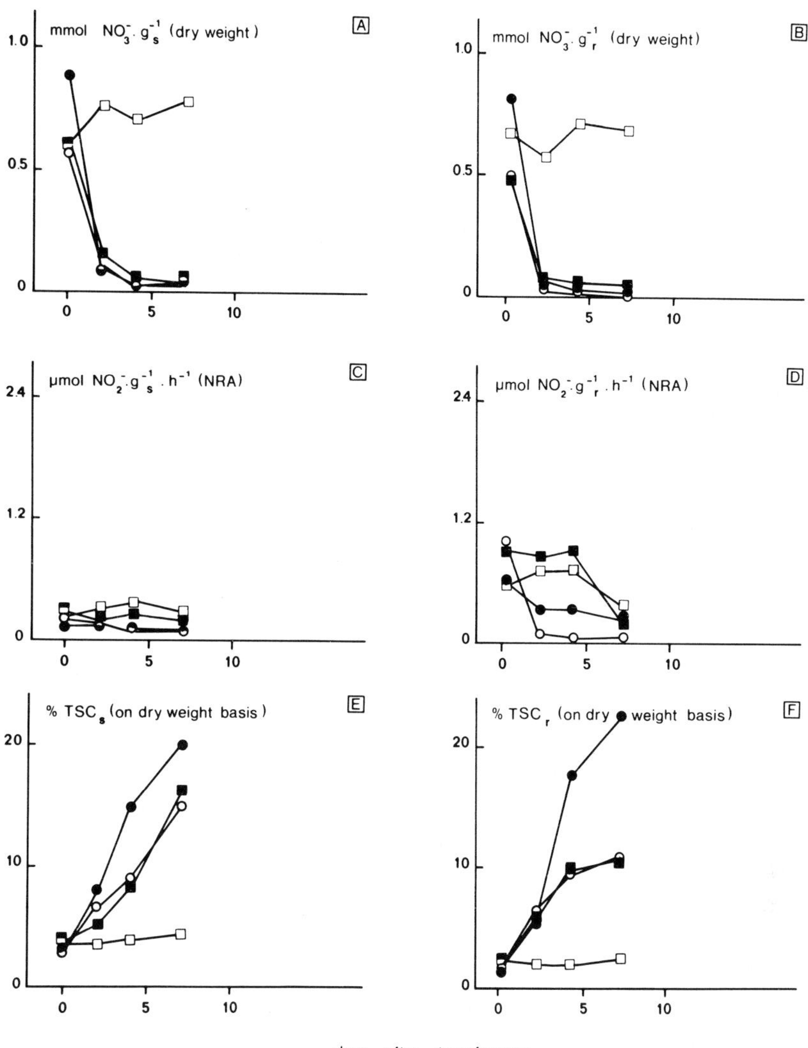

Fig. 2. Time course of nitrate content (A, B); NRA (C, D) and percentage TSC (E, F), after transference from $+NO_3^- + A$ to: $+NO_3^- + N_2$ (■); $-NO_3^- + A$ (○), $-NO_3^- + N_2$ (●) and $+NO_3^- + A$ = control, s = shoot, r = root.

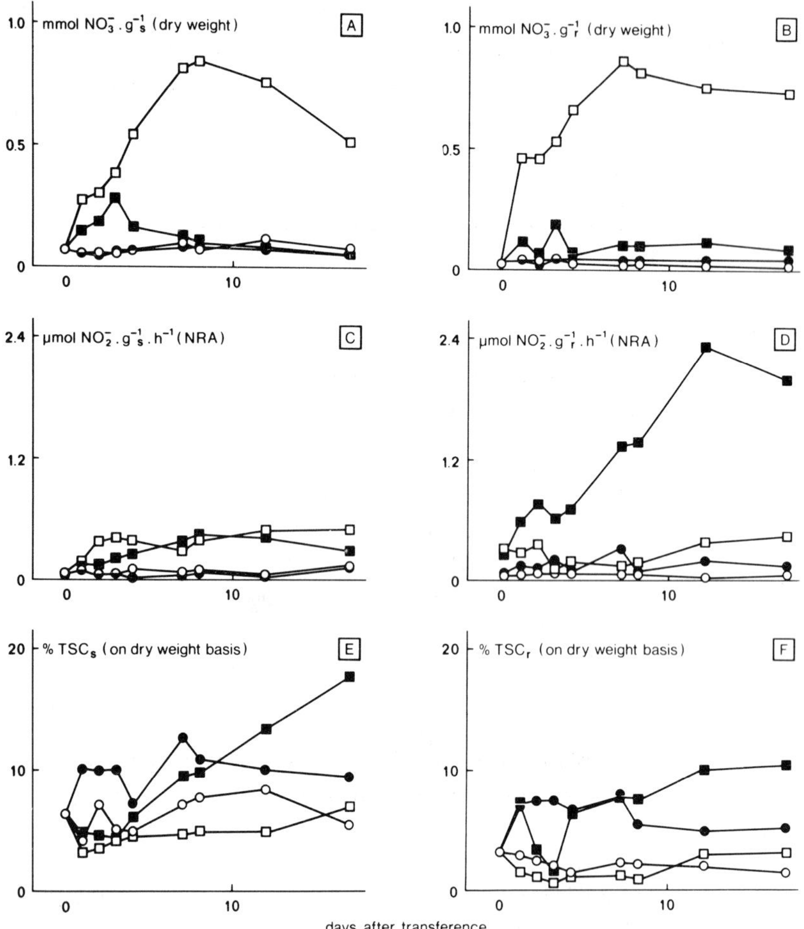

Fig. 3. Time course of nitrate content (A, B), NRA (C, D) and %-age TSC (E, F), after transference from $-NO_3^-$ + A to: $+NO_3^-$ + A (□); $+NO_3^-$ + N_2 (■), $-NO_3^-$ + N_2 (●) and $-NO_3^-$ + A (○) = control, s = shoot, r = root.

REFERENCES

1. Brouwer, R. 1962 Nutritive influences on the distribution of dry matter in the plant. Neth. J. Agric. Sci. 10, 399–408.
2. Lambers, H., Steingröver, E. and Smakman, G. 1978 The significance of oxygen transport and of metabolic adaptation in flood tolerance of Senecio species. Physiol. Plant. 43, 277–281.
3. Lee, R. B. 1980 Sources of reductant for nitrate assimilation in non-photosynthetic tissue: a review. Plant, Cell and Environment 3, 65–90.

75. Relationship between the root and above ground parts of different sunflower inbreds regarding the content of some mineral elements

M. R. SARIĆ

Faculty of Natural Sciences, Institute of Biology, Novi Sad, Yugoslavia

and

D. ŠKORIĆ

Faculty of Agriculture, Institute of Field and Vegetable Crops, Novi Sad, Yugoslavia

In the past, the problem of genotypic specificity was studied from various points of view at different levels and with large differences in methodology. A large number of results is available now, pointing out differences in the content of nutrient elements of individual genotypes [1].

The studies conducted so far have been concerned mostly with the problem of genotypic specificity in relation to mineral nutrition of corn, wheat, and much less frequently with other plant species. We have found relatively few papers dealing with sunflower [2, 3, 4].

It may be assumed that plant species which have a noticeably larger number of varieties will also have larger differences among them regarding the specific requirements for mineral nutrition. Considering the fact that sunflower genotypes are much fewer than corn or wheat genotypes as well as that papers on genotypic specificity of sunflower for mineral nutrition are scarce, we decided to study this problem with sunflower.

RESULTS AND DISCUSSION

Genotypic specificity for mineral nutrition was studied on 20 sunflower inbreds which differed only slightly in vegetation period. The inbreds were grown for 25 days in water cultures and the plants were examined for the weight of dry matter and contents of N, P, K, Ca and Mg in root, stem and leaves.

The sunflower inbreds varied considerably in all parameters investigated. Dry matter weight of roots ranged from 0.66 to 1.03, stems 1.40–2.61 and leaves from 1.69 to 2.83 mg/plant. The largest leaf area was 12.60 the smallest 7.57 cm^2/plant.

The results obtained indicate the existence of pronounced genotypic specifi-

R. Brouwer et al. (eds.), Structure and Function of Plant Roots, 399–401.

Table 1. Relationships between minimum and maximum contents of N, P, K, Ca and Mg in plant organs of 20 sunflower inbreds expressed in percentage

Plant organ	N	P	K	Ca	Mg
Root	77	42	46	76	60
Stem	75	64	63	49	68
Leaf	87	53	62	41	49

cities regarding both the average contents of these elements and their contents in different plant organs. The average content of N per plant organs varied from 4.380 to 4.937, P, 760–1.185, K, 4.366–6.000, Ca, 569–1064 and Mg from 353 to 530 mg/100 g of dry matter. Relationships between inbred lines of sunflower are very different depending on the organ and element studied, as illustrated by variations between minimum and maximum contents (Table 1).

The results indicate the existence of high genotypic specificity for the content of individual ions in the inbreds. However, the inbreds showed specific differences, e.g. in their distribution of nitrogen and especially phosphorus and potassium. These results show that the genetic specificity in mineral nutrition is reflected not only in the contents of certain ions but also in their distribution in certain plant organs (Table 2).

Such differences in the distribution of these elements in individual plant organs presumably occurred due to the specific nature of the processes of their uptake, movement, and metabolism. The inbred 46 takes up and retains phosphorus in

Table 2. Distribution of nitrogen, phosphorus, and potassium per sunflower plant organ (in mg/100 gr of dry matter)

Plant organ	Elements					
	N		P		K	
	Inbreds no.					
	26	48	46	50	28	25
Leaf	6981	6888	910	1250	4400	5200
Stem	3401	3983	833	950	6400	6300
Root	4301	3447	1258	877	5400	4750
Average	4894	4773	1000	1025	5400	5415

the root whereas the inbred 50 transports phosphorus quickly into the leaves retaining a small quantity in the root. It may be assumed that the inbred 50 has a more intensive phosphorus metabolism. It is therefore necessary to determine not only the total content of an element per plant but also the contents per individual plant organs as well as the chemical form of the element, either free or bound. To solve this problem, it would be useful to resort to grafting among the inbreds to establish the importance of the root, i.e., the scion for the uptake, movement, and distribution of ions. We are continuing our studies in this direction.

Our results suggest that in sunflower, the plant species with a relatively small number of varieties, i.e. hybrids or genotypes, the variations in the contents of individual elements are similar to the variations found with other species which have many more varieties, as wheat and corn.

It might be assumed that there is a relationship between potassium, calcium, and magnesium in plant tissue, i.e., the inbred with less calcium will have more magnesium or potassium. However, our results do not indicate a regularity in the contents of these three elements. It should also be emphasized that there were no correlations between dry matter weights of the root, stem, and leaf or between leaf area and the content of the elements examined.

Finally, these results show that there are a number of criteria suitable for the evaluation of genetic specificity of mineral nutrition of plants which must be considered when conducting similar studies.

REFERENCES

1. Sarić, M. 1980 Some problems and criteria of evaluation of the existence of genetic specificity in relation to mineral nutrition. Introductory lecture in the session 'Effects of genotypes on mineral nutrition'. Federation of European Societies of Plant Physiology, II Congress, Santiago de Compostela, 27 July–1 August 1980.
2. Mahdok, O. P. and Walker, R. B. 1969 Magnesium nutrition of two species of sunflower. Plant Physiol. 44, 1016–1022.
3. Foy, C. D., Orellana, R. G., Schwartz, J. W. and Fleming, A. L. 1974 Responses of sunflower genotypes to aluminium in acid soil and nutrient solution. Agron. J. 66, 293–296.
4. Diaz M. de la Guardia, Sáiz, J. A. de Omenaca, Pérez, E. B. Torres and Montes, F. Asusti 1980 Differentes respuestas de las lineas, hibridos y variedades de girasol a la nutricion con bajo nivel de calsio. IX Conferencia internacional del girasol. Madrid.

76. Response of barley roots to infection by the parasitic fungus *Erysiphe graminis* DC.

C. PAULECH, F. FRIČ, P. MINARČIC, S. PRIEHRADNÝ and G. VIZÁROVÁ

Institute of Experimental Biology and Ecology of the Slovak Academy of Sciences, 885 34 Bratislava, Czechoslovakia

Powdery mildew of barley is an ectoparasitic fungus attacking the above ground organs only. Physiological and biochemical processes of infected plants are significantly affected by the fungus [6]. Its harmful influence is also manifested in the root system and during pathogenesis. Root dry matter decrease is greater than that of above-ground organs [7]. This phenomenon is caused by lowered translocation of photoassimilates from infected leaves to roots [1]. The root dry matter reduction is accompanied by anatomical and morphological changes too. Lateral root growth inhibition, and some reduction of root elongation occurs [9]. Mitotic cell division of the apical root meristems is significantly lowered at the time of fructification. However, at the end of the incubation time, and at the beginning of the parasite's fructification a transient increase in intensity of mitotic cell division of apical root meristems has been noted [5]. This phenomenon is accompanied by changes of free cytokinin levels in the roots. In the first phase of pathogenesis (2nd to 4th day) after inoculation the content of endogenous cytokinins increased. At this period, an inhibition of elongation of the roots could be observed, together with the inhibition of growth, and forming lateral roots. Changes within the central cylinder of the roots were also noticed: the diameter of the central vessel was smaller and the cells of the central cylinder were crowded, and also smaller. The cell walls in the central cylinder and the cortical cells were thickened and more lignified. At the time of fructification in the roots of healthy and diseased plants, no differences were observed in the content of free endogenous cytokinins [10, 11].

Root respiration of diseased plants is slightly increased shortly after inoculation as compared with the healthy plants. In the later phase of pathogenesis respiration is, on the contrary, significantly decreased. The root respiration decrease is not due to shortage of respiration substrate but to functional deficiency of certain mitochondria in root cells [2].

Reduced root growth of the diseased plants is also accompanied by a lower water uptake and transpiration. The transpiration in the early stage of pathogen-

R. Brouwer et al. (eds.), Structure and Function of Plant Roots, 403–405.

esis is more affected than the water uptake by the roots. Owing to this effect the hydration of the plant organs is better than that of healthy plants. This stage lasts till the period of advanced fructification of the fungus when the water balance gets worse [8].

During the powdery mildew pathogenesis of barley the uptake of phosphate by the roots as well as its incorporation into organic compounds and their transport into above ground plant organs is also reduced [3, 4].

The changes found in intensity of physiological process in the roots of diseased plants are closely connected with the permeability changes of root cell membranes. Their permeability increases during infection. The ^{14}C-labelled photoassimilates as well the ^{32}P-labelled phosphate ions were exuded more intensively from the roots of diseased plants than those from healthy ones [3].

It appears that the parasitic fungus studied, living solely on the above-ground assimilation plant organs, rapidly damages the root structure and function after the inoculation. Metabolical changes in the root system increase the physiological disturbances of the above-ground plant organs caused directly by the ectoparasitic fungus.

REFERENCES

1. Friĉ, F. 1975 Translocation of ^{14}C-labelled assimilates in barley plants infected by powdery mildew (Erysiphe graminis f. sp. hordei Marchal). Phytopathol. Z. 84, 88–95.
2. Frič, F. and Čiamporová, M. 1975 Barley root respiration during Powdery mildew (Erysiphe graminis f. sp. hordei Marchal) pathogenesis (in Slovak). Acta Inst. bot. Acad. Sci. slovacae ser. B. 1, 17–26.
3. Frič, F. 1978 Absorption and translocation of phosphate in barley plants infected with Powdery mildew. Phytopath. Z. 91, 23–32.
4. Frič, F. and Vrátny, P. 1980 Physiological processes influenced by Powdery mildew in infected and uninfected organs of diseased barley (in Slovak). Zbor. ref. 3. zjazdu SBS, Zvolen, pp. 307–314.
5. Minarčic, P. and Paulech, C. 1975 Influence of powdery mildew on mitotic cell division of apical root meristems of barley. Phytopath. Z. 83, 341–347.
6. Paulech, C., Frič, F., Haspelová-Horvatovičová, A., Priehradný, S. and Vizárová, G. 1975 Influence of powdery mildew (Erysiphe graminis hordei Marchal) upon physiological processes of barley. Polnohospodárska veda 1, No. 4, pp. 1–204. Slovak Acad. Sci., Bratislava Czechoslovakia.
7. Paulech, C. 1969 Einfluss des Getreidemehltaupilzes Erysiphe graminis DC auf die Trockensubstanzmenge und auf das Wachstum der vegetativen Pflanzenorgane. Biológia (Bratislava) 24, 709–719.
8. Priehradný, S. 1979 Changes in water balance by powdery mildew infected susceptible barley cultivar. Acta Phytopathologica Acad. Sci. Hungaricae 14, 351–361.
9. Vizárová, G., Paulech, C. and Minarčic, P. 1975 Influence of increased endogenous cytokinin level upon the growth of barley roots and their morphology (in Slovak). Acta Univ. Agric. 23, 949–955.

10. Vizárová, G. and Minarčic, P. 1974 The influence of powdery mildew upon the cytokinines and the morphology of barley roots. Phytopath. Z. 81, 49–55.
11. Vizárová, G. and Kováčová, M. 1980 Investigations of free cytokinins in barley roots during pathogenesis of mildew by thin-layer chromatography (in Slovak) Biológia (Bratislava) 35, 727–732.

77. Translocation of growth regulators from roots in relation to the stem apical dominance in pea (*Pisum sativum* L.) seedlings

STANISLAV PROCHÁZKA

Department of Botany and Plant Physiology, University of Agriculture, 662 65 Brno, Czechoslovakia

Since it has been found that roots are a source of cytokinins [2], [8] and gibberellins [1, 7] and that exogenous applications of cytokinins and gibberellins can promote bud growth [5], [9] several authors have studied the relationships between the transport of growth regulators from roots and the development of lateral buds [3]. The role of growth regulators synthetized in roots and translocated into the above-ground part of plants in relation to the apical dominance of the stem has not yet been completely explained. The condition of a bud releasing itself from a correlative inhibition is changing very quickly and involves two stages, viz. (a) the release from the correlative inhibition and (b) the subsequent growth of the lateral bud [6].

Considering these facts we have studied the transport of ^{14}C-benzylaminopurine (^{14}C-BA), ^{14}C-gibberellic acid (^{14}C-GA), and $KH_2{}^{32}PO_4$ (^{32}P) from roots into lateral buds of intact, decapitated, and decapitated + IAA-treated 5-day-old pea seedlings. During all the experiments the plants were kept in darkness at 20 °C and at a relative humidity of 80 per cent.

When studying the transport of ^{14}C-BA (Table 1) from roots into lateral buds within 12 h after decapitation (i.e. in the period following shortly after the release of lateral buds from the inhibition) a slightly higher ^{14}C-activity was found in lateral buds and/or surrounding tissues of decapitated plants. There were no differences in activities after the application of other labelled compounds on

Table 1. Distribution of ^{14}C in pea seedlings (dpm/part of plant) after the application of ^{14}C-BA to roots. ^{14}C-BA was applied immediately after decapitation. Plants were gathered after 12 h

Part of plant	Intact		Decapitated + lanolin		Decapitated +0.007% IAA		Decapitated +0.5% IAA	
	$\bar{x}$	s_x	$\bar{x}$	s_x	$\bar{x}$	s_x	$\bar{x}$	s_x
Apex of epicotyl	230	33	732	234	327	116	200	161
Lateral buds	87	10	100	25	80	8	85	24

R. Brouwer et al. (eds.), Structure and Function of Plant Roots, 407–409.

Table 2. Distribution of ^{14}C in pea seedlings (dpm/part of plant) after the application of ^{14}C-BA to roots. ^{14}C-BA was applied for 12 hours immediately after decapitation. Plants were gathered after another 12 hours

Part of plant	Intact		Decapitated + lanolin		Decapitated +0.007% IAA		Decapitated +0.5% IAA	
	$\bar{x}$	s_x	$\bar{x}$	s_x	$\bar{x}$	s_x	$\bar{x}$	s_x
Apex of epicotyl	3.799	499	5.681	950	8.075	2.021	638	277
Lateral buds	114	10	283	76	103	37	91	9

Table 3. Distribution of ^{14}C in pea seedlings (dpm/part of plant) after the application of ^{14}C-GA to roots 24 h after decapitation. ^{14}C-GA was applied for 12 h and plants were gathered after another 12 h

Part of plant	Intact		Decapitated + lanolin		Decapitated +0.007% IAA		Decapitated +0.5% IAA	
	$\bar{x}$	s_x	$\bar{x}$	s_x	$\bar{x}$	s_x	$\bar{x}$	s_x
Apex of epicotyl	1.233	59	294	34	1.129	7	1.182	65
Lateral buds	51	10	167	25	43	5	68	21

Table 4. Distribution of ^{32}P in pea seedlings (dpm/part of plant) after the application of ^{32}P to roots 24 h after decapitation. ^{32}P was applied for 12 h and plants were gathered after another 12 h

Part of plant	Intact		Decapitated + lanolin		Decapitated +0.007% IAA		Decapitated +0.5% IAA	
	$\bar{x}$	s_x	$\bar{x}$	s_x	$\bar{x}$	s_x	$\bar{x}$	χ_{υ}
Apex of epicotyl	237	39	25	21	244	18	268	59
Lateral buds	63	51	122	51	27	11	23	1

intact and decapitated plants and for that reason the results are not tabulated.

When studying the transport of ^{14}C-BA, ^{14}C-GA, and ^{32}P (Tables 2, 3, 4) 24 h after decapitation (i.e. during the period of the subsequent growth of lateral buds) an increased ^{14}C-activity was observed in lateral buds of decapitated plants. An increased activity from ^{14}C-BA was observed at first (Table 2) followed by those from ^{14}C-GA and ^{32}P (Tables 3, 4) in lateral buds of decapitated plants.

Our experiments suggest that apices of intact plants are able to attract substances of regulatory and nutritive nature from roots [4]; however, this regulatory ability of the apex can be partly simulated by exogenous indole-3-acetic acid (IAA). IAA synthetized in the stem apex probably amplifies the apical dominance *via* its direct effect upon the growth of lateral buds as well as its ability to attract other regulators and nutrients. Shortly after the release of lateral buds from the correlative inhibition cytokinins are transported from roots into lateral buds and surrounding tissues. Only thereafter the further growth of lateral buds is associated with an intensive flow of gibberellins and nutrients from roots into these buds. It can therefore be concluded that gibberellins and nutrients from roots preferentially affect the elongation growth of buds.

REFERENCES

1. Jones, R. L. and Phillips, I. D. J. 1966 Organs of gibberellin synthesis in lightgrown sunflower plants. Plant. Physiol. 4, 1381–1386.
2. Kende, H. 1964 Preservation of chlorophyll in leaf sections by substances obtained from root exudate. Science 145, 1066–1067.
3. Morris, D. A. and Winfield, P. J. 1972 Kinetin transport to axillary buds of dwarf pea (Pisum sativum L.). J. Exp. Bot. 23, 346–355.
4. Phillips, I. D. J. 1969 Apical dominance in physiology of plant growth and development. In: Wilkins, M. B. (ed.), pp. 165–202. London: McGraw-Hill.
5. Sachs, T. and Thimann, K. V. 1964 Release of lateral buds from apical dominance. Nature 201, 939–940.
6. Sachs, T. and Thimann, K. V. 1967 The role of auxins and cytokinins in the release of buds from dominance. Amer. J. Bot. 54, 136–144.
7. Sebanek, J. 1966 The effect of amputation of the epicotyl on the level of endogenous gibberellins in the roots of pea seedlings. Biol. Plant. 8, 470–475.
8. Short, K. C. and Torrey, J. G. 1972 Cytokinins in seedling roots of pea. Plant Physiol. 49, 155–160.
9. Wickson, M. and Thimann, K. V. 1958 The antagonism of auxin and kinetin in apical dominance, II. The transport of IAA in pea stems in relation to apical dominance. Physiol. Plant. 13, 539–554.

Subject index